www.brookscole.com

www.brookscole.com is the World Wide Web site for Thomson Brooks/Cole and is your direct source to dozens of online resources.

At *www.brookscole.com* you can find out about supplements, demonstration software, and student resources. You can also send e-mail to many of our authors and preview new publications and exciting new technologies.

www.brookscole.com
Changing the way the world learns®

INTRODUCTION TO

Mathematical Modeling Using Discrete Dynamical Systems

INTRODUCTION TO

Mathematical Modeling Using Discrete Dynamical Systems

Frederick R. Marotto
Fordham University

THOMSON

BROOKS/COLE

Australia • Canada • Mexico • Singapore • Spain
United Kingdom • United States

THOMSON

BROOKS/COLE

Introduction to Mathematical Modeling Using Discrete Dynamical Systems
Frederick R. Marotto

Publisher: *Bob Pirtle*

Assistant Editor: *Stacy Green*

Editorial Assistant: *Katherine Cook*

Technology Project Manager: *Earl Perry*

Marketing Manager: *Tom Ziolkowski*

Marketing Assistant: *Jennifer Velasquez*

Marketing Communications Manager: *Bryan Vann*

Signing Representative: *Lisa Cappazzolo Beckman*

Project Manager, Editorial Production: *Cheryll Linthicum*

Art Director: *Vernon T. Boes*

Print Buyer: *Doreen Suruki*

Permissions Editor: *Kiely Sisk*

Production Service: *Hearthside Publishing Service, Laura Horowitz*

Text Designer: *Roy R. Neuhaus*

Copy Editor: *Melissa Moore*

Illustrator: *Matrix Art, Jade Myers*

Cover Designer: *Larry Didona*

Cover Image: *Michael Boys/CORBIS; AgeFotostock*

Cover Printer: *Phoenix Color Corp*

Compositor: *Interactive Composition Corporation*

Printer: *R.R. Donnelley/Crawfordsville*

Thomson Higher Education
10 Davis Drive
Belmont, CA 94002-3098
USA

For more information about our products, contact us at:
Thomson Learning Academic Resource Center
1-800-423-0563

For permission to use material from this text or product, submit a request online at http://www.thomsonrights.com.

Any additional questions about permissions can be submitted by email to thomsonrights@thomson.com.

Library of Congress Control Number: 2005926312

ISBN 0-495-01417-6

Contents

Preface   ix

1 C H A P T E R

Mathematical Modeling and Dynamical Systems   1

1.1   Modeling Reality   2
1.2   Discrete Dynamical Systems   11

2 C H A P T E R

Linear Equations and Models   20

2.1   Some Linear Models   21
2.2   Linear Equations and Their Solutions   28
2.3   Homogeneous Equations and Their Applications   39
2.4   Solutions of Non-Homogeneous Equations   49
2.5   Applications of Non-Homogeneous Equations   58
2.6   Dynamics of Linear Equations   68
2.7   Empirical Models and Linear Regression   78

3 C H A P T E R

Nonlinear Equations and Models   87

3.1   Some Nonlinear Models   88
3.2   Autonomous Equations and Their Dynamics   96

3.3   Cobwebbing, Derivatives and Dynamics   110

3.4   Some Mathematical Applications   127

3.5   Periodic Points and Cycles   141

3.6   Parameterized Families   155

3.7   Bifurcation and Period-Doubling   164

3.8   Chaos   176

4   C H A P T E R

Modeling with Linear Systems   189

4.1   Some Linear Systems Models   190

4.2   Linear Systems and Their Dynamics   197

4.3   Some Vector and Matrix Arithmetic   208

4.4   Stability and Eigenvalues   218

4.5   Repeated Real Eigenvalues   229

4.6   Complex Numbers and Their Arithmetic   236

4.7   Complex Eigenvalues   244

4.8   Non-Homogeneous Systems   254

5   C H A P T E R

Modeling with Nonlinear Systems   265

5.1   Nonlinear Systems and Their Dynamics   266

5.2   Linearization and Local Dynamics   276

5.3   Bifurcation and Chaos   287

5.4   Fractals   302

Appendix   315

Answers to Odd-Numbered Exercises   327

Bibliography   339

Index   341

Preface

The past several decades have witnessed the birth of some entirely new branches of applied mathematics, as well as equally dramatic advances in computer technology that have revolutionized the manner in which everyday work is conducted in the mathematical sciences and engineering. However, these recent developments appear to have made little impact on many undergraduate mathematics curricula. To some extent this is due to a general and quite reasonable reluctance to part with traditional mathematical methods that have proven so effective over the past few centuries, in favor of relatively new and untested ones. There is also the issue of a proper context in which to introduce these new mathematical ideas or take advantage of the computer technology that is currently available.

It is believed that an elementary mathematical modeling course giving equal emphasis to both traditional and contemporary topics, as well as to both analytical and computational tools of investigation, would be an excellent way of addressing these and other issues presently of interest to a number of mathematics departments across the country, especially those affiliated with smaller liberal arts colleges. Although many such departments have already come to this conclusion, it appears that few mathematical modeling textbooks presently available take this approach.

Goals of This Text

In an attempt to meet these needs, the aims of this text are therefore as follows:

(i) To introduce beginning and intermediate students of mathematics, the natural sciences and the social sciences to some powerful mathematical modeling techniques, both standard analytical and modern computational, that are accessible at their level;

(ii) To provide through a diversity of applications convincing evidence of the usefulness and therefore relevance of certain topics in a typical undergraduate mathematics major that are often viewed these days as being excessively theoretical or abstract, e.g., sequences, limits, linear algebra, complex variables, etc.; and

(iii) To expose such undergraduates, very likely for the first time in their academic lives, to some fascinating areas of contemporary mathematical discovery.

Today, the unique vehicle by which each of these goals can be accomplished must certainly be *discrete dynamical systems*. The rationale is that with minimal mathematical background one may quickly progress in this field from the traditional study of exponential growth and decay that simple linear equations always exhibit, to an investigation of recently discovered chaotic dynamics often associated with nonlinear systems, all within the span of a single introductory course.

Why Mathematical Modeling and Discrete Dynamical Systems?

At present, an elementary mathematical modeling course based upon discrete dynamical systems may be a wise addition to an undergraduate curriculum for the following reasons: First, it is worth stressing to mathematics majors and minors, as early and as frequently as possible, the wide range of applicability of their chosen field of study, thereby encouraging their continued pursuit of the subject. A modeling course does just that. Also of benefit is impressing upon them, through the introduction of some exciting areas of current research, that, although it has a long history stretching back to antiquity, mathematics is nonetheless a vibrant and evolving discipline. Perhaps no better choice in demonstrating this point is *dynamical systems,* which was recently given its own subject classification by the American Mathematical Society due to the high volume of activity in the field during the past few decades. And, since some of the most intriguing phenomena associated with dynamical systems, i.e., *chaos, strange attractors* and *fractals,* can be observed by numerical and graphical means only, such a course makes the perfect arena for introducing at least one of the popular mathematical software packages currently available. Finally, to allow students from other disciplines, such as the physical and biological sciences, economics, business, etc., to participate in this endeavor, the course must be not only applied but also elementary and comprehensive, since these students' mathematical backgrounds may be limited, and their programs of study are not likely to provide sufficient time for further exploration of these ideas. As pointed out earlier, an introductory course in discrete dynamical systems would satisfy this need as well.

The alternative approach to deterministic modeling, which has historically been *differential equations,* will no doubt remain the preference for students of physics and engineering, especially at large technology-based institutions. The natural laws underlying most processes in these fields are fairly well understood and best expressed using classic differential equations models that allow description and prediction with incredible accuracy. But this is not generally the case for biological and social processes, where many fundamental principles are either unknown at present or at least inexpressible using a few simple equations. Consequently, it makes little sense in those cases to insist upon differential equations modeling, which requires a more sophisticated mathematical background for its study, as well as several undergraduate courses in the field before arriving at contemporary issues, when much simpler iterative models work just as well.

Outline of the Text

Most of the material comprising this text has been successfully tested several times as the curriculum for a recently developed mathematical modeling course at Fordham University. One should find that it is fairly comprehensive, not only with regard to subject matter, but also logistically, i.e., an ample supply of exercises at varying levels of difficulty with many answers provided, and numerous suggested computer projects with specific instructions for their completion. This should make outside sources unnecessary.

After a brief introductory chapter that discusses the nature of mathematical modeling in general and discrete dynamical systems in particular, Chapters 2 and 3 cover one-dimensional linear and nonlinear equations respectively, and Chapters 4 and 5, two-dimensional linear and nonlinear systems, respectively. In keeping with its introductory character, there are few formal theorems and fewer rigorous proofs, although some sort of justification is given for nearly all important concepts. Deriving formulas for exact solutions is emphasized only for the simplest linear models of Chapter 2. Since this text is intended for a course in *discrete dynamical systems* and not *difference equations,* beginning with Chapter 3, determining the dynamics of solutions becomes the primary focus. Although exact solutions are obtained for several linear systems in Chapter 4, this is done primarily to show how unnecessarily tedious that process is when compared with determining a system's equally useful dynamics.

Each of Chapters 2 through 5 begins with the construction of a host of simple iterative models of the type that are to be investigated later in that chapter. This provides some motivation for the analysis that is to follow, and ensures that applications will not be sacrificed if time runs short. Not all models need be discussed, however. An instructor may pick those of maximum interest to the class.

Nearly all these models are for illustrative purposes only and, consequently, greatly over-simplified. Since this text is intended for beginning undergraduates from potentially many disciplines, it is the acquisition of mathematical skills that was deemed most important, and not the models themselves. Interested students should be encouraged to pursue more legitimate models elsewhere, such as in the research literature or advanced courses of their disciplines. Perhaps individual or team projects could be assigned in which discrete models from specific applied areas are constructed and analyzed.

Each section of every chapter ends with a set of exercises requiring only a pencil and paper to complete, although a simple calculator would often be helpful as well. These are usually followed by several optional but recommended computer projects for which a computer and appropriate software are needed. Most of these projects might be more appropriately classified as *computer exercises* since they are not very lengthy. However, by combining several together, some challenging projects can be designed. This is especially true with regard to analytically and numerically investigating the dynamics of parameterized families in Chapters 3 and 5.

Answers to most odd-numbered exercises, those not involving graphs or proofs, appear in the back of the book. The Appendix explains how most computer projects can be done using either a spreadsheet program such as *Microsoft Excel* or the remarkably powerful software package *Mathematica*. Although computer programs are given there for completing the projects by either method, the former has several practical advantages. First, spreadsheet programs are already available on virtually all personal computers these days, which makes trips to a computer lab or the purchase and installation of special software unnecessary. Second, spreadsheets are much easier to master, and in fact, many students may already be familiar with their use. This saves a good deal of class time that might otherwise be spent on technical computer issues.

A Student Solutions Manual containing fully worked-out solutions to all odd-numbered exercises is available on a separate CD. Also, a Complete Solutions Manual with fully worked-out solutions to every exercise in the text appears on the associated instructor website, along with some suggestions for individual and team projects that might be assigned throughout the semester.

Suggested Course Curricula

Although more material is included here than could possibly be covered in one semester at a typical college or university, a number of topics are optional: sums, products, linear regression, two-state Markov processes and some other applications in Chapter 2; the Newton and Euler methods in Chapter 3; and three-state Markov processes in Chapter 4. A single course based upon this text may therefore take on one of several forms, depending upon the approach of the instructor and the background of the students involved.

For students whose level of sophistication is differential calculus, which was assumed as this text was being written, the course would likely include most of the first three or four chapters. This is an outside estimate, especially if some class time must be taken out to familiarize students with the particular computer hardware and software they may be expected to use. Under this plan it is doubtful that there would be enough time for more than a brief mention of the ideas from Chapter 5, some of which could therefore be introduced simultaneously with analogous material from Chapter 4. Other, more complex, issues from that final chapter, such as bifurcation, chaos and fractals, might then be explored only numerically in some computer projects.

Alternatively, an instructor in this situation might instead choose to shave several weeks off the coverage of Chapter 4 by first briefly showing how the eigenvalues of a 2 by 2 matrix are computed, and then demonstrating with a few examples the validity of the stability and rotation criteria for autonomous linear systems that are based upon these eigenvalues. With this approach much of Chapter 4, in particular most of Sections 4.4, 4.5 and 4.7, containing some of the text's most tedious material, could be eliminated. By trimming down Chapter 4 this way, as well as skipping the optional material mentioned earlier, additional time would be made available for the more contemporary topics considered in the latter part of Chapter 5.

For more mathematically sophisticated students, an instructor may wish to devote less time to linear iterative equations and their applications, and more to nonlinear equations and two-dimensional systems. Consequently, much of Chapter 2 could be briefly skimmed or altogether skipped. Perhaps the only parts of Chapter 2 that are truly essential for what follows are Sections 2.1, 2.4 and 2.6. If this is the plan, Chapters 3 and 4 could certainly be covered in their entirety, especially if students are already familiar with linear algebra concepts, in which case most of the material from Chapter 4 involving eigenvectors and eigenvalues may already be known. Chapter 5 could then be covered, probably in its entirety, especially if multivariate calculus has already been taken.

Prerequisites

With regard to prerequisites, high-school algebra and some differential calculus are sufficient for the investigation of one-dimensional dynamics in the first half of the book, which should make that material immediately accessible by many freshmen. Since vectors in the plane and 2 by 2 matrices are needed in Chapters 4 and 5, an introduction to these topics is given in Section 4.3, just before they are first used. Limiting the discussion to two-dimensional dynamics makes an entire course in linear algebra unnecessary. For students who have already taken such a course, this material may be used for a quick review if needed. A similar introduction to complex numbers appears in Section 4.6, just before complex eigenvalues are considered.

Partial differentiation is taken up in several places, first in the context of linear regression in Section 2.7, and then again in Section 5.2, where stability criteria are developed for nonlinear systems. In the former case, since only the simplest types of functions are considered, taking the step from derivative to partial derivative is easily accomplished without the need for an entire course in multivariate calculus. Alternatively, partial derivatives can be avoided altogether in Section 2.7, by using heuristic arguments rather than careful derivation to arrive at the normal equations. The linearization techniques and stability theorems of Section 5.2 are recommended, however, only for those who possess more substantial multivariate calculus skills (or are willing to develop them here). Others will have to settle for just the numerical investigations of nonlinear dynamics in the plane that are undertaken in Chapter 5.

No previous programming experience is necessary for the suggested computer projects. Basic familiarity with the keyboard and mouse, which nearly all students have these days, is the only skill assumed. The Appendix gives enough information so that even someone new to spreadsheet programs would be able to complete nearly all of the computer projects using *Microsoft Excel*®. More sophisticated computer users may wish to try *Mathematica*®, also discussed in the Appendix.

Apologies and Acknowledgments

Finally, apologies should be given for any errors that might appear in this text, and for its omission of many important topics, especially with regard to one-dimensional dynamics, e.g., Sharkovskii ordering, Schwarzian derivative, Feigenbaum numbers, symbolic dynamics, etc. To give an introductory course a realistic chance of getting through the text in one semester, some difficult content decisions had to be made, especially in Chapter 3. Apologies, as well as gratitude, should also be conveyed to those Fordham University students who were either willing or unwilling participants in the author's experiments while this book was being written.

There are a number of other individuals whose contributions should be gratefully acknowledged, in particular, the reviewers, for their many helpful suggestions and corrections:

David Cooke, Hastings College
Bernard Fusaro, Florida State University, Talahassee
Christopher Hee, Eastern Michigan University
John Koelzer, Rockhurst University
Barry Spieler, Birmingham-Southern College
Bart Stewart, United States Military Academy

The author is also extremely appreciative of the skill and dedication exhibited by the editors and other talented professionals associated with Brooks/Cole and with Hearthside Publishing Services, in putting this text together. And a special thanks to the author's wife, Beverly, for her love, support and patience during this time.

Frederick Marotto

1

Mathematical Modeling and Dynamical Systems

In this brief introductory chapter we discuss the nature and goals of mathematical modeling, and introduce some of the terminology used both to describe the general features of such models and to classify them. We also begin our analysis of the particular class of mathematical models that are to be the primary focus of this book, by constructing several elementary examples of discrete dynamical systems.

1.1   Modeling Reality

Our understanding of the world we inhabit is largely the result of the mental images we form of the complex processes unfolding before us and the simplified models our minds construct when attempting to comprehend their underlying principles. If such an abstract model of reality is quantitative in nature, we might call it a **mathematical model.**

Mathematical models need not involve very sophisticated mathematics. Counting, arithmetic, and indeed the very notion of *number* itself, constitute mathematical modeling tools that have certainly proven to be fundamentally important to all cultures since the dawn of recorded history. Geometry and the later development of algebra and graphing have provided additional techniques, not only for describing important spatial properties of the three-dimensional world we live in, but also for more conveniently investigating natural and scientific phenomena, both quantitatively with the use of equations, and visually with the use of graphs. And calculus has given us a mathematical language for dealing with processes that change continuously over time or space, which has contributed enormously to our understanding of the laws of the physical universe.

The explosion in scientific knowledge and technology over the past few centuries has been both a contributing factor in, and also to some extent a direct consequence of, the development of more contemporary forms of mathematical modeling, which has further enhanced our understanding of the world around us. The goal of this text is to provide an introduction to some of these more recent mathematical ideas.

Dynamical Systems

A frequent motivation for those who construct or use mathematical models is the desire to make predictions: *What time will the sun rise tomorrow morning? How many people will have the flu next winter? How much will a certain investment be worth a year from now?* To help answer such questions, numerous types of models have been developed over the past several centuries for the investigation of natural and social processes that evolve over time. Today, these models are commonly referred to as **dynamical systems.**

Dynamical systems models often fall into one of two general categories: **deterministic** models, which attempt to make predictions with 100% certainty, and **stochastic** models, whose goal is to present a range of possible outcomes, each with its own associated probability of occurring. Deterministic models are ordinarily employed when the number of quantities involved in the process being modeled is relatively small and all the underlying scientific principles are fairly well understood. In this case only a few variables are usually needed, and the equations that express the relationships between them can often be stated very succinctly and with great precision.

For example, suppose we wish to predict how fast an object will be falling t seconds after it is dropped from a certain initial height. It was first observed by Galileo and has long been accepted in physics that any object falling freely undergoes the same constant acceleration $-g$, approximately -32 ft/sec^2 or -9.8 m/sec^2 near the surface of the earth, as long as the effects of air-resistance are negligible. From calculus, one learns that acceleration

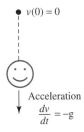

$v(0) = 0$

Acceleration

$$\frac{dv}{dt} = -g$$

FIGURE 1.1

is the first derivative of velocity $v(t)$, which therefore means that this velocity function must satisfy

$$\frac{dv}{dt} = -g \tag{1}$$

And if the object is indeed just dropped rather than thrown, then its initial velocity at time $t = 0$ must be $v(0) = 0$ (see Fig. 1.1).

This constitutes a simple mathematical model for the single changing quantity $v(t)$. It is easy to see that this model is deterministic since, for any particular $t \geq 0$, we are attempting to predict the exact value of $v(t)$. With a little calculus a solution of the simple **differential equation** (1) that also satisfies $v(0) = 0$ can easily be found:

$$v(t) = -gt \tag{2}$$

With this solution it is now possible to predict with great accuracy (except for air-resistance) the velocity of the falling object at any time $t \geq 0$ until it hits the ground.

Contrast this situation with that of attempting to make predictions when important underlying factors are not completely known. For example, suppose we toss a coin and try to predict whether the outcome will be Heads or Tails. Although the coin must certainly be subject to some well-known laws of physics, so that its complete path could be computed in theory, so many factors might influence that path, and therefore the final outcome, that it would be impractical to catalog them all. One would have to know in advance such quantities as: the exact direction that the coin was tossed and its initial velocity; the angle of the coin at that time; its initial angular velocity or spin, and the axis of rotation; its height when released and when it lands; etc. And even if such quantities were available, writing down the set of differential equations that govern the coin's motion would still be rather tedious, not to mention actually trying to solve them.

For this reason, calculating the exact outcome of a coin toss would seldom be attempted in practice. Instead, an altogether different approach is used and a different means of describing the outcome is employed. Rather than taking on the impossible task of accounting for every factor that might influence the result of a coin toss, all these factors are *merged* together into an *average* effect that they have on the outcome. This average effect can only be known by observing many repeated tosses of the coin.

With results recorded of countless coin tosses throughout human history, it has been concluded that this average effect causes a symmetrically balanced or *fair* coin to turn up Heads half the time and Tails the other half of the time. (We do not consider here the unlikely possibility that the coin lands on its edge!) Another way to state this is

$$P(H) = 0.5 \quad \text{and} \quad P(T) = 0.5 \tag{3}$$

where $P(H)$ and $P(T)$ stand for the probabilities that Heads and Tails occur respectively.

Equation (3) is referred to as the **probability distribution** for the set of outcomes of the coin toss problem. Rather than indicating precisely what the outcome of any given toss will be, instead every possible outcome is listed there, each with a probability of occurring. This makes the coin toss problem a stochastic one.

In this text we will be concerned primarily with deterministic dynamical systems, although an important class of stochastic models — Markov processes — will be considered in Chapters 2 and 4. For centuries deterministic models have been used primarily in the physical sciences and engineering, and have proven to be quite successful there, while stochastic modeling has been used extensively in the social sciences. This is largely due to our having long ago established many laws of the physical universe, such as those involving gravitational, electromagnetic, and thermodynamic forces, and our ability to accurately express them using relatively few variables and equations — usually differential equations such as (1).

But in the social sciences: business, economics, psychology, sociology, etc., a vast array of unknowable factors enter into most processes being investigated. Many of these depend upon the thoughts, emotions and behavior patterns of many interacting individuals, which makes them difficult to quantify or even identify. Consequently, simple and precise deterministic modeling of social processes is for the most part unfeasible at present. Stochastic models, based primarily upon empirically observed data and statistical analysis, are the best that can be expected in these cases. Despite this, because they provide some simple and interesting modeling applications, we will be investigating in the following chapters a variety of deterministic models from population biology, finance, micro-economics and other social sciences as well.

Testing and Refining a Model

When a mathematical model is first constructed, in order to keep it simple, comprehensible, and therefore perhaps solvable, only what are considered to be the most significant factors affecting the process being modeled may actually be included in the design. To keep the model from becoming too complex, or *intractable,* out of necessity all other influences must often be ignored. But how does one know which factors are truly important and which can safely be excluded?

One way to answer this is to test the model by comparing its predictions with what we actually observe. If agreement between the two is sufficiently close to satisfy the modeler, then the model may be accepted as it stands. But if there is substantial disagreement between a model's predictions and the reality it proposes to represent, then the model will need to be refined. That is, different or additional features will need to be designed into it.

For example, in the falling object problem considered earlier, experimentation could easily verify that the solution given by (2) adequately describes the velocity $v(t)$ of virtually any smooth and streamlined object for small values of time t. But if an object is dropped from a great height and falls through the earth's atmosphere for a relatively long period of time, this solution for $v(t)$ becomes increasingly inaccurate. Air-resistance, which has a tendency to slow an object's velocity, now becomes a significant factor. This explains why

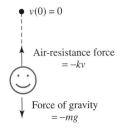

$v(0) = 0$

Air-resistance force
$= -kv$

Force of gravity
$= -mg$

FIGURE 1.2

a skydiver never falls faster than a certain limiting velocity, even though (2) mistakenly predicts that $v(t) \to -\infty$ as $t \to \infty$.

If model (1) is refined by the inclusion of air-resistance, physical principles such as Newton's law of motion transform it into

$$m\frac{dv}{dt} = -kv - mg \tag{4}$$

where m is the object's mass, $-mg$ is the force of gravity and $-kv$ is the air-resistance force, which depends upon the size, shape, and other characteristics of the falling object (see Fig. 1.2). Of course, while this model is more accurate than model (1), its solution for $v(0) = 0$, which can be shown with some calculus to be

$$v(t) = \frac{mg}{k}(e^{-kt/m} - 1) \tag{5}$$

still may not describe the object's exact velocity. Many factors have been left out of this new design as well, some of which may prove to be significant when trying to arrive at the true velocity. So, further testing and refinement may be needed.

This cycle of testing and refinement is common when mathematical models are being developed. Of course, as can readily be seen if (4) is compared with (1), each refinement is likely to make the model more complex and therefore more difficult to work with. Consequently, there must often be a trade-off between the tractability of a model and its accuracy.

Continuous and Discrete Time

One might assume that, whenever modeling a process that evolves over time, the dynamical system used should always be one that attempts to describe the state of that process over the entire time interval of interest. In the falling object problem considered earlier, for example, solving either differential equation (1) or (4) yields a velocity function $v(t)$, given by (2) or (5) respectively, which is defined for all real time t in the interval $t \geq 0$ (until the object hits the ground). Either of these velocity functions $v(t)$ can be evaluated for any such choice of t, and the resulting quantity used as an approximation to the true velocity at that time.

However, this type of **continuous-time** modeling, which is most often implemented with the use of a differential equation, is not always the best approach. There are numerous situations in which attempting to describe the state of a process for all real-time values is undesirable, due to the complexity that would be associated with the resulting model. In cases like these, a **discrete-time** model may prove to be a wiser choice. In this type of dynamical system, instead of trying to describe a process for *all* real values of time over some interval, the state of the process is determined for only a collection of separate or discrete-time values, which are often equally spaced on the time axis and perhaps infinite in number.

Discrete-time models, or **discrete dynamical systems** as they are commonly known today, often lend themselves to analysis and quantitative solution much more readily than do their corresponding continuous-time versions. To demonstrate this point, suppose we try to trace the path of a bouncing ball over both continuous and discrete time. With each bounce a ball generally loses some of the energy it had just before that bounce. For simplicity, let us assume that the ball loses $1/4$ of its energy with each bounce, and again there is no air-resistance. This means that the velocity of the ball each time it leaves the ground, the maximum height that it then rises to, and the total time of flight until it next hits the ground, all keep decreasing with each bounce. Consequently, in a continuous-time model, although the differential equation satisfied by the height $s(t)$ of the ball between any two consecutive bounces is not difficult to write, deriving its exact solution for all $t \geq 0$ is nevertheless quite tedious.

To see this, first recall from calculus that the acceleration of the ball between consecutive bounces is the second derivative of its height $s(t)$. Since, as we saw earlier, all objects falling freely and without air-resistance have an acceleration of $-g$, then we must have

$$\frac{d^2s}{dt^2} = -g \qquad (6)$$

which is a second-order differential equation for $s(t)$.

To find $s(t)$ for all $t \geq 0$, the differential equation (6) must be solved first for $t_0 \leq t \leq t_1$, where $t_0 = 0$ and t_1 represent the time when the ball first hits the ground (see Fig. 1.3). This solution can easily be verified to have the form

$$s(t) = -\tfrac{1}{2}gt^2 + v_0 t + s_0 \quad \text{for} \quad t_0 \leq t \leq t_1$$

where v_0 and s_0 are constants that can be determined by making use of the initial height and velocity of the ball, which we assume are known. Once this is done, the value of t_1, which depends upon them, can also be computed.

To next find $s(t)$ for $t_1 \leq t \leq t_2$, where t_2, as shown in Figure 1.3, represents the second time when the ball hits the ground, (6) must be solved again. The solution now is

$$s(t) = -\tfrac{1}{2}gt^2 + v_1 t + s_1 \quad \text{for} \quad t_1 \leq t \leq t_2$$

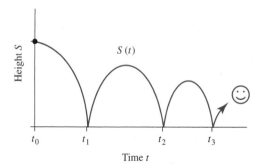

FIGURE 1.3

where v_1 and s_1 are constants to be determined, which depend not only upon v_0, s_0 and t_1, but also on the ball's having lost $1/4$ of its energy during the first bounce. Once v_1 and s_1 are found, t_2 can then be computed.

Similarly, the height of the ball from t_2 until time t_3 when it next hits the ground, again shown in Figure 1.3, must be

$$s(t) = -\tfrac{1}{2}gt^2 + v_2 t + s_2 \quad \text{for} \quad t_2 \le t \le t_3$$

where v_2 and s_2 are again unknown constants. These, along with t_3, would have to be determined using previously found data, and taking into consideration the further loss of the ball's energy after the second bounce.

In general, we see that the height of the ball for all continuous-time $t \ge 0$ is given by

$$s(t) = -\tfrac{1}{2}gt^2 + v_n t + s_n \quad \text{for} \quad t_n \le t \le t_{n+1}$$

where t_n represents the nth time that the ball hits the ground. Here, all three values t_n, v_n and s_n would need to be computed in turn for $n = 1, 2, 3, \ldots$, to have a complete description of $s(t)$ for all $t \ge 0$.

At this point it should be clear that obtaining $s(t)$ for all real time $t \ge 0$ would require a rather substantial set of calculations. We next see how this compares with a discrete dynamical system that models essentially the same process. Before constructing that model, it may be worth asking ourselves: *Since we are not attempting to describe the ball's height for all continuous-time, what is it exactly that we are hoping to glean from the model?*

Although this question might be answered in any number of ways, here we choose the following: *We would like to determine the maximum height that the ball reaches after each bounce.* Finding just these maximum heights seems reasonable since common sense allows us to fill in the rest. After achieving its maximum height each time, the ball falls to the ground and bounces back up again to its next maximum height. It was the attempt to determine the height of the ball for all *in-between* times that led to the continuous-time model that we are presently trying to avoid. Consequently, we would now be willing to accept a less detailed description of a bouncing ball in the hope of obtaining a much simpler model.

It may be worth pointing out, however, that among the details that are to be sacrificed in constructing this discrete model is the set of time-values at which the ball reaches those maximum heights. These time-values are not equally spaced on the time axis. Rather, the amount of time that elapses between successive bounces, and hence between successive maxima, decreases with each bounce. If this set of time-values is deemed important, a separate discrete model for them might be constructed using some additional physical principles.

Now that we have decided to view the bouncing ball problem as a discrete process involving just a sequence of maximum heights, we next construct a model for them. Suppose we let S_0 represent the initial height of the ball when dropped (with an initial velocity of 0). Since we are assuming as before that the ball loses $1/4$ of its energy with each bounce and there is no air-resistance, the ball must therefore retain $3/4$ of its energy from one maximum height to the next. From physics this means that it must rise back up to a maximum height of

$$S_1 = \tfrac{3}{4}S_0$$

after the first bounce (see Fig. 1.4). For the same reason, the ball must rise to a maximum height of

$$S_2 = \tfrac{3}{4}S_1$$

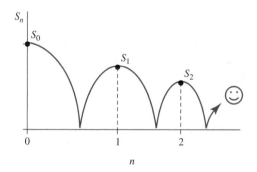

FIGURE 1.4

after the second bounce, and a maximum height of

$$S_3 = \tfrac{3}{4} S_2$$

after the third, as shown in Figure 1.4. It is easy to see that in general each maximum height S_n is related to the next S_{n+1} by the equation

$$S_{n+1} = \tfrac{3}{4} S_n \qquad (7)$$

where $n = 0, 1, 2, \ldots$.

The **iterative equation** (7) represents a discrete-time model for our bouncing ball. With S_0 known, one can **iterate** this equation repeatedly to determine a unique sequence of iterates $S_1, S_2, S_3, \ldots$. For example, if $S_0 = 4$ ft then, referring to (7),

$$n = 0 \text{ gives} \quad S_1 = \tfrac{3}{4} S_0 = \tfrac{3}{4} \cdot 4 = 3,$$

$$n = 1 \text{ gives} \quad S_2 = \tfrac{3}{4} S_1 = \tfrac{3}{4} \cdot 3 = 2.25,$$

$$n = 2 \text{ gives} \quad S_3 = \tfrac{3}{4} S_2 = \tfrac{3}{4} \cdot 2.25 = 1.6875,$$

etc.

On the other hand, if $S_0 = 6$ ft then

$$n = 0 \text{ gives} \quad S_1 = \tfrac{3}{4} S_0 = \tfrac{3}{4} \cdot 6 = 4.5,$$

$$n = 1 \text{ gives} \quad S_2 = \tfrac{3}{4} S_1 = \tfrac{3}{4} \cdot 4.5 = 3.375,$$

$$n = 2 \text{ gives} \quad S_3 = \tfrac{3}{4} S_2 = \tfrac{3}{4} \cdot 3.375 = 2.53125,$$

etc.

Any complete sequence of such iterates S_n, for $n = 0, 1, 2, \ldots$, constitutes a **solution** of (7). It should be apparent that (7) actually has an infinite number of solutions, depending upon the value chosen for S_0.

One should note how much simpler it is to generate a discrete-time solution S_n of the bouncing-ball problem than it was to obtain a continuous-time solution $s(t)$. Unlike continuous-time models, whose frequent reliance upon differential equations means that precise solutions may be difficult to come by, very accurate solutions of discrete dynamical systems such as (7) are always readily available, especially if some type of computing device can be employed to actually carry out the calculations. The relative ease of generating solutions this way, through **direct numerical iteration,** explains to some extent the recent popularity of discrete dynamical systems in our present era of extremely fast yet rather inexpensive computer hardware.

One should also note that (7) is not the only way to express the relationship between successive iterates S_0, S_1, S_2, etc. For example, one might instead write

$$S_n = \tfrac{3}{4} S_{n-1} \tag{8}$$

where $n = 1, 2, 3, \ldots$. Assuming that S_0 still represents the given initial height of the ball, then letting $n = 1$ in (8) yields

$$S_1 = \tfrac{3}{4} S_0$$

which is the same value of S_1 that (7) gave. Similarly, letting $n = 2$ in (8) yields

$$S_2 = \tfrac{3}{4} S_1$$

as before. Each subsequent iterate can be successively computed from (8) by letting $n = 3, 4, 5, \ldots$. In the end this generates the exact same solution S_0, S_1, S_2, $\ldots$ that (7) did earlier.

Although (8) does provide an alternate means of representing the problem, we prefer the more commonly used style of (7). We henceforth adopt this style exclusively whenever writing such an iterative equation. We also prefer to describe the process of solving equations like (7) as **iteration,** although such equations are sometimes referred to as **recursive,** and the process of solving them as **recursion.**

With a small modification to (7) one can obtain a discrete model for any bouncing ball that retains the fraction r of its energy after each bounce, where $0 \le r \le 1$ and air-resistance is again assumed to be negligible. Since we be will referring to this model occasionally, we summarize it as follows:

BOUNCING BALL MODEL

The equation

$$S_{n+1} = r\, S_n$$

where $0 \le r \le 1$ models the height of a bouncing ball that retains the fraction r of its energy after each bounce. S_0 is the initial height of the ball when dropped, and each subsequent S_n for $n = 1, 2, 3, \ldots$ represents the maximum height after the nth bounce.

As we shall soon see, constructing discrete models such as this, determining their solutions and investigating the types of behavior they can exhibit are to be the central issues of this book.

Exercises 1.1

In Exercises 1–11, if trying to predict the quantity given, decide which would be the better approach, a deterministic model or a stochastic one.

1. The outcome when a pair of dice is rolled.

2. The velocity of an object at any given moment if its exact position is known at all times.

3. The distance that an object will travel during the first 2 minutes if its exact velocity is known at all times.

4. The amount of time the next shopper will have to wait on line at a supermarket checkout.

5. The time sunrise occurs for a certain spot on the earth's surface on a certain day of the year.

6. The amount of rainfall this year for a certain town.

7. A virus population after 5 hours if it was originally 10^6 and is known to double in size every hour.

8. The length of time it will take to pay off a loan of $5000 with a fixed interest rate of 8% when the monthly payment is $100.

9. The amount of interest one must pay over 20 years on a mortgage with a variable interest rate.

10. The price 6 months from now of a stock that is presently selling at $35 per share.

11. The grade on your next exam.

12. Use (8) to compute S_1, S_2 and S_3 beginning with (a) $S_0 = 4$; (b) $S_0 = 6$. Compare your results with those found earlier for (7).

13. Determine which of the following iterative processes are equivalent to (7):

    (a) $S_{k+1} = \frac{3}{4}S_k$ for $k = 0, 1, 2, \ldots$

    (b) $S_{n-1} = \frac{3}{4}S_n$ for $n = 1, 2, 3, \ldots$

    (c) $S_{m-1} = \frac{4}{3}S_m$ for $m = 1, 2, 3, \ldots$

14. In the continuous-time bouncing ball problem discussed in this section, suppose the ball is dropped from a height of 4 feet at time $t = 0$. Determine the height of the ball $s(t)$ from $t = 0$ until the second time it hits the ground by computing: $v_0, s_0, t_1, v_1, s_1, t_2$. *Note:* Since the ball retains 3/4 of its energy after the first bounce, from physics we must have $\lim_{t \to t_1^+} v(t) = -\frac{\sqrt{3}}{2} \lim_{t \to t_1^-} v(t)$.

In Exercises 15–18, construct a Bouncing Ball Model by determining r and S_0 using the given information.

15. The ball is dropped from an initial height of 6 feet and loses 40% of its energy with each bounce.

16. The ball is dropped from an initial height of 5.5 feet and its maximum height after each bounce is 10% less than its previous maximum height.

17. $S_0 = 5$ and $S_1 = 3.5$ 18. $S_1 = 4.5$ and $S_2 = 2$

In Exercises 19 and 20 use the Bouncing Ball Model to compute S_1, S_2, ..., S_5 for the given value of r and S_0.

19. $r = 1/2$ and $S_0 = 6$ 20. $r = 0.8$ and $S_0 = 5$

21. If one were to construct a discrete model for the height $s(t)$ of the center point of a circular drum at time t after it is struck, what meaning would likely be associated with the values of $S_0, S_1, S_2, \ldots$?

22. If one were to construct a discrete model for the angle $\Theta(t)$ that a swinging pendulum makes with the vertical at time t, what meaning would likely be associated with the values of $\Theta_0, \Theta_1, \Theta_2, \ldots$?

1.2   Discrete Dynamical Systems

As pointed out in the previous section, deterministic models need not be implemented using continuous-time differential equations. In fact, throughout the remainder of this text we will seldom encounter that sort of equation, and never really need to solve one. Rather, we will be considering processes that can quite successfully be modeled using discrete dynamical systems, which requires no introduction whatsoever of differential equations nor even integral calculus. One should note, however, that our work will involve a good deal of algebra, both linear and nonlinear, as well as a bit of differential calculus from time to time.

In this section we offer an overview of the types of mathematical models we will be studying in this text and of the analytic tools we will be developing for their investigation. We begin by constructing several discrete models based upon well-known financial processes, and then preview some additional models of greater variety and complexity, to be explored more fully in later chapters. Although the Bouncing Ball Model of the previous section already provides a nice example of an iterative process, bouncing balls are generally of little interest to anyone outside the world of competitive sports! The focus here will therefore be on some rather important areas of human endeavor (presumably of greater significance than bouncing balls), in which discrete dynamical systems play an indispensable role.

Simple Interest

Let us first consider the growth of an investment through **simple interest.** Suppose an investor deposits a certain principal P_0 into a savings account or other financial instrument that pays simple interest at a certain fixed rate r. Banks and other financial institutions usually quote the interest rate r on a per year basis, for example, $r = 5\% = 0.05$ per year or $r = 3.25\% = 0.0325$ per year. If this is the case, then after one year the account will have earned

$$I = r P_0$$

in interest, and so the investor will then have a total of

$$P_1 = P_0 + I$$

in that account. If left untouched for another year, since the interest is simple, another $I = r P_0$ will be added to the previous year's balance P_1, which yields

$$P_2 = P_1 + I$$

in the account. Similarly, the total will be

$$P_3 = P_2 + I$$

at the end of the third year.

If this process continues, then we can see that each year the account balance will increase by adding precisely the same amount of simple interest I to the previous year's balance. Another way to say it, if the balance after the nth year is P_n, then the balance after the following year P_{n+1} can be determined from P_n using the iterative equation

$$P_{n+1} = P_n + I$$

where $n = 0, 1, 2, \ldots$. From this it should be apparent that such financial growth constitutes a discrete dynamical system. We summarize these ideas as follows:

SIMPLE INTEREST MODEL

The equation

$$P_{n+1} = P_n + I \tag{1}$$

models the growth of an investment earning simple interest. P_0 is the original principal, and each subsequent P_n for $n = 1, 2, 3, \ldots$ represents the total value of the investment at the end of n years. The amount of interest each year is $I = r P_0$, where r is the fixed annual interest rate.

While (1) aptly describes how an investment grows from one year to the next under simple interest, direct numerical iteration of that equation would constitute a rather inefficient method of answering such questions as: *What will be the future value of the account after 5 years, after 10 years,* etc.?, since it would require computing all previous balances along the way. For example, to determine P_{10} by iterating (1), we must compute in turn

$$P_1 = P_0 + I, \quad P_2 = P_1 + I, \quad P_3 = P_2 + I, \quad \ldots$$

until we finally arrive at $P_{10} = P_9 + I$.

What would be desirable in this case is an equation that allows any particular P_n to be evaluated directly, without the need to compute $P_1, P_2, \ldots, P_{n-1}$ first. Such an equation is called the **formula for the exact solution** of the discrete model, and is usually stated by giving the value of the subscripted variable being modeled, in this case P_n, directly as a function of the variable n. It can be easily verified that the solution of the Simple Interest Model (1) is

$$P_n = P_0 + nI$$

for $n = 0, 1, 2, \ldots$. Technically speaking, this equation actually represents the entire infinite set of all possible solutions of the iterative equation (1), depending on the value that is assumed for P_0. Since $I = r P_0$, this **Simple Interest Formula** may also be written as $P_n = P_0 + nr P_0$, or

$$P_n = (1 + rn)P_0 \tag{2}$$

By substituting the appropriate value of n into this formula, any particular year's balance P_n can be obtained immediately, without needing to compute any preceding year's balance. For example, since the original principal P_0 and interest rate r are known quantities, then P_{10} can be quickly determined using

$$P_{10} = (1 + 10r)P_0$$

which does not require the use of the values $P_1, P_2, P_3, \ldots, P_9$. How to derive formulas for exact solutions of elementary types of discrete dynamical systems, such as this one, will be a primary concern of the next chapter.

Compound Interest

Although simple interest scenarios have their place in the world of finance, most investments are actually made with the expectation of **compound interest,** which over the long term causes an investment to grow much more rapidly. Compounding of interest means that periodically all simple interest earned thus far is added to the principal, and the next interest computation is based upon this new principal. This situation too can best be modeled with the use of a discrete dynamical system.

To accomplish this, for simplicity first suppose that the period of compounding is 1 year. Using (2) with $n = 1$, an original principal P_0 earning interest at a rate of r per year will therefore yield

$$P_1 = (1 + r)P_0$$

after one year. Since interest is to be compounded, this P_1 becomes the new principal upon which the next interest computation is based. In other words, the following year's balance P_2 is determined by computing 1 year's simple interest using P_1 as the principal, i.e.,

$$P_2 = (1 + r)P_1$$

Similarly,

$$P_3 = (1 + r)P_2, \quad P_4 = (1 + r)P_3, \quad P_5 = (1 + r)P_4, \quad \ldots$$

This leads to the iterative equation

$$P_{n+1} = (1 + r)P_n$$

Of course, interest is seldom compounded on an annual basis. Monthly or perhaps quarterly are much more common time periods or **terms** used for compounding. To accommodate such a change, only two modifications need be made with regard to the equation above: The first is the adjustment of the interest rate to conform to the appropriate compounding period. For monthly compounding we divide the yearly rate r by 12 to find the monthly rate $i = r/12$. On the other hand, if compounding is done quarterly, r must instead be divided by 4, which yields the quarterly rate $i = r/4$. A similar calculation must be done for other compounding periods.

The only other modification we need to make is in our interpretation of n. Rather than have n represent years, we now let n measure the total number of terms. For example, n may be the total number of months or quarters during which the account earns interest, depending upon the compounding period. With these two modifications we arrive at the following discrete dynamical system:

COMPOUND INTEREST MODEL

The equation

$$P_{n+1} = (1 + i)P_n$$

models the growth of an investment earning interest at a rate of i per term, compounded once per term. P_0 is the original principal, and each subsequent P_n for $n = 1, 2, 3, \ldots$ represents the total value of the investment at the end of n terms.

In the next chapter it will be seen that the solution of the Compound Interest Model, or, alternatively, its entire infinite set of solutions, is given by the **Compound Interest Formula**

$$P_n = (1 + i)^n P_0$$

which describes a rather different type of growth when compared with the solution (2) of the Simple Interest Model.

Preview of Other Models

Discrete dynamical systems have much wider application in the world of finance than just as a means of computing simple and compound interest. For example, in the next chapter we will be deriving discrete models for the growth of savings annuities and for paying off interest-bearing loans. Also in that chapter, as well as subsequent ones, a wide range of iterative models from the natural and social sciences will be developed, primarily from the fields of population biology and micro-economics.

Few of these upcoming models, however, are likely to have a form as simple as those we have seen so far. For instance, in Chapter 3 we will see a model for the spread of an infectious disease through a fixed population:

$$I_{n+1} = I_n - r\,I_n + s\,I_n \left(1 - \frac{I_n}{N}\right)$$

and another for the price changes over time of a commodity when subject to the laws of supply and demand:

$$P_{n+1} = \frac{a}{P_n} + b\,P_n + c$$

In Chapter 4 we begin the investigation of discrete processes in which several different quantities of interest are not only changing over time but also interact with one another. Such processes must be modeled using a **system** of iterative equations, such as

$$P_{n+1} = r_1\,P_n - s_1\,Q_n$$
$$Q_{n+1} = s_2\,P_n + r_2\,Q_n$$

which describes the populations of two interacting species, a **predator** and its **prey.** And in Chapter 5 we will discuss systems of iterative equations having even greater complexity, including a now famous one first considered by B. Mandelbrot:

$$x_{n+1} = x_n^2 - y_n^2 + a$$
$$y_{n+1} = 2\,x_n\,y_n + b$$

which has been credited with giving birth to the newly created field of mathematics known as **fractal geometry.** Fractals, which adorn countless posters, calenders, screen-savers, etc., are among the most bizarre and fascinating images ever generated by a computer. Today, there exist literally thousands of books, articles and web pages devoted to the creation and analysis of fractals, or to the appreciation of their beauty. Several examples of fractal images are shown in Figures 1.5 and 1.6.

FIGURE 1.5

FIGURE 1.6

Quantitative Solution vs. Qualitative Behavior

It may come as a surprise to many readers that, despite the discussion earlier in this section concerning the benefit of having a formula for the exact solution of an iterative process, we will not be attempting to obtain such formulas for most of the upcoming models just mentioned. Due to the complexity of nearly all the models we encounter after Chapter 2, formulas for their exact solutions are either completely unknown at present, or, at the very least, somewhat tedious to make use of. Thus, another form of analysis will often need to be undertaken.

The type of investigation that we actually will be engaged in much of the time involves determining the **dynamics** of a model. This means that, rather than obtaining its exact quantitative solutions either through direct numerical iteration or using a formula, we will instead try to predict in advance the general **qualitative behavior** that all, or at least most, of its infinite set of solutions must have. Hopefully, the dynamics we are able to obtain this way will be sufficient to allow useful conclusions to be drawn with regard to the model, and meaningful predictions made with regard to the underlying process being modeled. These predictions might involve answering such questions as: *Will a population thrive and flourish, or will it become extinct? Will a disease be quickly eliminated from a population, or will it run rampant and perhaps never be completely eradicated? Will the price of some item increase, decrease, stabilize at some positive level or oscillate wildly?*

To help explain why the determination of this sort of qualitative behavior is often adequate when investigating real-world processes, we first describe several situations in which exact solutions are essential. Consider any of the financial models we have mentioned; for example, the Compound Interest Model $P_{n+1} = (1+i)P_n$. Before an investment is made, it would certainly be understood, and in fact be expected, by any investor that an initial deposit P_0 will grow without bound over time, as long as the interest rate i is positive. Many of them may even be familiar with the nature of the compounding process and how their savings will grow more quickly this way rather than under simple interest.

But those general features alone are not likely to satisfy an investor's curiosity regarding the value of the investment. What is needed is the precise account balance P_n at any time n. No investor would settle for anything less than a complete accounting of the investment at all times over the months and years that follow. A description of the qualitative dynamics would fail in this case to match the importance of knowing the exact solution, i.e., the Compound Interest Formula $P_n = (1+i)^n P_0$, which is why this formula has such importance in the financial world. Similar remarks could be made with regard to any of the other financial models mentioned earlier.

It should be clear, therefore, that formulas for exact solutions of models such as these are crucial. However, elementary financial models tend to be the exception rather than the rule. The reason for this, one might argue, is that the underlying financial scenarios themselves are artificial creations, specifically designed by humans for both their usefulness and simplicity of calculation. They are *unnatural* and completely controlled situations in which no outside or unknown forces are allowed to influence the outcomes. These financial models are not just approximations to the real-world processes they represent, but more than that, they describe those processes *exactly*. It should therefore come as no surprise that formulas for exact solutions in these cases are indispensable.

However, the characteristics just described for these elementary financial scenarios are not shared with processes that arise in most other fields of endeavor. When constructing models in the natural and social sciences, including even financial models when risky investments such as stocks, options, etc. are involved, attempts are seldom made to build in all the factors that might influence the processes being investigated. As described in the previous section, even if all these factors were known, which is rare, their inclusion would likely complicate a model well beyond what is possible to represent mathematically or to even comprehend.

This is certainly the case for the various population models we will be considering in the following chapters. When applied to any but the most elementary forms of life, such models must intentionally disregard a whole host of specific characteristics of a population that might influence its growth. Rather than trying to incorporate all these numerous and often unknown features within the model, only the most important can be included. Attempting to do more would only complicate the model and perhaps render it useless as a tool for making meaningful predictions.

Economic models, too, must in general ignore a vast array of social and psychological factors that might influence such quantities as the price, supply and demand for a particular commodity. Just as there exist many unknown biological and ecological forces that affect the growth of a population, a large number of socio-economic forces have yet to be uncovered before reliable economic prediction can be made.

A similar situation exists even within the physical sciences, which have historically been the most successful within the natural and social sciences at modeling our world. Although many of the laws of the physical universe are known with great precision, which has led to the development of much of our modern industrial and technological society, a number of areas within the physical sciences still are not so well understood. For example, meteorology and weather prediction, as well as fluid dynamics, still await simple and accurate modeling techniques.

In short, most models from the natural and social sciences don't even attempt to describe our world exactly. Rather, they approximate reality as best they can. So, expending an inordinate amount of time and effort to find a formula for the exact quantitative solution of a problem that is itself only an approximation to reality, assuming such a formula could ever be found, would be somewhat pointless. This is especially true if there exists an alternate and more immediate way of drawing useful conclusions from the model. The trick is to realize that, even though the approximate model may differ quantitatively from the true process being modeled, the two may nevertheless share many of the same dynamics. Instead of an exact solution, therefore, it is this qualitative behavior that one should often go in search of.

Asymptotic Behavior

One of the most common ways of determining the dynamics of a model, oftentimes even when a formula for its exact solution is not available, is through the analysis of its long-term or **asymptotic behavior.** However, what is meant by *long-term behavior* is not necessarily as it may sound. Ordinarily, one might interpret this phrase as referring to that which occurs after a lengthy period of time as measured on a human time scale, i.e., after perhaps years, decades or centuries. *So why,* one might wonder, *should we be concerned with what happens so far in the future when it's the more immediate behavior that we're more likely to be interested in knowing?*

What **long-term behavior** actually refers to in an applied setting, however, does not necessarily have to occur after that long of a period of time. For example, when an ordinary incandescent light bulb is switched on, it does not immediately glow at full strength. Rather, there is a steady increase in brightness over a fraction of a second until the bulb's total luminosity is reached. The long-term behavior of a light bulb is to glow at full strength, but this is achieved in less than a second!

In general, long-term or asymptotic behavior refers to what one ultimately observes after perhaps a brief **transient** period. The actual length of that transient period depends greatly upon the application. In some cases, like the light bulb, it is quite short, and the asymptotic behavior is consequently very important since those are the dynamics that one observes most of the time. In other applications, such as the population dynamics of a highly developed species, the transient period may indeed be decades or centuries. But even so, asymptotic behavior may still be worth determining in these cases, as it often provides an explanation of the dynamics one is currently observing, and thereby allows long-term prediction to potentially be made.

Consequently, much of our effort after Chapter 2 will involve investigating the asymptotic dynamics of the discrete models we construct. The types of mathematical tools we will make use of in that endeavor may include elementary algebra and graphing, differential calculus, matrix algebra, complex numbers and partial derivatives. These will be accompanied, of course, by the regular dependence upon computer-generated solutions and their graphs, which often indicate the range of dynamics possible for a model and suggest the direction that our analysis should take.

Exercises 1.2

For which of the quantities described in Exercises 1–11 would ongoing evaluation be highly recommended, and for which would it likely suffice to have knowledge of just the long-term, asymptotic behavior?

1. The angular velocity of the blades of a fan after it is switched on.

2. The position of a tornado that has just formed.

3. The price of a certain volatile stock just purchased by a day-trader.

4. The price of a certain blue-chip stock just purchased for someone's retirement portfolio.

5. The number of unemployed workers after interest rates are lowered.

6. Someone's muscle tone and strength after starting an exercise program.

7. The temperature of a sick child who has just developed a fever.

8. The temperature of water in a tea kettle after being placed on a hot stove.

9. The level of toxins present in a long-polluted commercial harbor after a cleaning process is begun.

10. The level of toxins present in the drinking water after a water-main break.

11. The compression strength of concrete after it is poured.

12. Suppose $1000 is deposited into an account earning simple interest at an annual rate of 3%. (a) Construct the corresponding Simple Interest Model. (b) How much will be in the account after 4 years?

13. Suppose a deposit is made into an account earning simple interest and after 2 years the balance is $27,600. (a) If the annual interest rate is 4%, how much was the original deposit? (b) If the original deposit was $25,000, what is the annual interest rate?

14. Unlike most formulas for exact solutions of discrete processes, the Simple Interest Formula $P_n = (1 + rn)P_0$ can be used for continuous-time or non-integer values of n. If $r = 7\%$ and $P_0 = \$5000$, compute each of the following: (a) The balance after $2\frac{1}{2}$ years. (b) The balance after 9 months.

15. For continuous-time n, as described in the previous exercise, how long will it take an investment to double in value if it earns 8% simple interest per year?

16. Suppose an account earns compound interest at an annual rate of 6.6%. Determine the interest rate i if compounding is done: (a) monthly; (b) quarterly; (c) bi-weekly.

17. Suppose $8000 is deposited into an account that earns 6% annual interest compounded monthly. (a) Construct the corresponding Compound Interest Model. (b) Compute the account balance after each of the first 3 months.

18. Suppose a deposit is made into an account that earns interest that is compounded quarterly, and the account balance after 3 months is $5481. (a) If the original deposit was $5416, what is the annual interest rate? (b) If the annual interest rate is 6%, what was the original deposit?

19. In the Bouncing Ball Model $S_{n+1} = r S_n$ constructed in the previous section: (a) If $0 \le r < 1$, determine the asymptotic behavior of the height S_n as $n \to \infty$, for any $S_0 \ge 0$. (b) If $r = 1$, how would you describe the asymptotic behavior of S_n as $n \to \infty$, for any $S_0 \ge 0$?

20. Under some circumstances the interest rate i in the Compound Interest Model can be negative. Although this cannot happen for a savings account, it can with riskier investments such as stocks in a declining market. (a) If $i = -10\%$ and $P_0 = \$1000$, compute P_1, P_2 and P_3. (b) For any i satisfying $-1 \le i < 0$, determine the asymptotic behavior of P_n as $n \to \infty$, for any $P_0 \ge 0$.

21. Equation (5) of Section 1.1 states that the velocity $v(t)$ of a falling object that is subject to air resistance is given by $v(t) = \frac{mg}{k}(e^{-kt/m} - 1)$, where $m, g, k > 0$. (a) Find the asymptotic behavior of any such $v(t)$ by computing $\lim_{t \to \infty} v(t)$. (b) For some skydiver, if $m = 5.5$, $k = 1.28$ and $g = 32$, determine the *terminal velocity*, i.e., the fastest that the skydiver ever falls.

2

Linear Equations and Models

In this chapter we undertake a detailed investigation of the most elementary type of discrete dynamical system: linear iterative equations of one variable, which are among the few for which it is always possible to derive formulas for exact solutions. We first introduce some of the classic theory of linear iteration, along with corresponding applications from the natural and social sciences. Then, toward the end of the chapter, we make a transition to the more contemporary dynamical systems approach to the analysis of these iterative processes, which will be needed in the investigations of nonlinear equations and systems of equations in the chapters that follow.

SECTION

2.1   Some Linear Models

Due to their simplicity, linear iterative equations are often the first type of dynamical system considered when modeling scientific and natural processes. The Bouncing Ball Model and the models of simple and compound interest introduced in the previous chapter are all based upon this type of iteration scheme.

Each of these models is referred to as **linear** since its corresponding equation expresses a linear relationship between the subscripted variables under consideration. For example, in the Bouncing Ball Model $S_{n+1} = r S_n$, if we replace S_n by x, and S_{n+1} by y, then

$$S_{n+1} = r S_n \quad \text{becomes} \quad y = rx$$

Since r is a constant, or **parameter,** that does not depend upon $x = S_n$ or $y = S_{n+1}$, this means that $y = rx$ is a linear equation involving the two variables x and y only, or equivalently, y is a linear function of x. The model is therefore classified as linear. Similarly, in the Simple Interest Model $P_{n+1} = P_n + I$, if we replace P_n by x, and P_{n+1} by y, then

$$P_{n+1} = P_n + I \quad \text{becomes} \quad y = x + I$$

Since $I = r P_0$ is a parameter representing the constant amount of interest each year, then $y = x + I$ also says that y is a linear function of x, and so the model is again linear.

The reader may wish to check that the Compound Interest Model $P_{n+1} = (1 + i)P_n$ with parameter i is also linear. As we shall see later, recognizing and classifying an iterative model this way is quite important for its analysis.

The following models offer more complex examples of linear iteration than we have seen so far. Although their solutions will be taken up later, introducing them at this stage, it is hoped, will provide both a direction to and some additional motivation for the detailed investigation we will be undertaking in this chapter.

Financial Models

Besides linearity, the two investment models we have described thus far, simple and compound interest, share another characteristic. They both involve making just a single deposit P_0, and so all future growth of the investment is a result of interest alone. But this is not the only type of investment scenario possible. It is quite common for an investor to contribute regularly to an account already earning compound interest, creating what is called an **ordinary annuity.** For example, many retirement savings plans, such as the well known 401(k) plans, work this way. In such a situation, the account balance is likely to increase much more quickly since growth now emanates from two sources: compound interest as well as additional deposits.

To develop a model for the future value of this type of annuity, we begin with the same assumptions that led to the previous compound interest model and incorporate one additional feature: Suppose that the fixed amount d is deposited into the account at the end of each term. To arrive at an iterative equation for the next account balance P_{n+1} in terms of the present one P_n, this now means that we must take $(1 + i)P_n$ from the previous compound interest equation and add the deposit d to it. This leads to the following model.

ANNUITY SAVINGS MODEL

The equation

$$P_{n+1} = (1+i)P_n + d$$

where $d > 0$ models the growth of an investment earning compound interest at a rate of i per term, with an additional deposit of d at the end of each term. P_0 is the original principal, and each subsequent P_n for $n = 1, 2, 3, \ldots$ represents the total value of the investment at the end of n terms.

The reader may verify that this savings model is again linear with parameters i and d.

A small modification to this equation leads to another that is equally important in the financial industry. Suppose that, instead of making a deposit into the account at the end of each term, a withdrawal is made. This is mathematically equivalent to the situation in which money is borrowed with the intent of paying back the entire principal along with any interest owed, by making a regular series of equal payments to the lender. In this case the lender is acting as the investor, who deposits a principal with the borrower and then makes a withdrawal from this account, the actual loan payment, at the end of each term. We generally assume that each loan payment is large enough to cover all the interest that has accrued during the previous term, plus at least some of the principal. Thus the unpaid balance, or principal that is still owed, will decrease with each payment.

To model this type of loan process, suppose the amount P_0 is borrowed at the fixed interest rate of i per term. If this money is held by the borrower for one term, then $(1+i)P_0$ is the total amount of principal plus interest owed to the lender. But if the borrower then pays the amount d to the lender, then this amount becomes

$$P_1 = (1+i)P_0 - d$$

which is the new unpaid balance or principal after one term. Similarly, after the next several terms these balances will be

$$P_2 = (1+i)P_1 - d, \quad P_3 = (1+i)P_2 - d, \quad P_4 = (1+i)P_3 - d, \quad \ldots$$

which gives rise to the following linear model.

LOAN PAYMENT MODEL

The equation

$$P_{n+1} = (1+i)P_n - d$$

where $d > 0$ models the unpaid balance on a loan with an interest rate of i per term and a payment of d at the end of each term. P_0 is the original amount borrowed, and each subsequent P_n for $n = 1, 2, 3, \ldots$ represents the unpaid balance at the end of n terms.

One may note the similarity between the linear equations corresponding to these last two financial models, with the only difference being the sign preceding the constant d.

Population Growth Models

Another field in which linear iterative processes play a key role is the study of population dynamics. Suppose ecologists wish to model the population growth of some living species displaying a certain growth characteristic common to many varieties of bacteria and viruses, and often times even to plant, insect and animal populations. Namely, the population size multiplies at a constant rate. That is, if we simplify the growth process by viewing the population as changing suddenly and at fixed time intervals only, each of which we may think of as one **generation,** then the population size of the next generation may be obtained by multiplying the present population size by a certain fixed constant.

To see how such a growth process may occur, suppose we make the following assumptions: First, the number of *births* during any generation is proportional to that generation's population size. In the case of bacteria and viruses undergoing cell division, this is rather common, but it may also be true for more highly developed species. We additionally assume that the number of *deaths* during each generation is also proportional to the population size — again a reasonable assumption for most species. We call the constant of proportionality in the former case b, the **birth rate,** and in the latter d, the **death rate.** To have a meaningful population model, it is clear that both b and d must have non-negative values. Common sense might tell us that d must also satisfy $d \leq 1$, since no more than 100% of a population can die.

A model of this ecological process can now be developed as follows: If we call P_n the population of the nth generation, then the number of births during that generation will be bP_n and the number of deaths dP_n. The population of the next generation P_{n+1} may then be determined by taking P_n, adding the number of births and subtracting the number of deaths. That is,

$$P_{n+1} = P_n + bP_n - dP_n = (1 + b - d)P_n$$

Oftentimes for the sake of simplicity, we absorb the b and d into the single parameter $r = 1 + b - d$, called the **growth rate,** and write

$$P_{n+1} = rP_n$$

Assuming that the initial population P_0 is known, we now have a means of predicting the population of all future generations, presuming of course that the growth rate remains constant from one generation to the next. These ideas are summarized as follows.

LINEAR POPULATION MODEL

The equation

$$P_{n+1} = rP_n$$

models the population growth of a species whose growth rate r is a non-negative constant. This growth rate may be computed as $r = 1 + b - d$, where b is the birth rate and d the death rate, which satisfy $b \geq 0$ and $0 \leq d \leq 1$.

A principle that is inherently assumed in this model is that the group whose population we are investigating is somewhat isolated from outside influences. As a result, the population of any generation depends only on the size of the one before. But, one might argue, this is seldom the case in the natural world. For example, members of the same species as that of the population under consideration, but who originated elsewhere, may regularly join the group being modeled, thereby contributing to an increase in its size. Alternatively, a fixed number of the group may be regularly removed, either by migrating permanently to another location or by being harvested, presumably by humans. Either of these results in a decrease in the population size.

The former case is referred to as **immigration,** and the latter as **migration** or **harvesting.** Each of these processes can be modeled by making a small modification to the simple population model above. If we assume that either immigration, migration or harvesting occurs at a level that remains constant for all generations, then we need to alter the equation $P_{n+1} = r P_n$ by merely adding or subtracting a constant. For immigration we add a positive constant k, representing the number joining the group at each step. For migration or harvesting, we subtract the constant k equal to the number who leave the group or are harvested at each step, or equivalently, we add the negative of that value. This yields the following model.

LINEAR IMMIGRATION/MIGRATION/ HARVESTING MODEL

The equation

$$P_{n+1} = r P_n + k$$

models the growth of a population with growth rate r, that is undergoing either immigration, migration or harvesting at the constant level k. For immigration $k > 0$, and for migration or harvesting $k < 0$.

The reader may again verify that, regardless of the sign of the parameter k, this equation is linear.

An Economic Model

One field in which mathematical modeling has always played a prominent role is economics, especially micro-economics. Linear iterative equations in particular provide a simple means of modeling some well known marketplace behavior. Suppose, for example, we wish to investigate the price dynamics of a certain commodity. If P_n represents the price at any given time step n, certain economic *forces* may cause this price to change. One such force is consumer demand. It is commonly accepted that the price will be pushed upward as the demand for that product increases, and downward as the demand decreases. That is, the next price P_{n+1} will increase or decrease from the present price P_n according to an equation such as

$$P_{n+1} = P_n + Demand\ Force$$

(Here, for simplicity we assume the supply remains constant.) Although this *Demand Force* can never really be known, we can arrive at a simple mathematical approximation to it by observing that: when the current demand D_n is large, the *Demand Force* acting on the price should be positive, which causes an increase in the price, and when this demand is small, the *Demand Force* should be negative, causing the price to decrease. This idea can be conveniently expressed using a linear equation of the form

$$Demand\ Force = a_1\ D_n - b_1$$

where a_1 and b_1 are non-negative constants.

Our price model may now be written

$$P_{n+1} = P_n + a_1\ D_n - b_1$$

Unfortunately, however, there is an extra variable D_n in the model. To eliminate D_n from this equation we make use of another well known economic principle: As the price of a product increases, the demand from consumers generally decreases. In other words, the demand is a decreasing (or at least non-increasing) function of the price. If we again use a linear equation to express this idea while assuming that the demand D_n reacts immediately to the price P_n, then we might say that

$$D_n = b_2 - a_2\ P_n$$

where a_2 and b_2 are non-negative constants.

Substituting this value of D_n into our price model yields

$$P_{n+1} = P_n + a_1\ (b_2 - a_2\ P_n) - b_1 = (1 - a_1 a_2)P_n + a_1 b_2 - b_1.$$

If for simplicity we call the parameters

$$a = 1 - a_1 a_2 \quad \text{and} \quad b = a_1 b_2 - b_1$$

then we arrive at the final version of our model.

LINEAR PRICE MODEL

The equation

$$P_{n+1} = a P_n + b$$

with $a \leq 1$ and any value b, models the price of a product at time n.

It may be worth pointing out that, although the iterative equation corresponding to this price model resembles several that have preceded it, such as those associated with the Annuity Savings Model and the Linear Immigration/Migration/Harvesting Model, there is one significant difference. While common sense would tell us that the constant representing the coefficient of P_n in each of those previous models ($1 + i$ and r, respectively) must logically be non-negative, this is not necessarily so in the Linear Price Model. The condition $a \leq 1$ means that the parameter a may take on negative values. As we shall see later, this implies that a somewhat different set of dynamics is possible for this model.

Other Models

There are, of course, many other models involving linear iteration that might be introduced here. For example, the oscillations of a spring or pendulum; the vibrations of a string or membrane held taut at the ends; the repeated filtering of an isolated body of water or air mass that was initially or is regularly being polluted; supply or demand levels for a certain product; as well as the dynamics associated with what are known as two-state Markov processes, which have application in any number of fields. The iterative equations corresponding to these models, however, share linear forms similar to those already discussed, and thus little would be gained from their introduction at this point. The reader is encouraged to develop some of these models in several homework exercises at the end of this section. Others, Markov processes in particular, will be investigated in later sections of this chapter.

Exercises 2.1

1. For each of the following financial models compute $P_1, \ldots, P_5$:

   (a) $P_{n+1} = 1.1P_n + 1000,\ P_0 = 0$ (b) $P_{n+1} = 1.01P_n - 1000,\ P_0 = 100{,}000$

2. For each of the following population models compute $P_1, \ldots, P_5$:

   (a) $P_{n+1} = 1.2P_n,\ P_0 = 1000$ (b) $P_{n+1} = 0.5P_n + 200,\ P_0 = 2800$

3. For each of the following price models compute $P_1, \ldots, P_5$:

   (a) $P_{n+1} = 65 - 2P_n,\ P_0 = 22$ (b) $P_{n+1} = 8000 - 0.9P_n,\ P_0 = 4000$

4. For the price model $P_{n+1} = 100 - P_n$: (a) If $P_0 = 40$ compute $P_1, \ldots, P_5$. (b) If $P_0 = 75$ compute $P_1, \ldots, P_5$. (c) Can you predict what will happen for any other value of P_0?

5. Construct the Annuity Savings Model that satisfies: (a) The initial balance is $800, and each month the account gets 1% interest and $150 is deposited. (b) The initial balance is $470, the account gets 6% annual interest compounded monthly and a deposit of $225 is made each month.

6. Construct the Annuity Savings Model that satisfies: (a) $i = 1\%$, $P_0 = \$1000$, $P_1 = \$1125$; (b) $P_0 = \$10{,}000$, $P_1 = \$12{,}400$, $P_2 = \$14{,}836$.

7. (a) In the Annuity Savings Model with $P_0 = \$200$ and $i = 1\%$, what deposit d will make $P_1 = 2P_0$? (b) In the Annuity Savings Model with $d = \$400$ and $P_0 = \$500$, what interest rate i will make $P_1 = 2P_0$?

8. Construct the Loan Payment Model that satisfies: (a) Payments of $450 are made quarterly on an original loan of $7800 with an interest rate of 2% per quarter. (b) $6500 is borrowed at an annual rate of 8% with payments of $300 per month.

9. Construct the Loan Payment Model that satisfies: (a) $d = \$75$, $P_0 = \$500$, $P_1 = \$450$; (b) $P_0 = \$10{,}000$, $P_1 = \$9800$, $P_2 = \$9594$.

10. (a) In the Loan Payment Model with $P_0 = \$25{,}000$ and $i = \frac{1}{2}\%$, what payment d will make all $P_n = P_0$? (b) In the Loan Payment Model with $P_0 = \$500$ and $d = \$275$, what interest rate i will make $P_1 = \frac{1}{2}P_0$?

11. Construct the Linear Population Model that satisfies: (a) The population is initially 75,000 and is increasing by 15% per generation. (b) The initial population is 125,000, and each generation the birth rate is 7% and the death rate is 5%.

12. Construct the Linear Population Model that satisfies: (a) The initial population is 3500 and the ratio of the next population to the present one is 6/5. (b) The initial population is 3500 and the population doubles every 3 generations.

13. Construct the Linear Immigration Model that satisfies: (a) The initial population is 5400, the growth rate is 1.07 per generation and immigration is occurring at a constant rate of 175 per generation. (b) $P_0 = 7500$, $P_1 = 8000$, $P_2 = 8200$.

14. Construct the Linear Harvesting Model that satisfies: (a) The population, initially 100,000, is decreasing by 5% per generation and harvesting is occurring at a constant rate of 1500 per generation. (b) $P_0 = 75,000$, $P_1 = 77,000$, $P_2 = 79,500$.

15. Suppose a population, initially 275,000, has a birth rate of 8% and a death rate of 6.5% per generation. Construct the model that also satisfies: (a) Immigration occurs at 3500 per generation. (b) Migration occurs at 6300 per generation.

16. (a) For the Linear Price Model $P_{n+1} = 18 - 0.8P_n$, find the largest value of P_0 for which $P_n \geq 0$ for all $n \geq 0$. (b) For the Linear Price Model $P_{n+1} = 18 - 1.5P_n$, is there a value of P_0 for which $P_n \geq 0$ for all $n \geq 0$? If so, find one.

17. Suppose that a contaminated body of water is being cleaned by a filtering process. Each week this process is capable of filtering out a certain fraction a of all the pollutants present at that time. Here, a is a constant that satisfies $0 < a < 1$. (a) If T_0 tons of pollutants are initially in that body of water, write a general iterative equation for T_n, the amount of pollutants present n weeks later. (b) If each week 5% of all pollutants present can be removed, find the value of a and write the iterative equation for this process.

18. Suppose that a contaminated body of water is being cleaned by a filtering process. Each week this process is capable of filtering out a certain fraction a of all the pollutants present at that time, but another b tons of pollutants seep in. Here, a and b are constants that satisfy $0 < a < 1$ and $b > 0$. (a) If T_0 tons of pollutants are initially in that body of water, write a general iterative equation for T_n, the amount of pollutants present n weeks later. (b) If each week 10% of all pollutants present can be removed but another 2 tons seep in, find the values of a and b and write the iterative equation for this process.

19. Suppose that after a warm object is placed in a cooler room, the temperature difference between the object and the room is always a certain fraction a of that temperature difference a minute earlier. (This is a consequence of Newton's Law of Cooling.) Here, a is a constant that satisfies $0 < a < 1$. (a) If D_0 is the initial temperature difference, write a general iterative equation for D_n, the temperature difference n minutes later. (b) If $D_0 = 50°$ F and $D_1 = 48°$ F, find the value of a and write the iterative equation for D_n.

20. In the previous exercise assume that the temperature of the room is R and the initial temperature of the object is T_0. (a) Write a general iterative equation for the temperature T_n of the object n minutes after being placed in the room. (b) If $R = 70°$ F, $T_0 = 100°$ F and $T_1 = 95°$ F, write the iterative equation for T_n.

21. Suppose that when a pendulum is set in motion, it swings in such a way that the greatest positive (or negative) angle it makes on one side of the vertical is always a certain negative fraction $-a$ of the greatest negative (or positive) angle it previously made on the other side of the vertical. Here, a is a constant that satisfies $0 < a < 1$. (a) If Θ_0 is the initial angle of the pendulum when released and Θ_n is the nth greatest angle (positive or negative), write a general iterative equation for this process. (b) If $\Theta_0 = 10°$ and $\Theta_2 = 8°$, find the value of $-a$ and write the iterative equation for this process.

2.2   Linear Equations and Their Solutions

Although a number of important linear iterative models were introduced in the preceding section, relatively little was mentioned with regard to their solutions. In this and the next several sections we attempt to remedy that situation by introducing a variety of techniques commonly used to solve linear equations. We begin with a classification scheme that has important consequences for the solution process.

Autonomous and Non-Autonomous Equations

The discrete models introduced in the previous section are not only all linear, but they also share another property common to many dynamical systems. Namely, the variable representing time, or n in each of those previous models, does not appear *explicitly* in the equation. For example, in the Bouncing Ball Model $S_{n+1} = r S_n$, although the variable n is certainly used in each of the subscripts of the variable that measures the height of the ball, it does not appear anywhere else in the iterative equation. Another way to see it, using the technique described earlier for checking whether the equation is linear, if S_n is replaced with x, and S_{n+1} with y, then the variable n no longer appears at all in the resulting linear equation $y = rx$.

When this occurs, as it does for all the discrete iterative models introduced so far, a dynamical system is referred to as *autonomous*. Otherwise, it is called *non-autonomous*. It may already be apparent that most of our efforts in this chapter will be concentrated on those linear models that are autonomous. This prompts us to formulate a general description of the type of problem we are to consider most often:

$$x_{n+1} = a\, x_n + b \tag{1}$$

where a and b are any real constants.

It is easy to check that all the models we have been discussing are of this form. For example, letting

$$a = 1, \quad b = I \quad \text{and} \quad x_n = P_n, \quad \text{which of course implies} \quad x_{n+1} = P_{n+1}$$

yields the Simple Interest Model $P_{n+1} = P_n + I$. Alternately, with

$$a = r, \quad b = 0 \quad \text{and again} \quad x_n = P_n \quad \text{and} \quad x_{n+1} = P_{n+1}$$

we have the Linear Population Model $P_{n+1} = r P_n$. The reader may wish to verify that the other models we have considered are also of the form $x_{n+1} = ax_n + b$.

Although most of the common applications from the natural and social sciences are modeled by autonomous equations, in many circumstances non-autonomous ones are needed. The following examples demonstrate some of these.

<hr>

EXAMPLE 1

Suppose someone borrows $1000 at an interest rate of 12% per year paid monthly. Give the iterative equation that models the unpaid balance on the loan if: (a) $50 is paid back at the end of each month; (b) the borrower pays back nothing at the end of the first month, $10 at the end of the second month, \$20 at the end of the third month, \$30 at the end of the fourth month, etc.

Solution: (a) Since the loan is to be paid back monthly, we first need the monthly interest rate $i = 12\%/12 = 1\%$. We then use the Loan Payment Model from the previous section with $i = 0.01$ and $d = 50$, which gives $P_{n+1} = 1.01 P_n - 50$, where $P_0 = 1000$.

   (b) Although we still have $i = 0.01$ and $P_0 = 1000$, in this case we need to change the constant payment d to one that depends on n. When $n = 0$ the payment is \$0, and this increases by \$10 each month, which means the unpaid balances satisfy

$$P_1 = 1.01 P_0, \quad P_2 = 1.01 P_1 - 10, \quad P_3 = 1.01 P_2 - 20, \quad P_4 = 1.01 P_3 - 30, \quad \ldots$$

It is easily seen that the payment made at the end of the nth month is $10n$ dollars. This implies that $P_{n+1} = 1.01 P_n - 10n$ is the general form of the model.   ■

   Although the model from part (a) of this example is autonomous, the one from part (b) is not. This is due to the fact that n appears explicitly on the right-hand side of the equation

$$P_{n+1} = 1.01 P_n - 10n$$

Equivalently, the variable n still remains in the model after we write it in linear function form

$$y = 1.01x - 10n$$

In this case n appears in a term that is separated from the main variable x by a negative sign. But often in a non-autonomous equation these variables may be joined together in the same term, as the next example demonstrates.

EXAMPLE 2

Suppose a population is growing at a rate that is itself changing with each generation. In particular, suppose these growth rates are: $1, 1.1, 1.2, 1.3, \ldots$, respectively. Assuming that the initial population P_0 is known, give the iterative model for the growth of this population.

Solution: In this case we see that

$$P_1 = P_0, \quad P_2 = 1.1 P_1, \quad P_3 = 1.2 P_2, \quad P_4 = 1.3 P_3, \quad \ldots$$

In general this can be written $P_{n+1} = (1 + 0.1n) P_n$, which is the model for this process.   ■

   The population model from this last example is again non-autonomous, since its corresponding linear form is

$$y = (1 + 0.1n)x$$

The variable n appears this time in the coefficient of the primary variable x, rather than in a separate term.

   From these two examples one may be able to deduce what is true of all iterative equations for x_n that are linear but non-autonomous: Either the coefficient of x_n depends upon n, or a term separated from x_n by a plus or minus sign depends upon n, or both. To say that the coefficient of x_n depends upon n really means that it is a *function* of n, and so we could write this coefficient as $a(n)$. But for the purposes of an iterative equation, this function need only be defined on the non-negative integers $n = 0, 1, 2, \ldots$, which is the same domain for n that x_n has. For this reason, it is often preferable to write $a(n)$ as a_n, with the understanding that $a_0, a_1, a_2, \ldots$, are known real quantities. Similarly, the other term in the iterative equation that depends upon n may be called $b(n)$ or b_n.

All non-autonomous linear equations may therefore be written in the form

$$x_{n+1} = a_n x_n + b_n \tag{2}$$

It should be apparent that the special case of (2), in which all a_n are equal to the same constant a and all $b_n = b$ where b is also a constant, always yields an autonomous linear equation of the form of (1). This means that included within the form (2) are *all* linear equations, both autonomous and non-autonomous. Equation (2) therefore represents the most general form that a discrete linear equation can have. We formally state this as follows.

DEFINITION

The general form of a **linear equation** for x_n is given by (2), where all a_n and b_n are known real constants. When an initial value of x_0 is provided, a unique solution exists for all $n \geq 0$. If all $a_n = a$ and $b_n = b$ where a and b are constants, the equation is **autonomous.** Otherwise, it is **non-autonomous.**

EXAMPLE 3

Identify a_n and b_n for each of the following non-autonomous linear equations: (a) $x_{n+1} = \dfrac{3nx_n}{2n + 1}$; (b) $x_{n+1} = \dfrac{x_n - n}{2^n}$

Solution: (a) The equation $x_{n+1} = \dfrac{3nx_n}{2n + 1}$ may be written as $x_{n+1} = \dfrac{3n}{2n + 1}x_n$, and so it is immediately apparent that $a_n = \dfrac{3n}{2n + 1}$ and $b_n = 0$ for all n.

(b) We first separate the two terms on the right-hand side of the equation $x_{n+1} = \dfrac{x_n - n}{2^n}$ by writing it in the form $x_{n+1} = \dfrac{x_n}{2^n} - \dfrac{n}{2^n}$. From this it may be seen that $a_n = 1/2^n$ and $b_n = -n/2^n$. ∎

Computing Sums Iteratively

Besides the financial and population growth models introduced earlier, there are also many purely mathematical applications of non-autonomous linear equations that might be mentioned. Among the most important of these is the algorithm for computing the sum of a large collection of numbers: $c_1, c_2, c_3, \ldots, c_N$. Although the sum of these N values may be conveniently written

$$c_1 + c_2 + c_3 + \cdots + c_N = \sum_{k=1}^{N} c_k$$

the question arises: *How does one really compute this sum?* Or more commonly today: *How does one design an algorithm for calculating this sum that a computer could follow?*

Such questions can be answered by carefully observing what a person ordinarily does (often without realizing it) when computing such a sum. Usually, one begins with 0, and adds the first number c_1 to it, to obtain what might be called the first **partial sum.** Then, the next number c_2 is added to this first partial sum to obtain the second partial sum $c_1 + c_2$. This is followed by the addition of c_3, which yields the third partial sum $c_1 + c_2 + c_3$. If this process is continued, one will eventually arrive at the Nth sum, which will be equal to the entire sum $\sum_{k=1}^{N} c_k$.

With this in mind, it is possible to design an iterative process for computing this sum. Suppose we first let $S_0 = 0$, and then call the nth partial sum S_n for $1 \le n \le N$. That is, we let

$$S_1 = c_1 = S_0 + c_1$$
$$S_2 = c_1 + c_2 = S_1 + c_2$$
$$S_3 = c_1 + c_2 + c_3 = S_2 + c_3$$
$$\vdots$$
$$S_N = c_1 + c_2 + c_3 + \cdots + c_N = S_{N-1} + c_N$$

It is easy to see that in general these partial sums $S_n = \sum_{k=1}^{n} c_k$ satisfy

$$S_{n+1} = S_n + c_{n+1} \tag{3}$$

for $0 \le n \le N - 1$, which is a non-autonomous equation of the form (2).

Recognizing that the partial sums S_n satisfy a particular iterative equation provides several alternate means of computing the entire sum S_N. For example, one might store the values of $c_1, c_2, \ldots, c_N$ in one-dimensional array within a computer program, and then instruct the program to iterate (3) using the array values as the c_n. However, this would require actually typing the numerical values of all these c_n into the computer, which could be rather cumbersome if N is large. The same would of course be true if a spreadsheet program were used.

A better approach would be to try to describe the c_n using a general formula that could then be used by the program to automatically generate these values when (3) is iterated. This would mean that the actual numerical values of the c_n need never be typed into the computer at all, saving perhaps a great deal of time and energy. Unless the values of the c_n are random, a pattern for them can often be discerned and a simple formula developed, as the following examples show.

EXAMPLE 4

Construct an iterative equation for computing the sum

$$\frac{1}{2^0} + \frac{2}{2^1} + \frac{3}{2^2} + \frac{4}{2^3} + \frac{5}{2^4} + \cdots + \frac{25}{2^{24}}$$

Solution: If we first let $S_0 = 0$ and then define

$$S_1 = S_0 + \frac{1}{2^0}, \quad S_2 = S_1 + \frac{2}{2^1}, \quad S_3 = S_2 + \frac{3}{2^2}, \quad S_4 = S_3 + \frac{4}{2^3}, \quad \cdots$$

then it is not difficult to see that these partial sums all satisfy

$$S_{n+1} = S_n + \frac{n+1}{2^n}$$

Iterating this equation for $n = 0, \dots, 24$ will yield $S_{25} \approx 4.00$, which is the value of the entire sum. ∎

EXAMPLE 5

Construct an iterative equation for computing the sum

$$1 - \frac{1}{3} + \frac{1}{5} - \frac{1}{7} + \frac{1}{9} - \cdots + \frac{1}{1001}$$

Solution: Beginning again with $S_0 = 0$, this time we define

$$S_1 = S_0 + \frac{1}{1}, \quad S_2 = S_1 - \frac{1}{3}, \quad S_3 = S_2 + \frac{1}{5}, \quad S_4 = S_3 - \frac{1}{7}, \quad S_5 = S_4 + \frac{1}{9}, \quad \dots$$

To develop a general iterative equation for these partial sums, two observations must be made. First, the odd number in the denominator of the fraction that appears in each equation is always equal to one more than twice the subscript of the partial sum on the right-hand side of the equation. That is, the equations all have the form

$$S_{n+1} = S_n \pm \frac{1}{2n+1}$$

Next, the signs of the terms being added alternate between plus and minus, creating what is called an **alternating series.** A simple way to generate such a series is to multiply each term to be added (or subtracted) by $(-1)^n$. When n is odd, this makes the term negative, and when n is 0 or even, the term will be positive. This is precisely what we want here, so we write

$$S_{n+1} = S_n + \frac{(-1)^n}{2n+1}$$

This equation must be iterated for $n = 0, \dots, 500$, which will yield the value of the entire sum $S_{501} \approx 0.786$. ∎

Numerical Solution of Linear Equations

Distinguishing between autonomous and non-autonomous equations, as we have in this section, is not just an idle exercise. There is an important reason for making that distinction. Although the formula for the exact solution of any linear equation, whether autonomous or not, can always be written formally, solutions of non-autonomous problems nearly always involve sums and/or products that grow in size as n gets larger. Partial sums provide an example of this. In most cases the complexity of the formula for the exact solution of a non-autonomous equation renders it impractical as a means of actually evaluating the solution for large values of n.

To solve most non-autonomous problems and, as we shall later see, most autonomous ones when they are nonlinear, direct numerical iteration of the equation is often the best approach, provided of course a computer can be enlisted to actually carry out the calculations.

This is, after all, the reason why an iterative method was developed earlier for computing large sums.

Today, a variety of software is readily available that allows extensive iteration to be performed with ease. Several alternate means of accomplishing this are described in the Appendix of this book. Specific instructions and program segments are included there for those who wish to make use of either the popular spreadsheet program *Microsoft Excel,* or the more mathematically sophisticated *Mathematica.*

Regardless of what sort of software one prefers for generating numerical solutions of iterative equations, there are some specific issues that one must still decide upon, involving mostly how the resulting data is to be displayed. Sometimes, only a single iterate is needed, such as the future value of a financial investment at the end of a specific number of terms. Frequently, however, the solution x_n of an iterative model such as (1) or (2) is needed over an entire a range of steps, beginning with $n = 1$ (since x_0 was known in advance) and ending at perhaps $n = 20$, $n = 50$ or $n = 10{,}000$. In fact, this is usually the case when one is investigating such processes as population growth or price dynamics. To avoid what might prove to be temporary or transient anomalies in the data, and instead capture the true long-term or asymptotic behavior of a solution, a large number of iterates must be observed. Consequently, a table may be the most appropriate means of presenting this data.

EXAMPLE 6

Use a computer to generate the solution P_n of the population model from Example 2 with $P_0 = 500$. Display the results for $n = 0, \ldots, 20$ in a table.

Solution: The result of iterating $P_{n+1} = (1 + 0.1n)P_n$ starting with $P_0 = 500$ is shown in Table 2.1. Because of the population model interpretation of the equation, we have chosen

TABLE 2.1

n	P_n	n	P_n
0	500		
1	500	11	33522
2	550	12	70396
3	660	13	154872
4	858	14	356206
5	1201	15	854895
6	1802	16	2137237
7	2883	17	5556816
8	4901	18	15003403
9	8822	19	42009527
10	16761	20	121827629

to round off the entries in the table to the nearest whole numbers. From these first 20 iterates, one might (correctly) conclude that the population grows without bound. ∎

EXAMPLE 7

In the loan model of Example 1, which payment scenario, (a) or (b), will result in the loan being paid off sooner?

Solution: In this problem, we have no real way of knowing in advance how many iterations need to be performed in each case. So, one might begin by using a computer to iterate each a fixed number of times, and then look for the smallest value of n, if any, for which the unpaid balance satisfies $P_n \leq 0$. The loan is considered to be paid in full the first time P_n either equals 0 or has a negative value. If this does not occur within the range of iterates obtained, one should then, of course, increase the number of iterations until it does, or conclude that no such n exists. On the other hand, one might instead choose a more efficient approach, and design a computer program that will continue iterating until $P_n \leq 0$. An alternate means of terminating the program may also be needed, however, in case no such n exists.

Here we have chosen the former alternative. The result of iterating 24 times under each scenario, and rounding to 2 decimal places, is shown Table 2.2(a) and 2.2(b) respectively. We conclude that the loan will be paid off sooner under scenario (b), since P_n first becomes negative after 23 months under plan (a), but after only 16 months under (b). ∎

TABLE 2.2

n	P_n	n	P_n		n	P_n	n	P_n
0	1000.00				0	1000.00		
1	960.00	13	447.63		1	1010.00	13	328.77
2	919.60	14	402.10		2	1010.10	14	202.05
3	878.80	15	356.12		3	1000.20	15	64.07
4	837.59	16	309.69		4	980.20	16	−85.29
5	795.96	17	262.78		5	950.01	17	−246.14
6	753.92	18	215.41		6	909.51	18	−418.60
7	711.46	19	167.56		7	858.60	19	−602.79
8	668.57	20	119.24		8	797.19	20	−798.81
9	625.26	21	70.43		9	725.16	21	−1006.80
10	581.51	22	21.14		10	642.41	22	−1226.87
11	537.33	23	−28.65		11	548.83	23	−1459.14
12	492.70	24	−78.94		12	444.32	24	−1703.73

(a)   (b)

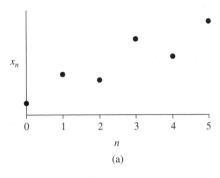
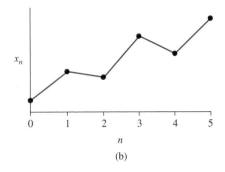

FIGURE 2.1

Time-Series Graphs

Oftentimes, much more can be inferred about a mathematical problem from a graph than from a table of numbers, especially if that table is somewhat lengthy. When one is more interested in the general *qualitative* behavior of a process being modeled, rather than the *quantitative* data it generates, graphing the solution may be preferable to tabulating its precise values.

For linear equations of the form (1) or (2), the only type of graph ordinarily needed is the **time-series graph.** This means that the points (n, x_n) are plotted in a two-dimensional coordinate system, using the horizontal axis for n and the vertical axis for the corresponding value of the solution x_n (see Fig. 2.1(a)). Since n often corresponds to time in an iterative problem, we are essentially plotting the solution as a function of time.

Because x_n is a function of n that is defined only over the discrete domain $n = 0, 1, 2, \ldots$, its time-series graph should always consist of just the discrete set of points (n, x_n) for $n = 0, 1, 2, \ldots$. However, one often chooses to connect consecutive points of the solution by straight line segments (see Fig. 2.1(b)). Not only does this help identify the location of these points, but also the relationship between them, and may ultimately contribute to the understanding of the overall qualitative behavior of solutions. Since that is perhaps the central goal in the field of dynamical systems, all future time-series graphs in this text will appear in this form.

EXAMPLE 8

Create a time-series graph for the first 10 data points (n, P_n) from Example 6.

Solution: The graph is shown in Figure 2.2. The rapidly increasing slopes of the line segments connecting consecutive points (n, P_n) are consistent with the premise that the growth rate of the population is increasing (or at least non-decreasing) with each step. ■

EXAMPLE 9

Suppose a population P_n changes from year to year according to the rule:

$$P_1 = 0.5P_0 + 300, \quad P_2 = 0.5P_1 - 100, \quad P_3 = 0.5P_2 + 300, \quad P_4 = 0.5P_3 - 100, \quad \ldots$$

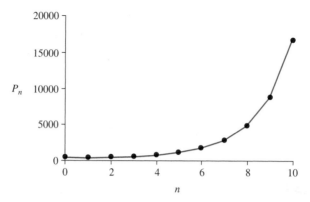

FIGURE 2.2

(a) Write this model in the form $P_{n+1} = a_n P_n + b_n$ by finding formulas for a_n and b_n for all $n \geq 0$. (b) Use a computer to perform 20 iterations of the resulting equation, starting with $P_0 = 1000$. Display P_n for $n = 0, \ldots, 20$ in a time-series graph.

Solution: (a) If written in the form $P_{n+1} = a_n P_n + b_n$, it is apparent that all $a_n = 0.5$, but the values of b_n alternate between -100 and $+300$:

$$b_0 = 300, \quad b_1 = -100, \quad b_2 = 300, \quad b_3 = -100, \quad \ldots$$

This alone, in fact, makes the model non-autonomous. To describe all these b_n using a single formula, we use a well known trick that says: To generate a sequence that alternates between two values, multiply half their difference by $(-1)^n$, and then either add this to or subtract this from their average, depending upon how the sequence begins. In this case this means we let

$$b_n = \frac{-100 + 300}{2} \pm \frac{300 - (-100)}{2}(-1)^n = 100 \pm 200(-1)^n$$

To decide whether the plus sign or the minus sign should be used here, we pick the one that makes the sequence correctly begin at $n = 0$ with $b_0 = 300$. Since the plus sign does this, we must have $b_n = 100 + 200(-1)^n$, which means the model may be written

$$P_{n+1} = 0.5P_n + 100 + 200(-1)^n$$

(b) One of the benefits of writing the model in the form of a single equation, as we just have, is that a computer can now be used to quickly iterate the problem for any prescribed number of steps. Figure 2.3 shows the result of doing so for $n = 0, \ldots, 20$ beginning with $P_0 = 1000$. This time-series graph seems to indicate that as n increases, the solution P_n eventually settles into a regular or **periodic** oscillation, alternating between two specific values. Upon closer inspection, these values are found to be approximately 67 and 333. ∎

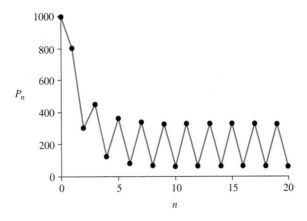

FIGURE 2.3

Exercises 2.2

1. Determine which of the following are linear:

   (a) $x_{n+1} = 3(2 - x_n) + 4$

   (b) $x_{n+1} = x_n + \dfrac{1}{x_n}$

   (c) $x_{n+1} = 5nx_n - n^2$

   (d) $x_{n+1} = n^2 - x_n^2$

2. Determine which of the following are autonomous:

   (a) $x_{n+1} = 4x_n - 3n$

   (b) $x_{n+1} + 6n = 2(x_n + 3n) - 4$

   (c) $(n + 1)x_{n+1} = nx_n + x_n - n - 1$

   (d) $x_{n+1} = x_n + (-1)^n$

3. Write each of the following non-autonomous equations in the form $x_{n+1} = a_n x_n + b_n$ and identify the discrete functions a_n and b_n.

   (a) $x_{n+1} = (5 - 2x_n)/n$

   (b) $x_{n+1} + n - 1 = nx_n - x_n$

   (c) $(n + 1)x_{n+1} = x_n + n + 1$

   (d) $3(x_{n+1} - x_n) = n$

4. Write each of the following autonomous equations in the form $x_{n+1} = ax_n + b$ and identify the constants a and b.

   (a) $x_{n+1} = 2(4 - x_n)/3$

   (b) $3x_{n+1} - 2x_n = 8$

   (c) $(x_{n+1} + x_n)/2 = 1$

   (d) $x_{n+1}/x_n = 0.9$

In Exercises 5–10 write each in the form $x_{n+1} = a_n x_n + b_n$, *and then compute and make a time-series plot of* $x_0, x_2, \ldots, x_5$ *using* $x_0 = 1$.

5. $x_1 = x_0$, $x_2 = 1.1x_1$, $x_3 = 1.2x_2$, $x_4 = 1.3x_3$, ...

6. $x_1 = x_0$, $x_2 = 2x_1 + 1$, $x_3 = 3x_2 + 2$, $x_4 = 4x_3 + 3$, ...

7. $x_1 = x_0 + 1$, $x_2 = x_1 - \frac{1}{2}$, $x_3 = x_2 + \frac{1}{4}$, $x_4 = x_3 - \frac{1}{8}$, ...

8. $x_1 = 1 - \dfrac{x_0}{2}$, $x_2 = 1 + \dfrac{x_1}{2}$, $x_3 = 1 - \dfrac{x_2}{2}$, $x_4 = 1 + \dfrac{x_3}{2}$, ...

9. $x_1 = 1 - x_0$, $x_2 = \frac{1}{2}x_1 - 2$, $x_3 = 4 - \frac{1}{4}x_2$, $x_4 = \frac{1}{8}x_3 - 8$, ...

10. $x_1 = 2x_0$, $x_2 = 1.5x_1 - 1$, $x_3 = 1.25x_2 + 2$, $x_4 = 1.125x_3 - 3$, ...

In Exercises 11–14 compute the partial sums $S_1, \ldots, S_5$.

11. $\displaystyle\sum_{k=1}^{100} k^2$ 12. $\displaystyle\sum_{k=1}^{10}(50 - 5k)$ 13. $\displaystyle\sum_{k=1}^{1000}\frac{k}{k+1}$ 14. $\displaystyle\sum_{k=1}^{100}\frac{(-1)^k}{2^k}$

In Exercises 15–22 construct an iterative method of computing the partial sums S_n *and identify which* N *would make* S_N *equal the entire sum.*

15. $\displaystyle\sum_{k=1}^{100}\frac{2k}{k^2+1}$ 16. $\displaystyle\sum_{k=1}^{10}\sqrt{k}$ 17. $\displaystyle\sum_{k=1}^{1000}\frac{(-1)^k}{2k-1}$ 18. $\displaystyle\sum_{k=1}^{100}\frac{-k}{2^k}$

19. $\sqrt{10} + \sqrt{30} + \sqrt{50} + \sqrt{70} + \sqrt{90} + \cdots + \sqrt{990}$

20. $1 + 4 + 9 + 16 + 25 + \cdots + 10{,}000$

21. $\dfrac{2}{1} - \dfrac{4}{3} + \dfrac{6}{5} - \dfrac{8}{7} + \dfrac{10}{9} - \cdots + \dfrac{100}{99}$

22. $\dfrac{1}{2} - \dfrac{2}{4} + \dfrac{3}{8} - \dfrac{4}{16} + \dfrac{5}{32} - \cdots + \dfrac{20}{2^{20}}$

23. Suppose someone opens an account that will get 4% annual interest compounded monthly with an initial deposit of $1000. Each month an additional deposit is made. Construct an iterative model for the account balance if the monthly deposits are:

    (a) $20, $25, $30, $35, $40, ...   (b) $1, $3, $9, $27, $81, ...

24. Suppose someone borrows $10,000 at an annual interest rate of 6% compounded monthly and makes monthly payments on the loan. Construct an iterative model for the monthly unpaid balance if the monthly payments are:

    (a) $1, $2, $4, $8, $16, ...   (b) $100, $200, $100, $200, $100, ...

25. Suppose a population is growing at a rate that is itself changing with time. Construct the iterative model for that population if the initial size is 100,000 and the growth rates of subsequent generations are:

    (a) 2, 1.5, 1.25, 1.125, ...   (b) 1, 1.5, 1.75, 1, 875, ...

26. Suppose a population is initially 5000, and is growing with an intrinsic growth rate of 1.02 per generation and also through immigration. Construct an iterative model for the population if the immigration levels of subsequent generations are:

    (a) 100, 120, 140, 160, ...   (b) 1000, 985, 970, 955, ...

27. Suppose a pond is initially contaminated with 100 tons of PCB's. Each month a filtering process removes 10% of the PCB's that are present, but additional PCB's seep into the pond. Find the iterative model for the monthly amount of PCB's in the pond if the monthly numbers of tons seeping into the pond are 2, 1.5, 1.25, 1.125, ...

▰▰▰▰ Computer Projects 2.2

In Projects 1 and 2 compute $x_1, \ldots, x_{50}$ *starting with* $x_0 = 0$, *and then draw a time-series graph of these iterates.*

1. The equation of Exercise 7.   2. The equation of Exercise 8.

In Projects 3–6 compute the value of each sum.

3. The sum from Exercise 11.
5. The sum from Exercise 20.

4. The sum from Exercise 17.
6. The sum from Exercise 22.

7. Use a computer to determine how long it will take to pay off the loan under each of the scenarios in Exercise 24.

8. Use a computer to determine how long it will take before the PCB level falls below 15 tons in Exercise 27.

2.3   Homogeneous Equations and Their Applications

Another classification scheme is used for linear equations that we have not yet mentioned. As a special case of the general linear equation

$$x_{n+1} = a_n x_n + b_n$$

formulated in the previous section, suppose all $b_n = 0$. This turns the problem into one of the form

$$x_{n+1} = a_n x_n \tag{1}$$

There is a name given to equations like this.

DEFINITION

If a linear equation can be written in the form (1), it is described as **homogeneous.** Otherwise it is **non-homogeneous.**

In this section, we consider homogeneous equations, their solutions and their numerous applications.

Solutions of Non-Autonomous, Homogeneous Equations

One reason for identifying a linear equation as homogeneous is that a formula for its exact solution will then be particularly easy to obtain. To find such a formula, we begin by writing out the first few steps of the iteration process. For $n = 0$, equation (1) says that

$$x_1 = a_0 x_0$$

For $n = 1$ this instead becomes

$$x_2 = a_1 x_1$$

Combining these two equations therefore gives us

$$x_2 = a_1 x_1 = a_1 (a_0 x_0)$$

and so x_2 may instead be written

$$x_2 = a_1 a_0 x_0$$

Next, equation (1) with $n = 2$ says that

$$x_3 = a_2 x_2$$

But because we have already found that $x_2 = a_1 a_0 x_0$, then

$$x_3 = a_2 x_2 = a_2(a_1 a_0 x_0)$$

which means that x_3 may be written

$$x_3 = a_2 a_1 a_0 x_0$$

From these observations, a general formula may be seen emerging. If the pattern that begins with

$$x_1 = a_0 x_0, \quad x_2 = a_1 a_0 x_0, \quad \text{and} \quad x_3 = a_2 a_1 a_0 x_0$$

continues (which it does), then we would expect in general that

$$x_n = a_{n-1} a_{n-2} \cdots a_0 x_0 \quad \text{or equivalently} \quad x_n = x_0 a_0 a_1 \cdots a_{n-1}$$

Since the product $a_0 a_1 \cdots a_{n-1}$ may be written more concisely as

$$a_0 a_1 \cdots a_{n-1} = \prod_{k=0}^{n-1} a_k$$

we state the following.

THEOREM *Non-Autonomous, Homogeneous Solution*

The general non-autonomous, homogeneous equation $x_{n+1} = a_n x_n$ has the exact solution

$$x_n = x_0 \prod_{k=0}^{n-1} a_k = x_0 a_0 a_1 \cdots a_{n-1} \text{ for all } n \geq 1. \qquad \blacksquare$$

EXAMPLE 1

Write a formula for the exact solution P_n of the population model from Example 2 of Section 2.2 if $P_0 = 500$.

Solution: In the previous section we found that the model in question may be written $P_{n+1} = (1 + 0.1n)P_n$. Since this non-autonomous equation is homogeneous, the above rule applies with $a_n = 1 + 0.1n$. So the solution may be written

$$P_n = P_0 \prod_{k=0}^{n-1} a_k = P_0 \prod_{k=0}^{n-1} (1 + 0.1n)$$

And since $P_0 = 500$, the exact solution for all $n \geq 0$ is

$$P_n = 500 \prod_{k=0}^{n-1} (1 + 0.1n) \qquad \blacksquare$$

The above rule and example demonstrate a point that was made in the previous section: Exact solutions of non-autonomous equations nearly always involve sums or products that grow in size as n gets larger. This is obviously the case for (1), whose solution, as we have just seen, is

$$x_n = x_0 \prod_{k=0}^{n-1} a_k = x_0 \, a_0 a_1 \cdots a_{n-1}$$

The number of calculations required to evaluate x_n using this formula is directly proportional to n. For example, five calculations are needed to evaluate x_5, but 100 are needed for x_{100}. In such cases using a computer to numerically iterate the equation is just as efficient as attempting to use the formula for the exact solution.

In some rare circumstances, however, the product involved in the solution of a homogeneous problem unexpectedly **telescopes,** dramatically simplifying the formula for the exact solution, and thereby making it quite useful. The following is an example of this.

EXAMPLE 2

Compute and, if possible, simplify the solution of $x_{n+1} = \left(1 + \dfrac{1}{n+1} \right) x_n$ with $x_0 = 1$.

Solution: Since this equation is homogeneous with $a_n = 1 + \dfrac{1}{n+1}$ and $x_0 = 1$, we have

$$x_n = x_0 \, a_0 a_1 \cdots a_{n-1} = 1 \cdot \left(1 + \frac{1}{1} \right) \cdot \left(1 + \frac{1}{2} \right) \cdot \left(1 + \frac{1}{3} \right) \cdot \ \cdots \ \cdot \left(1 + \frac{1}{n} \right)$$

for all $n \geq 1$. Although this is the solution, it can be simplified by first observing that

$$1 + \frac{1}{1} = \frac{2}{1}, \quad 1 + \frac{1}{2} = \frac{3}{2}, \quad 1 + \frac{1}{3} = \frac{4}{3}, \quad \cdots, \quad 1 + \frac{1}{n} = \frac{n+1}{n}$$

which means the solution may be written

$$x_n = \frac{2}{1} \cdot \frac{3}{2} \cdot \frac{4}{3} \cdot \ \cdots \ \cdot \frac{n+1}{n}$$

It is clear that the numerator in every term of this product will cancel with the denominator of the term that is immediately to the right — except, of course, for the right-most numerator $n + 1$. Since this is the only factor (besides 1) that remains, the solution simplifies to $x_n = n + 1$. ∎

Computing Products Iteratively

Unlike the product in the previous example, most do not cancel out completely, nor even simplify in any significant way. Computing most large products, like large sums, is usually a tedious task. But, as with sums, there is a way to perform the task iteratively.

According to the rule above, the solution of $P_{n+1} = a_n P_n$ when $P_0 = 1$ is

$$P_n = \prod_{k=0}^{n-1} a_k = a_0 a_1 \cdots a_{n-1}$$

for all $n \geq 1$. Knowing this provides a useful iterative approach to computing the product of any large collection of numbers $c_1, c_2, \ldots, c_N$. We simply let $a_n = c_{n+1}$ for $0 \leq n \leq N - 1$. This means the iterative equation

$$P_{n+1} = a_n P_n \quad \text{is equivalent to} \quad P_{n+1} = c_{n+1} P_n$$

and its solution

$$P_n = a_0 a_1 \cdots a_{n-1} \quad \text{is equivalent to} \quad P_n = c_1 c_2 c_3 \cdots c_n$$

Therefore, to compute the the entire product $P_N = c_1 c_2 c_3 \cdots c_N$ we need only iterate the equation for the **partial products** $P_n = c_1 c_2 c_3 \cdots c_n$, given by

$$P_{n+1} = c_{n+1} P_n \tag{2}$$

for $n = 0, \ldots, N - 1$, beginning with $P_0 = 1$. In the end this will yield the value of P_N. Since we have already seen that any linear equation can be easily iterated using a computer, we thus have a practical means of numerically evaluating large products.

EXAMPLE 3

Use an iterative equation and a computer to find the product

$$\frac{1}{2} \cdot \frac{3}{4} \cdot \frac{5}{6} \cdot \frac{7}{8} \cdot \quad \cdots \quad \cdot \frac{99}{100}$$

Solution: We can compute this product by first setting up an iterative equation for the partial products P_n in the form (2). This means we must find a general formula for the terms $c_1, c_2, c_3, \ldots$ to be multiplied. Since

$$c_1 = \frac{1}{2}, \quad c_2 = \frac{3}{4}, \quad c_3 = \frac{5}{6}, \quad \ldots, \quad c_{50} = \frac{49}{50}$$

then we see that each has the general form $c_n = \dfrac{2n - 1}{2n}$. To use (2) we need the value of c_{n+1}, which can be obtained by substituting $n + 1$ for n in the equation for c_n. That is,

$$\text{since} \quad c_n = \frac{2n - 1}{2n} \quad \text{then} \quad c_{n+1} = \frac{2(n + 1) - 1}{2(n + 1)} = \frac{2n + 1}{2n + 2}$$

The iterative equation for $P_n = c_1 c_2 \cdots c_n$ is therefore

$$P_{n+1} = \frac{2n + 1}{2n + 2} P_n$$

Using a computer to iterate this equation for $n = 0, \ldots, 49$ beginning with $P_0 = 1$ yields the value of the entire product $P_{50} \approx 0.079589$.   ■

EXAMPLE 4

Construct an iterative equation for computing $n!$ for any positive integer n, and use it to evaluate 12!.

Solution: We can create an iterative equation for computing $n! = 1 \cdot 2 \cdot 3 \cdot \cdots \cdot n$ by observing that, if we let $c_1 = 1, \; c_2 = 2, \; c_3 = 3, \ldots$, then the solution of (2) must be

$$P_n = c_1 c_2 c_3 \cdots c_n = 1 \cdot 2 \cdot 3 \cdot \cdots \cdot n = n!$$

Since this implies that $c_{n+1} = n + 1$, then (2) becomes

$$P_{n+1} = (n + 1)P_n$$

If we begin iterating using $P_0 = 1$, this will yield $P_n = n!$ for all $n \geq 1$. To compute 12! we iterate 12 times to obtain $P_{12} = 12! = 479,001,600$.   ■

Solutions of Autonomous, Homogeneous Equations

When a homogeneous equation is also autonomous, it must have the form

$$x_{n+1} = a \, x_n \tag{3}$$

where a is a constant. Since this may be viewed as a special case of (1) in which all $a_n = a$, its corresponding solution may be written

$$x_n = x_0 \prod_{k=0}^{n-1} a_k = x_0 \prod_{k=0}^{n-1} a = x_0 \, (a \cdot a \cdot \, \cdots \, \cdot a) = x_0 \, a^n$$

This gives us another formula.

THEOREM   *Autonomous, Homogeneous Solution*

The general autonomous, homogeneous equation $x_{n+1} = ax_n$ has the exact solution $x_n = x_0 a^n$ for all $n \geq 0$.   ■

Since (3) is the general form of many discrete models that arise in the natural and social sciences, including several we have been discussing, this rule is quite important. For example, it confirms that the Bouncing Ball Model

$$S_{n+1} = r \, S_n \quad \text{has the solution} \quad S_n = S_0 \, r^n$$

It similarly implies that the Linear Population Model

$$P_{n+1} = r \, P_n \quad \text{has the solution} \quad P_n = P_0 \, r^n$$

and that the Compound Interest Model

$$P_{n+1} = (1 + i) \, P_n \quad \text{has the solution} \quad P_n = P_0 \, (1 + i)^n$$

This last solution is known as the **Compound Interest Formula.**

EXAMPLE 5

Suppose a population, initially 8000, has a birth rate of 6% and a death rate of 2% per generation. How large will the population be after 10 generations?

Solution: Since $P_0 = 8000$ and $r = 1 + 0.06 - 0.02 = 1.04$, the population is governed by the equation $P_{n+1} = 1.04 P_n$, whose solution is $P_n = 8000(1.04)^n$. After 10 generations the population will therefore be $P_{10} = 8000(1.04)^{10} \approx 11,842$.   ■

EXAMPLE 6

When a population has a birth rate b that is lower than its death rate d, then that population will decrease over time and may eventually become extinct. For a population with $b = 0.1$, $d = 0.3$ and an initial size of 1000: (a) Find the equation for the population size P_n after n generations; (b) Determine how long it will take for the population to fall below 1, which we interpret as extinction.

Solution: (a) Since the growth rate is $r = 1 + 0.1 - 0.3 = 0.8$, the population must satisfy $P_{n+1} = 0.8 P_n$, whose solution is $P_n = P_0(0.8)^n$. Using $P_0 = 1000$, this becomes $P_n = 1000(0.8)^n$.

   (b) To determine when $P_n \leq 1$, we must solve

$$1000(0.8)^n \leq 1 \quad \text{or} \quad (0.8)^n \leq 0.001$$

Taking the natural logarithm of both sides gives

$$n \ln(0.8) \leq \ln(0.001), \quad \text{which implies} \quad n \geq \frac{\ln(0.001)}{\ln(0.8)} \approx 30.96$$

since $\ln(0.8)$ is negative. It will therefore take approximately 31 generations before extinction occurs.   ∎

EXAMPLE 7

If interest is compounded monthly, what annual rate will make an investment double in value in 8 years?

Solution: Since 8 years corresponds to 96 months, then to say that an original principal P_0 doubles in 8 years implies that $P_{96} = 2P_0$, where $P_n = P_0(1 + i)^n$ is the solution of the Compound Interest Model. The monthly interest rate i that will make P_0 double in 8 years must therefore satisfy

$$P_{96} = P_0(1 + i)^{96} = 2P_0 \quad \text{or} \quad (1 + i)^{96} = 2$$

Solving this for i yields

$$i = 2^{1/96} - 1 \approx 0.007246$$

To obtain the annual rate, we multiply this monthly rate i by 12 to obtain approximately $0.087 = 8.7\%$ per year.   ∎

Exponential Growth and Decay

The formula for solving equation (3) has certain implications with regard to how such a process must behave. Since the variable n appears only as an exponent on the right-hand side of $x_n = x_0 a^n$, this solution is called an **exponential function** of n. If $a > 0$, as it is for all of the autonomous, homogeneous models we have been discussing, this generally means that one of only two types of behaviors is possible. If $a > 1$ then a^n will increase by a factor of a with each step n, and consequently so too will x_n, as long as $x_0 > 0$. This type of growth is called **exponential growth.** On the other hand, if $0 < a < 1$, then solutions decrease by the factor a with each step. This is called **exponential decay.** Typical time-series graphs for the two cases are shown in Figures 2.4(a) and 2.4(b) respectively. For obvious reasons, we do not discuss the borderline case $a = 1$.

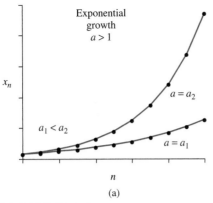
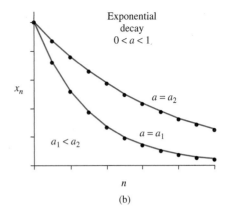

FIGURE 2.4

TABLE 2.3

n	P_n	n	P_n	n	P_n	n	P_n
0	10.00			0	10.00		
1	11.00	11	21.00	1	11.00	11	28.53
2	12.00	12	22.00	2	12.10	12	31.38
3	13.00	13	23.00	3	13.31	13	34.52
4	14.00	14	24.00	4	14.64	14	37.97
5	15.00	15	25.00	5	16.11	15	41.77
6	16.00	16	26.00	6	17.72	16	45.95
7	17.00	17	27.00	7	19.49	17	50.54
8	18.00	18	28.00	8	21.44	18	55.60
9	19.00	19	29.00	9	23.58	19	61.16
10	20.00	20	30.00	10	25.94	20	67.27

(a)   (b)

Exponential growth is about the *fastest* type of growth observed for natural and social processes (although it is not the fastest type of growth mathematically possible). For example, suppose we compare the yield over time when the principal $P_0 = \$10$ receives: *(i)* simple interest at a rate of 10% per year, versus *(ii)* interest at that same annual rate of 10%, but this time compounded annually as well. We saw earlier that the solution of the general Simple Interest Model is $P_n = P_0(1 + rn)$, and so the value of the investment after n years in case *(i)* is

$$P_n = P_0(1 + rn) = 10(1 + 0.10n) = n + 10$$

On the other hand, the solution of the general Compound Interest Model established in this section gives us a value of

$$P_n = P_0(1 + i)^n = 10(1 + 0.10)^n = 10(1.1)^n$$

after n years in case *(ii)*.

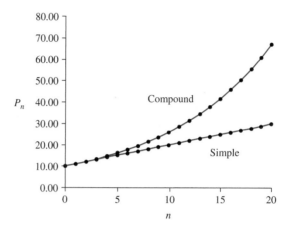

FIGURE 2.5

To see how these compare over time, suppose we compute the value of the investment under the two scenarios for each of the first 20 years. Table 2.3(a) shows the result of evaluating $P_n = n + 10$ for $n = 1, \ldots, 20$, and Table 2.3(b) the same for $P_n = 10(1.1)^n$. Their time-series graphs are shown together in Figure 2.5. Note how much faster the investment grows under the compound interest scenario *(ii)* rather than under the simple interest scenario *(i)*, even though the same interest rate is involved.

The simple interest investment P_n in case *(i)* undergoes **linear growth** since the equation $P_n = n + 10$ constitutes a linear function of n. Its graph is thus a straight line with a positive slope. But earning compound interest as in case *(ii)*, the investment $P_n = 10(1.1)^n$ grows exponentially. The graph here is not only increasing, but increasing at an increasing rate, or concave up. Over time, this will far surpass any linear growth function.

Not only does any exponential function of n increase *faster* than any linear function, but it also grows (it can be shown using limits) faster than any polynomial function, and most other functions involving only positive powers of n. Similarly, any exponential function a^n with $0 < a < 1$ decreases to 0 as n increases, faster than most functions consisting of only negative powers of n.

Due to an exponential function's very rapid rate of increase or decrease, a wide range of function values often results from a relatively narrow domain. This creates an awkward situation when graphing or otherwise dealing with the function. For example, try accurately plotting the points $(2, 10)$, $(4, 1000)$, $(6, 10^5)$ and $(8, 10^7)$ on the same graph.

For this reason, a **logarithmic scale** is sometimes used to help compress the range of function values that need to be considered in an application. Instead of dealing with the three points (x, y) given above, for example, suppose we compute the three new points $(x, \log_{10} y)$. This yields $(2, 1)$, $(4, 3)$, $(6, 5)$ and $(8, 7)$ respectively. These can be plotted on the same graph with ease. In fact, they now fall along the same straight line (see Fig. 2.6).

It is generally the case that, whenever dealing with measurements that vary widely, using a logarithmic scale compresses the data range. In the natural sciences, the Richter scale for measuring earthquake intensities, the pH scale for measuring acidity levels and the decibel scale for measuring sound levels are all commonly used logarithmic scales.

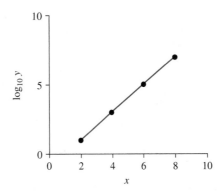

FIGURE 2.6

Exercises 2.3

For each of the products given in Exercises 1–6, if $P_0 = 1$ construct an iterative equation for computing the partial products P_n, identify which N would make P_N equal the entire product, and compute $P_1, \ldots, P_5$.

1. $\displaystyle\prod_{k=1}^{100} \frac{2k}{3k+4}$

2. $\displaystyle\prod_{k=1}^{10} \left(1 - \frac{1}{2^k}\right)$

3. $\displaystyle\prod_{k=1}^{20} \frac{(-1)^k}{5^k+1}$

4. $\dfrac{2}{5} \cdot \dfrac{4}{9} \cdot \dfrac{6}{13} \cdot \dfrac{8}{17} \cdot \; \cdots \; \cdot \dfrac{20}{41}$

5. $\dfrac{50}{3} \cdot \dfrac{-100}{9} \cdot \dfrac{150}{27} \cdot \dfrac{-200}{81} \cdot \; \cdots \; \cdot \dfrac{-500}{3^{10}}$

6. $\left(1 - \dfrac{1}{2}\right) \cdot \left(1 + \dfrac{2}{4}\right) \cdot \left(1 - \dfrac{3}{8}\right) \cdot \left(1 + \dfrac{4}{16}\right) \cdot \; \cdots \; \cdot \left(1 + \dfrac{20}{2^{20}}\right)$

In Exercises 7–10 the solution x_n with $x_0 = 1$ is a product that telescopes to a relatively simple expression. In each case find that expression.

7. $x_{n+1} = \dfrac{n+1}{n+2} x_n$

9. $x_{n+1} = \dfrac{n+3}{n+1} x_n$

8. $x_{n+1} = \sqrt{\dfrac{n+2}{n+1}} x_n$

10. $x_{n+1} = -\dfrac{n+2}{n+1} x_n$

In Exercises 11–14 find a formula for the exact solution x_n for the given x_0.

11. $x_{n+1} = 1.5x_n, \; x_0 = 10$

13. $x_{n+1} = -1.75x_n, \; x_0 = 3000$

12. $x_{n+1} = 2x_n/3, \; x_0 = 5$

14. $x_{n+1} = -x_n/2, \; x_0 = 1$

15. Suppose that a ball is dropped from a height of 6 feet and after each bounce its maximum height is 5/8 of its previous maximum height.

   (a) How high will it go after the 5th bounce?

   (b) Make a time-series graph of its maximum height after each of the first five bounces.

16. Suppose that a savings account pays 3% annual interest compounded monthly. If $5000 is initially deposited: (a) How much will be in the account after 4 years? (b) When will the account balance reach $7500?

17. Suppose that $10,000 is initially deposited into an account paying interest at an annual rate of 8%. How much more will be in the account after 5 years if it gets compound rather than simple interest, if compounding is done: (a) quarterly? (b) monthly?

18. Suppose that an investment of $1000 grows to $1500 in 2 years getting interest that is compounded monthly. (a) What is the interest rate? (b) How much interest (in dollars) does the investment earn during the first year?

19. If a population is growing by 5% per generation: (a) How many generations will it take for an initial population of 1000 to grow to 1600? (b) What growth rate will make this happen in four generations?

20. Suppose that every hour the number of cells of a certain virus doubles. How long will it take for the number of cells to increase by 10,000 if initially there are: (a) 2000 cells? (b) 20,000 cells?

21. Suppose that after an object having an initial temperature of 150° F is placed in a 70° F room, the temperature difference between the object and the room is always 90% of that temperature difference a minute earlier. (a) How long will it take for the temperature of the object to fall below 80° F? (b) How long will it take for the temperature of the object to fall from 80° F to 75° F?

22. Suppose that when a certain radioactive substance decays, its mass M_n on day n satisfies $M_{n+1} = a M_n$. Assuming its mass is measured to be 250 grams initially and 100 grams on day 7: (a) Find the formula for the mass M_n. (b) What is the approximate half-life of the substance, i.e., the time its takes to lose half its mass?

23. Suppose that a lake is initially contaminated with 50 tons of PCB's. Each month a cleaning process is capable of filtering out 8% of all the PCB's present. (a) How many tons of PCB's will be in the lake after 1 year? (b) How long will it take before the PCB level falls below 1 ton?

24. Suppose that when a certain pendulum is set in motion, it swings in such a way that the greatest positive (or negative) angle it makes on one side of the vertical is always 95% of the greatest negative (or positive) angle it previously made on the other side of the vertical, which always occurs 3 seconds earlier. If the initial angle it makes with the vertical when let go is 20°: (a) How many times will the pendulum cross the vertical before the magnitude of the angle it achieves is 1° or less? (b) Approximately how long will it take for this to happen?

25. Suppose that when a certain skyscraper is disturbed from rest, its top sways back and forth in such a way that its maximum deflection from the vertical axis of the building to either side is always 98% of its previous maximum deflection on the other side of that axis, which always occurs 8 seconds earlier. After a strong wind gust, if the top of the building is 2 feet from that vertical axis: (a) How many times will the top of the building cross the vertical axis before the amplitude of the oscillation falls below 3 inches? (b) Approximately how long will it take for this to happen?

26. In the previous exercise, assuming it always takes 8 seconds for the top of the building to go from the vertical axis to its maximum deflection and back again: (a) How many

times will the top of the building cross the vertical axis before its average velocity falls below 1 inch per second? (b) Approximately how long will it take for this to happen?

27. For each of the following sets of points (x_n, y_n), plot the graph of $(x_n, \log_{10} y_n)$. From this graph determine whether the growth or decay appears to be exponential.

   (a) $(1, 40), (2, 90), (3, 160), (4, 250), (5, 360)$;
   (b) $(1, 4), (2, 10), (3, 26), (4, 68), (5, 170)$;
   (c) $(2, 2), (4, 1.6), (6, 1.28), (8, 1.02), (10, 0.82)$.

▬▬▬ Computer Projects 2.3

1. Evaluate the product given in Exercise 4.

2. Evaluate the product given in Exercise 6.

In Projects 3–6 compute and draw a time-series graph for the first 20 iterates of each solution.

3. The model from Exercise 15.

4. The model from Exercise 20.

5. The model from Exercise 24.

6. The model from Exercise 25.

7. (a) Compare the growth of the simple and compound interest investments described in Exercise 17(a) by plotting together their time-series graphs for each quarter of the first 5 years. (b) Repeat this using instead the monthly compound interest scenario of Exercise 17(b).

8. Each of the following sets of points (x_n, y_n) indicates either exponential growth or decay. Use a computer to plot the graphs of $(x_n, \log_c y_n)$ for: (i) $c = 2$; (ii) $c = 10$; (iii) $c = e$. What can you conclude about the base c used when determining exponential behavior?

   (a) $(0, 1), (5, 4), (10, 16), (15, 64), (20, 256), (25, 1024), (30, 4096), (35, 16, 384)$
   (b) $(2, 64), (4, 32), (6, 16), (8, 8), (10, 4), (12, 2), (14, 1), (16, 0.5), (18, 0.25), (20, 0.125)$

SECTION
2.4   Solutions of Non-Homogeneous Equations

Having now developed formulas for the exact solutions of all homogeneous linear equations, both autonomous and non-autonomous, we set our sights next on those that are non-homogeneous. However, we shall not be considering equations that are both non-autonomous and non-homogeneous, i.e., those having the general form

$$x_{n+1} = a_n x_n + b_n \tag{1}$$

As pointed out earlier, the complexity of the formula for the exact solution of (1) renders that formula impractical as a means of either computing particular iterates x_n of the solution for large values of n, or of drawing useful conclusions concerning the overall dynamics of the process being modeled. Numerical solution is generally the best approach in this case. For the sake of completeness, however, we do state the formula for the exact solution of (1):

$$x_n = x_0 \prod_{k=0}^{n-1} a_k + \sum_{k=0}^{n-2} b_k \prod_{i=k+1}^{n-1} a_i + b_{n-1}$$

though its proof is left as an exercise.

As seen in Section 1 of this chapter, most of the common linear iterative models from the natural and social sciences are actually autonomous, and consequently have the form

$$x_{n+1} = a\, x_n + b \tag{2}$$

where a and b are constants. Although we shall not be investigating (1), the simpler autonomous equation (2) does, fortunately, lend itself to analysis. Having treated in the previous section the homogeneous case of (2), in which $b = 0$, the main goal of this section will therefore be to solve the full non-homogeneous problem (2) with $b \neq 0$. In anticipation of this, some mathematical background may first be in order.

Geometric Sequences and Series

Earlier in this chapter the issue of computing large sums was considered. In general, finding a sum such as

$$\sum_{k=1}^{n} c_k = c_1 + c_2 + \cdots + c_n$$

for most collections of numbers $c_1, c_2, \ldots, c_n$, is a rather tedious task when n is large. In some circumstances, however, the terms being added have a certain form that allows the entire sum to be determined immediately, through some means other than actually adding terms one-by-one.

Such is the case when the terms are sequential powers of the same quantity a. Suppose we wish to add the numbers

$$1,\ a,\ a^2,\ a^3, \ldots, a^{n-1}$$

where a is any real number and n is a positive integer. The sum of any sequence having this form is called a **geometric series.** This sum S may be written

$$S = 1 + a + a^2 + a^3 + \cdots + a^{n-1} = \sum_{k=0}^{n-1} a^k$$

since $a^0 = 1$ for any a. To compute S, its terms could be individually added. For example, if $a = 2$ then

$$1 + 2 + 2^2 + 2^3 + 2^4 = 1 + 2 + 4 + 8 + 16 = 31$$

But there is a another way that proves to be much faster when n is large.

To derive the method, suppose we first multiply the sum S by a. This gives

$$a\,S = a\,(1 + a + a^2 + a^3 + \cdots + a^{n-1}) = a + a^2 + a^3 + a^4 + \cdots + a^n$$

If we then subtract S from this result, we have

$$a S - S = a + a^2 + a^3 + a^4 + \cdots + a^n - 1 - a - a^2 - a^3 - \cdots - a^{n-1}$$

Here, it should be apparent that almost all the terms cancel except for those involving the lowest and highest powers of a, i.e., 1 and a^n. This means that

$$a S - S = a^n - 1$$

If we think of this as an algebraic equation for the unknown sum S that we are looking for, then it is possible to easily solve the equation to find S. Factoring S on the left-hand side of the equation gives

$$(a - 1) S = a^n - 1, \quad \text{and then dividing by } a - 1 \text{ yields} \quad S = \frac{a^n - 1}{a - 1}$$

This is the sum of the series.

What we have just constructed is a way to compute the sum of any geometric series without adding terms one-by-one. The method is summarized below.

THEOREM   *Geometric Series*

The sum of the sequence of numbers $1, a, a^2, a^3, \ldots, a^{n-1}$, for any real a and any positive integer n, may be computed by

$$1 + a + a^2 + a^3 + \cdots + a^{n-1} = \frac{a^n - 1}{a - 1}$$ ∎

EXAMPLE 1

Compute the value of $1 + 2 + 4 + 8 + 16 + \cdots + 2^{24}$.

Solution: Since the sum constitutes a geometric series with $a = 2$ and $n = 25$, it must equal

$$\frac{a^n - 1}{a - 1} = \frac{2^{25} - 1}{2 - 1} = 2^{25} - 1 = 33,554,431$$ ∎

One may notice that the above formula does not appear to work when $a = 1$, since division by 0 is not allowed. However, if we view things the right way, the formula will still be valid. Suppose we interpret the case $a = 1$ as

$$\lim_{a \to 1} \frac{a^n - 1}{a - 1}$$

Using L'Hopital's rule for evaluating limits when there is an indeterminate form gives

$$\lim_{a \to 1} \frac{a^n - 1}{a - 1} = \lim_{a \to 1} \frac{n\, a^{n-1}}{1} = n$$

This is the correct value of the geometric series when $a = 1$.

Infinite Geometric Series

In many occasions in mathematics the sum of an infinite number of terms is needed. For example, when computing the value of a function using a Taylor series or when trying to solve a certain type of differential equation, the answer may be available only as an **infinite series** of the form

$$c_1 + c_2 + c_3 + \cdots = \sum_{k=1}^{\infty} c_k$$

The value of this sum is generally computed using

$$\sum_{k=1}^{\infty} c_k = \lim_{n \to \infty} \sum_{k=1}^{n} c_k$$

if the limit exists. In general one must be concerned whether such an infinite sum **converges** to a usable finite value, or **diverges** to infinity. For example, the infinite series

$$\frac{1}{1} + \frac{1}{2} + \frac{1}{3} + \frac{1}{4} + \cdots \quad \text{diverges, while} \quad \frac{1}{1^2} + \frac{1}{2^2} + \frac{1}{3^2} + \frac{1}{4^2} + \cdots \quad \text{converges}$$

Various criteria exist for distinguishing convergence from divergence in any particular case. For an infinite geometric series, that criteria is especially easy to determine. We have already seen that for any value of a when n is finite,

$$1 + a + a^2 + a^3 + \cdots + a^{n-1} = \sum_{k=0}^{n} a^k = \frac{a^n - 1}{a - 1}$$

So, if wish to evaluate the infinite sum

$$1 + a + a^2 + a^3 + \cdots = \sum_{k=0}^{\infty} a^k = \lim_{n \to \infty} \sum_{k=0}^{n} a^k$$

then it is reasonable to expect that we need only take the limit as $n \to \infty$ of the sum of the first n terms of the finite geometric series. That is,

$$\sum_{k=0}^{\infty} a^k = \lim_{n \to \infty} \sum_{k=0}^{n} a^k = \lim_{n \to \infty} \frac{a^n - 1}{a - 1}$$

For the case of $|a| < 1$, as n gets larger the value of $|a^n|$ gets smaller, ultimately approaching 0. This means that

$$\lim_{n \to \infty} \frac{a^n - 1}{a - 1} = \frac{0 - 1}{a - 1} = \frac{1}{1 - a}$$

which is the sum of the infinite geometric series in this case. When $|a| \geq 1$, this limit does not exist, indicating that the series either diverges (for $a \geq 1$ or $a < -1$), or at least does not converge (for $a = -1$). We state our observations formally.

THEOREM   *Infinite Geometric Series*

If $|a| < 1$ the infinite geometric series converges and may be computed by

$$1 + a + a^2 + a^3 + \cdots = \frac{1}{1 - a}$$

If $|a| \geq 1$ the series does not converge to a finite value.   ∎

EXAMPLE 2

Compute the value of

$$\frac{4}{1} + \frac{4}{5} + \frac{4}{5^2} + \frac{4}{5^3} + \cdots$$

Solution: If 4 is factored from each term, the result is the infinite geometric series with $a = 1/5$. The sum must therefore equal

$$4 \left(1 + \left(\frac{1}{5}\right) + \left(\frac{1}{5}\right)^2 + \left(\frac{1}{5}\right)^3 + \cdots \right) = 4 \left(\frac{1}{1 - 1/5} \right) = 4 \left(\frac{1}{4/5} \right) = 5 \quad ∎$$

Solutions of Autonomous, Non-Homogeneous Equations

With this background we are now in a position to derive a relatively simple formula for the exact solution of the non-homogeneous equation (2) for any real a and b. We begin by writing out the first few steps of (2) for $n = 0$ and $n = 1$ respectively:

$$x_1 = a\,x_0 + b \quad \text{and} \quad x_2 = a\,x_1 + b$$

Substituting the first of these into the second gives us

$$x_2 = a\,x_1 + b = a\,(a\,x_0 + b) + b \quad \text{or} \quad x_2 = a^2 x_0 + a\,b + b$$

Next, since $x_3 = ax_2 + b$, substitution also gives us

$$x_3 = a\,(a^2 x_0 + a\,b + b) + b \quad \text{or} \quad x_3 = a^3 x_0 + a^2 b + a\,b + b$$

Continuing this way, we see that it is always possible to write x_n in terms of just a, b and x_0. For example, the next two would be

$$x_4 = a^4 x_0 + a^3 b + a^2 b + a\,b + b \quad \text{and} \quad x_5 = a^5 x_0 + a^4 b + a^3 b + a^2 b + a\,b + b$$

Looking at the above equations for $x_1, x_2, \ldots, x_5$, the beginning of a pattern and a general formula may be seen. It appears that every iterate of (2) can be written in the form

$$x_n = a^n x_0 + a^{n-1} b + \cdots + a^3 b + a^2 b + a\,b + b$$

This is equivalent first to

$$x_n = a^n x_0 + b\,(a^{n-1} + \cdots + a^3 + a^2 + a + 1)$$

and if we reverse the order of addition, then also to

$$x_n = a^n x_0 + b\,(1 + a + a^2 + a^3 + \cdots + a^{n-1})$$

Although this is a formula for the exact solution of (2), we can do better. Recognizing that part of what we have here constitutes a geometric series, we can nicely abbreviate the manner in which it is written. Using the above theorem, we can replace

$$1 + a + a^2 + a^3 + \cdots + a^{n-1} \quad \text{with} \quad \frac{a^n - 1}{a - 1}$$

which yields a much more concise way of writing the solution:

$$x_n = a^n x_0 + b\,\frac{a^n - 1}{a - 1} \tag{3}$$

We formally state these observations as follows.

THEOREM   *Autonomous, Non-Homogeneous Solution*

The general autonomous, non-homogeneous equation $x_{n+1} = ax_n + b$ for any real a and b has the exact solution (3) for all $n \geq 0$.   ■

One may note that for the special case in which $b = 0$, the formula (3) reduces to $x_n = x_0 a^n$, which is consistent with the solution we previously found for this homogeneous case.

EXAMPLE 3

Find the exact solution of $x_{n+1} = 0.75x_n - 2$ if $x_0 = 50$.

Solution: In this case, $a = 0.75$ and $b = -2$. Using $x_0 = 50$, the solution may be written

$$x_n = 50(0.75)^n + (-2)\frac{(0.75)^n - 1}{0.75 - 1} = 50(0.75)^n + 8((0.75)^n - 1)$$

which simplifies to $x_n = 58(0.75)^n - 8$.   ■

As a result of this theorem, it is now possible to solve all of the non-homogeneous models introduced in Section 1. Several examples appear below. Some additional and more complex applications of non-homogeneous equations will be discussed in the next section.

EXAMPLE 4

Suppose $P_{n+1} = 30 - 0.5P_n$ models the price of some item n months after the beginning of the year. If $P_0 = 18$, find the approximate price 12 months later.

Solution: Here $a = -0.5$, $b = 30$ and $P_0 = 18$. The solution is therefore

$$P_n = 18(-0.5)^n + 30\frac{(-0.5)^n - 1}{-0.5 - 1} = 18(-0.5)^n - 20((-0.5)^n - 1)$$

or $P_n = 20 - 2(-0.5)^n$. The price 12 months later must be $P_{12} = 20 - 2(-0.5)^{12} \approx$ \$20.00.   ■

EXAMPLE 5

Suppose a population has a growth rate of 1.2 per generation, but is being harvested at a rate of 100 per generation. If its present size is 700, how long will it take before the population exceeds 1400?

Solution: The model describing this process can be written $P_{n+1} = 1.2P_n - 100$, where $P_0 = 700$. The population after n generations will therefore be

$$P_n = 700(1.2)^n - 100\frac{(1.2)^n - 1}{1.2 - 1} = 200(1.2)^n + 500$$

To determine when the population exceeds 1400 we must solve

$$200(1.2)^n + 500 \geq 1400 \quad \text{which is equivalent to} \quad (1.2)^n \geq 4.5$$

Taking the natural logarithm of both sides gives

$$n \ln 1.2 \geq \ln 4.5 \quad \text{and so} \quad n \geq \frac{\ln 4.5}{\ln 1.2} \approx 8.25$$

Since this implies that $P_8 < 1400 < P_9$, it will therefore take nine generations for the population to exceed 1400.   ∎

Change of Variable

There is another way to derive the formula (3) as the solution of $x_{n+1} = ax_n + b$, using a completely different argument that avoids any discussion whatsoever of series. The technique involves a **change of variable,** or a well-chosen substitution that results in a simplification of the equation to be solved. For discrete problems the goal would be to replace the original equation for x_n with a simpler one involving a different variable, perhaps u_n. After solving that simpler equation by finding a formula for the exact solution for u_n, the solution for x_n can then be obtained by reversing the original substitution.

Uncovering a change of variable that simplifies the problem to be solved is not often possible, and even when it is, a fair amount of guesswork is usually needed. In the present case, let us assume that the non-homogeneous equation $x_{n+1} = ax_n + b$ can be simplified by using the substitution

$$x_n = u_n + \frac{b}{1 - a}, \quad \text{which implies} \quad x_{n+1} = u_{n+1} + \frac{b}{1 - a}$$

The justification for using this particular change of variable is best shown by first substituting these new expressions for x_n and x_{n+1} into the linear equation that is to be solved, and then observing the outcome. With these substitutions $x_{n+1} = ax_n + b$ becomes

$$u_{n+1} + \frac{b}{1 - a} = a\left(u_n + \frac{b}{1 - a}\right) + b = a u_n + \frac{a b}{1 - a} + b$$

Subtracting $b/(1 - a)$ from both sides then yields

$$u_{n+1} = a u_n + \frac{a b}{1 - a} - \frac{b}{1 - a} + b = a u_n + \frac{a b - b}{1 - a} + b$$

$$= a u_n + \frac{(a - 1)b}{1 - a} + b = a u_n - b + b$$

$$= a u_n$$

What we discover from this is that the substitution

$$x_n = u_n + \frac{b}{1-a} \quad \text{converts} \quad x_{n+1} = a\,x_n + b \quad \text{into} \quad u_{n+1} = a\,u_n$$

Although the problem is not yet solved, that is now an easy task. Since $u_{n+1} = au_n$ is homogeneous, we must have $u_n = u_0 a^n$. Using the reverse substitution

$$u_n = x_n - \frac{b}{1-a} = x_n + \frac{b}{a-1}$$

the solution for x_n can then be found. That is, letting

$$u_n = x_n + \frac{b}{a-1} \quad \text{converts} \quad u_n = u_0 a^n \quad \text{into} \quad \left(x_n + \frac{b}{a-1}\right) = \left(x_0 + \frac{b}{a-1}\right) a^n$$

since the value of u_0 must be $x_0 + b/(a-1)$. Subtracting $b/(a-1)$ from both sides of that last equation yields

$$x_n = \left(x_0 + \frac{b}{a-1}\right) a^n - \frac{b}{a-1} = a^n x_0 + \frac{b\,a^n}{a-1} - \frac{b}{a-1} = a^n x_0 + b\,\frac{a^n - 1}{a-1}$$

which is the same solution for x_n found earlier using the geometric series.

Solving the non-homogeneous problem this way may appear redundant since we had already solved it earlier. But this latter method involving a change of variable is worth remembering since it will resurface when dealing with more complex iterative processes in the chapters that follow.

EXAMPLE 6

Solve $x_{n+1} = -x_n + 5$ with $x_0 = 4$, by converting to a homogeneous equation using a change of variable.

Solution: Since $a = -1$ and $b = 5$, the appropriate change of variable is

$$x_n = u_n + \frac{b}{1-a} = u_n + \frac{5}{1-(-1)} = u_n + 2.5$$

The resulting homogeneous equation for u_n is

$$u_{n+1} = a u_n = -u_n \quad \text{whose solution is} \quad u_n = u_0(-1)^n$$

Using the reverse substitution $u_n = x_n - 2.5$, gives

$$(x_n - 2.5) = (x_0 - 2.5)(-1)^n \quad \text{or} \quad x_n = 2.5 + (x_0 - 2.5)(-1)^n$$

With $x_0 = 4$ this becomes $x_n = 2.5 + 1.5(-1)^n$.   ∎

Exercises 2.4

For each of the geometric series given in Exercises 1–4, identify a *and* n, *and determine the sum* S.

1. $1 + \dfrac{3}{4} + \dfrac{9}{16} + \dfrac{27}{64} + \cdots + \dfrac{3^{10}}{4^{10}}$

2. $1 + 10^{-1} + 10^{-2} + 10^{-3} + \cdots + 10^{-20}$

3. $1 - 1.5 + 2.25 - 3.375 + \cdots + (1.5)^{12}$

4. $1 - 0.2 + 0.04 - 0.008 + 0.0016 - \cdots - (0.2)^{15}$

Evaluate each of the infinite sums given in Exercises 5–9 or say if it diverges.

5. $1 + 0.9 + 0.81 + 0.729 + 0.6561 + \cdots$

6. $1 + 1.1 + 1.21 + 1.331 + 1.4641 + \cdots$

7. $-5 + \dfrac{25}{4} - \dfrac{125}{16} + \dfrac{625}{64} - \dfrac{3125}{256} + \cdots$

8. $-3 + \dfrac{3}{1.1} - \dfrac{3}{1.21} + \dfrac{3}{1.331} - \dfrac{3}{1.4641} + \cdots$

9. $3 - 1.5 + 0.75 - 0.375 + 0.1875 - \cdots$

In Exercises 10–15 find a formula for the exact solution x_n for the given x_0.

10. $x_{n+1} = 3x_n - 8,\ x_0 = 6$

11. $x_{n+1} = 0.75x_n + 1,\ x_0 = 2$

12. $x_{n+1} = \dfrac{x_n}{4} + 5,\ x_0 = 3$

13. $x_{n+1} = 3 - 2x_n,\ x_0 = 1$

14. $x_{n+1} = -\tfrac{1}{2}(x_n + 1),\ x_0 = 1$

15. $x_{n+1} = 2 - x_n,\ x_0 = 0$

16. Suppose that each month $200 is deposited into a savings account with an original balance of $1500, that pays 4% annual interest compounded monthly. (a) What will the account balance be 2 years later? (b) How much interest will the account earn during that time?

17. Suppose someone with a 401(k) plan presently worth $50,000 wishes to retire 10 years from now. If the account is estimated to earn 8% annual interest compounded quarterly: (a) How much will be in the account at retirement time if each quarterly deposit is $2000? (b) How much should each quarterly deposit be to have $250,000 in the account at retirement time?

18. Suppose someone borrows $10,000 at an annual interest rate of 9% to be paid back monthly. (a) If $150 is paid each month, what will the unpaid balance be after 3 years? (b) How much interest is paid during this time?

19. Suppose that a population is presently 4500 and is growing by 20% per generation. Also, harvesting is occurring at a constant rate of 800 per generation. (a) What will the population equal after 10 generations? (b) After how many generations will the population reach 10,000?

20. Suppose that a population is presently 150,000 and is growing by 2% per generation. Immigration has also been occurring at a constant rate of 2500 per generation. (a) What will the population equal five generations from now? (b) What must have been the population five generations ago?

21. Suppose that in 1995 the price of a certain item was $80, and its approximate price n years later, P_n, satisfied $P_{n+1} = 0.95P_n + 7.25$ for $0 \le n \le 9$. (a) What was the approximate price in the year 2000? (b) If this trend continues what will the price be in 2010?

22. Suppose that at its initial offering a stock is priced at $25, and n weeks later its price P_n satisfies $P_{n+1} = 60 - 1.2P_n$. (a) Compute $P_1, \ldots, P_5$. (b) Assuming the model breaks down as soon as the price becomes negative, determine when this happens.

23. Suppose that a lake is initially contaminated with 10 tons of PCB's. Each month a cleaning process is capable of filtering out 10% of all the PCB's present, but another $1/2$ ton of PCB's seeps into the lake. (a) How many tons of PCB's will be in the lake after 1 year? (b) Determine the lowest PCB level that the lake will ever have under these conditions.

In Exercises 24–27 find a change of variables that will convert the given equation for x_n into a homogeneous one for u_n. Solve that new equation in terms of u_0 and then the original equation in terms of x_0.

24. $x_{n+1} = 2x_n - 2$

25. $x_{n+1} = \frac{1}{2}x_n + 10$

26. $x_{n+1} = 1.75x_n + 2$

27. $x_{n+1} = 1.5 - 0.25x_n$

28. Derive the general formula for the exact solution of a linear equation that is both non-autonomous and non-homogeneous: $x_{n+1} = a_n x_n + b_n$.

Computer Projects 2.4

For homogeneous equations $x_{n+1} = ax_n$ the ratio of successive iterates always satisfies $|x_{n+1}/x_n| = a$. For non-homogeneous equations this is not so, but in some circumstances this ratio approaches a constant as $n \to \infty$. For each of the following compute $|x_{n+1}/x_n|$ for $n = 0, \dots, 19$ with several choices of x_0. Describe your conclusions.

1. $x_{n+1} = 2x_n + 1$

2. $x_{n+1} = 2 - 1.5x_n$

3. $x_{n+1} = \frac{1}{3}x_n + 4$

4. $x_{n+1} = 3 - \frac{1}{4}x_n$

5. The model from Exercise 21.

6. The model from Exercise 22.

SECTION

2.5   Applications of Non-Homogeneous Equations

Using the formula for the exact solution, the autonomous, non-homogeneous equations considered in the examples of the previous section were dealt with in a rather straightforward manner. In this section we explore several additional types of applications that are of greater complexity than those introduced earlier.

Financial Formulas

In light of our previous findings, it is now possible to derive some important mathematical formulas commonly used in the world of finance. Recall first the Annuity Savings Model in which someone deposits the same amount d at the end of every term into an account that is getting compound interest at a rate of i per term. In Section 2.1 it was determined that the iterative equation

$$P_{n+1} = (1 + i)P_n + d$$

describes the growing value P_n of that investment over time. Since this is of the form $x_{n+1} = ax_n + b$ with $a = 1 + i$ and $b = d$, it can be solved using the formula for the exact solution:

$$P_n = P_0 \, a^n + b \, \frac{a^n - 1}{a - 1} = P_0 (1 + i)^n + d \, \frac{(1 + i)^n - 1}{(i + 1) - 1} \quad \text{or}$$

$$P_n = P_0 \, (1 + i)^n + d \, \frac{(1 + i)^n - 1}{i}$$

Because one is more likely to use this formula when first planning a long-term savings strategy beginning with little or no funds, the special case in which the original principal P_0 equals 0 is worth special mention. In this case, each deposit d is sometimes referred to as the *rent R*, since it is often a regular monthly payment. With $b = R$ and $P_0 = 0$, the above equation for P_n becomes the **Annuity Savings Formula**

$$P_n = R \, \frac{(1 + i)^n - 1}{i}$$

which is also sometimes written

$$P_n = R \, s_{n\rceil i} \quad \text{where} \quad s_{n\rceil i} = \frac{(1 + i)^n - 1}{i}$$

If one wants to know how much to deposit each term to end up with a specific balance P_n at the end of n terms, the above equations could be solved for R, giving the reverse Annuity Savings Formula

$$R = P_n \, \frac{i}{(1 + i)^n - 1} \quad \text{or} \quad R = P_n \, \frac{1}{s_{n\rceil i}}$$

■ **EXAMPLE 1**

Suppose that at the end of each month someone deposits $100 into an account earning interest at an annual rate of 6% compounded monthly. (a) How much will this investment be worth after 10 years? (b) For how long would these deposits need to be made before the total value of the investment reaches $20,000?

Solution: (a) Since $n = 10 \cdot 12 = 120$, we are looking for P_{120}. Using $R = 100$ and $i = 0.06/12 = 0.005$ we have

$$P_{120} = R \, \frac{(1 + i)^{120} - 1}{i} = 100 \, \frac{(1.005)^{120} - 1}{0.005} \approx 16{,}387.93$$

So, approximately $16,387.93 will be in the account after 10 years.

(b) In this case, we are looking for the value of n that will make $P_n = 20{,}000$. Since $R = 100$ and $i = 0.005$ we must therefore solve

$$P_n = 100 \, \frac{(1.005)^n - 1}{0.005} = 20{,}000$$

for n. Since $100/0.005 = 20{,}000$, then dividing by 20,000 yields

$$(1.005)^n - 1 = 1 \quad \text{or} \quad (1.005)^n = 2$$

Taking the natural logarithm of both sides then gives

$$n \ln 1.005 = \ln 2 \quad \text{and so} \quad n = \frac{\ln 2}{\ln 1.005} \approx 138.98$$

It will therefore take approximately 139 months or 11.6 years for the value of the investment to reach $20,000. ∎

Another financial model developed in Section 2.1 that is both autonomous and non-homogeneous is the Loan Payment Model

$$P_{n+1} = (1+i)P_n - d$$

where P_0 is the initial principal borrowed at interest rate i per term, d is the amount paid back to the lender at the end of every term and P_n is the unpaid balance after n terms. Since the only change from the Annuity Savings Model is the replacement of $b = d$ with $b = -d$, the solution must now be

$$P_n = P_0(1+i)^n - d\frac{(1+i)^n - 1}{i}$$

This formula is generally needed to calculate the payments for an **amortized** loan, i.e., a loan that is repaid by making a regular series of equal payments, each of which includes all the interest that has accrued during the previous term, plus some of the principal. Over time, the unpaid balance P_n therefore gradually decreases to 0. A mortgage, in which a buyer borrows enough to purchase a home or other real estate, and a car loan are common examples of this.

An amortized loan is always taken out for a specified period of time — perhaps 4 years for a car loan, or 20 to 30 years for a home mortgage. This means that the amount of each payment d must be computed so that the unpaid balance P_n will equal 0 for that one particular value of n corresponding to the duration of the loan. Assuming the interest rate i is fixed for the entire life of the loan, the above formula can be used to find the relationship between the original amount borrowed P_0 and payment d.

Since we want the unpaid balance P_n to equal 0 for some particular n, then substituting $P_n = 0$ into the equation above yields

$$0 = P_0(1+i)^n - d\frac{(1+i)^n - 1}{i} \quad \text{or} \quad d\frac{(1+i)^n - 1}{i} = P_0(1+i)^n$$

This equation can easily be solved for P_0, indicating how much can be borrowed in order to have the loan payment equal some given d, which is once again designated as the rent R. On the other hand, it is more commonly solved for $d = R$, which will then indicate how much the loan payment is for a given P_0. These give the two **Loan Payment Formulas:**

$$P_0 = R\frac{(1+i)^n - 1}{i(1+i)^n} \quad \text{and} \quad R = P_0\frac{i(1+i)^n}{(1+i)^n - 1}$$

respectively. These are sometimes written as

$$P_0 = R\, a_{n\rceil i} \quad \text{and} \quad R = P_0\frac{1}{a_{n\rceil i}} \quad \text{where} \quad a_{n\rceil i} = \frac{(1+i)^n - 1}{i(1+i)^n}$$

EXAMPLE 2

(a) For an amortized loan of $25,000 with an interest rate of 4.5% paid monthly for 5 years, what is the monthly payment? (b) At most how much can be borrowed under these conditions to have a maximum monthly payment of $400?

Solution: (a) Using the second Loan Payment Formula above with $P_0 = 25,000$, $i = 0.045/12 = 0.00375$ and $n = 5 \cdot 12 = 60$ yields

$$R = P_0 \frac{i(1+i)^n}{(1+i)^n - 1} = 25,000 \frac{0.00375(1.00375)^{60}}{(1.00375)^{60} - 1} \approx 466.08$$

So, the payment is approximately $466.08 per month.

(b) Again with $i = 0.00375$ and $n = 60$, but this time using the first Loan Payment Formula above and $R \leq 400$, gives us

$$P_0 = R \frac{(1+i)^n - 1}{i(1+i)^n} \leq 400 \frac{(1.00375)^{60} - 1}{0.00375(1.00375)^{60}} \approx 21,455.75$$

which means that at most $21,455.75 can be borrowed under these conditions.   ∎

Two-State Markov Processes

Another important area in which linear iterative equations play a major role is the study of **Markov processes.** For the most elementary type of such a process, in which only two discrete variables are involved, the model can always be expressed using a single autonomous, non-homogeneous equation for one of them. Based upon our findings in the previous section, this implies that a formula for the exact solution of any two-state Markov process can always be obtained.

To demonstrate the ideas involved, consider the following Markov process for the distribution of a fixed population:

> Suppose that during each decade 10% of all people living in a certain city move to the suburbs around that city, while 15% of all people living in those suburbs move into that city. The rest of those populations remain where they were. Also, assume that at the beginning of some decade 50,000 people live in that city and 50,000 in its suburbs, and this total population of 100,000 remains constant over time. Find equations for the city and suburban populations for each subsequent decade.

Here, we are interested in tracking over time two different but interconnected quantities: the city and suburban populations. These are the two **states** of this Markov process. To help clarify things, one often begins by summarizing the **transitions** between these states through the use of a **transition diagram,** also called a **state diagram.** In such a diagram, each state is drawn as one *node* in a network of nodes that are connected by arrows. The arrows indicate the direction and magnitude of the transitions between the states. These magnitudes are often interpreted as probabilities, which makes the model a *stochastic* one.

A transition diagram for the city/suburbs problem is shown in Figure 2.7. The arrow from *City* to *Suburbs* is labeled with 0.10, representing the fraction of city residents who move to the suburbs during each decade. The arrow from *Suburbs* to *City* is labeled with 0.15 for a similar reason. Although it was not explicitly mentioned that, from one decade to the next, 90% of city dwellers and 85% of suburban dwellers remain where they are, these are implied by the statement of the problem. It is therefore appropriate to have an arrow from *City* to itself labeled with 0.90, and one from *Suburbs* to itself labeled with 0.85. Note that the magnitudes of all arrows or transitions originating from a state must always add up to 1, since 100% of the population of any state must always be accounted for.

The next steps in solving a Markov process involve first choosing the discrete variables that will be used to quantify the states, and then constructing an appropriate iterative model

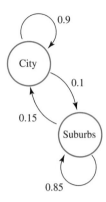

FIGURE 2.7

for them. Here, since we are interested in tracking two different populations, two different discrete variables are needed, at least at first. Therefore, suppose we let C_n represent the city population at the end of the nth decade, and S_n the suburban population at that time. To develop an iterative model for these two populations, equations must be derived that indicate how the populations at the end of the next decade, C_{n+1} and S_{n+1}, can be determined from C_n and S_n.

For the first of these variables, this is easily accomplished by observing that the next city population C_{n+1} consists entirely of two groups: *(i)* those city residents who still remain there from the nth decade, plus *(ii)* those who were residing in the suburbs at the end of the nth decade but recently moved to the city. Since 90% of city residents remain there from one decade to the next, then there must be $0.90C_n$ people in group *(i)*. Also, since 15% of suburban dwellers move to the city during every decade, then $0.15S_n$ must be in group *(ii)*. The equation for C_{n+1} can therefore be written as

$$C_{n+1} = 0.90C_n + 0.15S_n \tag{1}$$

Although it is not really needed, an iterative equation for S_n can similarly be derived. The suburban population S_{n+1} is the result of those who move to the suburbs from the city after the nth decade, $0.1C_n$, plus those who remain in the suburbs from the nth decade, $0.85S_n$. This means that

$$S_{n+1} = 0.10C_n + 0.85S_n \tag{2}$$

Equations (1) and (2) taken together constitute what is known as a **discrete linear system.** Although a general treatment of such systems must wait until Chapter 4, the special nature of a two-state Markov process allows its reduction to a single linear equation involving just one of those discrete variables. Although the choice of which variable to use is somewhat arbitrary, to be consistent we will always choose the first.

To obtain an equation for C_n alone, recall that the total population remains fixed at 100,000. This means that for every $n \geq 0$

$$C_n + S_n = 100,000 \quad \text{or} \quad S_n = 100,000 - C_n$$

Substituting this value of S_n into (1) yields

$$C_{n+1} = 0.90C_n + 0.15S_n = 0.90C_n + 0.15(100,000 - C_n)$$

which can be simplified to

$$C_{n+1} = 0.75C_n + 15{,}000$$

This is now an iterative equation for C_n alone. Solving this autonomous, non-homogeneous equation yields

$$C_n = C_0(0.75)^n + 15{,}000 \frac{(0.75)^n - 1}{0.75 - 1} = C_0(0.75)^n - 60{,}000((0.75)^n - 1)$$

Since we are told that the city population is initially $C_0 = 50{,}000$, this becomes

$$C_n = 50{,}000(0.75)^n - 60{,}000((0.75)^n - 1) = 60{,}000 - 10{,}000(0.75)^n$$

The conclusion is that $C_n = 60{,}000 - 10{,}000(0.75)^n$ is the approximate city population n decades after the process begins. To determine S_n, rather than construct and solve another iterative equation, this suburban population can instead be computed using

$$S_n = 100{,}000 - C_n = 100{,}000 - (60{,}000 - 10{,}000(0.75)^n)$$

which simplifies to $S_n = 40{,}000 + 10{,}000(0.75)^n$. Both populations are therefore known for all $n \geq 0$.

In general, any two-state Markov process can always be solved this way. Suppose that Figure 2.8 represents a general transition diagram for any such process. Note that, since we call the magnitude of the transition from the first state X to itself p, then the one from X to the other state Y must equal $1 - p$ to have the sum of the magnitudes of all transitions originating from X add up to 1. For a similar reason, the magnitudes of those transitions originating from the second state Y are called q and $1 - q$. We assume that p and q satisfy

$$0 \leq p \leq 1 \text{ and } 0 \leq q \leq 1, \quad \text{which implies} \quad 0 \leq 1 - p \leq 1 \quad \text{and} \quad 0 \leq 1 - q \leq 1$$

If we now let x_n and y_n represent the respective populations of the two states X and Y after n transitions, then x_n can be modeled at first by the iterative equation

$$x_{n+1} = p\, x_n + q\, y_n$$

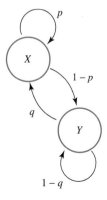

FIGURE 2.8

But if we also assume that the total population size remains fixed at some positive value N for all $n \geq 0$, then we must have

$$x_n + y_n = N, \quad \text{which means} \quad y_n = N - x_n$$

and so a model can be developed for x_n alone. That is,

$$x_{n+1} = p\, x_n + q\, y_n = p\, x_n + q\,(N - x_n)$$

which simplifies to

$$x_{n+1} = (p - q)\, x_n + q\, N$$

For all $n \geq 0$, the solution for x_n is therefore

$$x_n = x_0\,(p - q)^n + q\, N\, \frac{(p - q)^n - 1}{p - q - 1}$$

The solution for y_n can be obtained either by similarly constructing and then solving an iterative equation for it, or more simply by just using $y_n = N - x_n$ for all $n \geq 0$.

EXAMPLE 3

Suppose that researchers are training a group of 60 dolphins to communicate with humans. They notice that each week 20% of untrained dolphins become trained, but 10% of trained dolphins revert to untrained status. Assuming that originally all 60 dolphins were untrained, approximately how many will be trained and how many untrained after 6 weeks?

Solution: A transition diagram for this process appears in Figure 2.9. Suppose we let T_n and U_n represent the number of trained and untrained dolphins, respectively, after n weeks. Since each week 90% of trained dolphins remain trained and 20% of untrained dolphins become trained, then

$$T_{n+1} = 0.9T_n + 0.2U_n$$

Also, since $T_n + U_n = 60$, which implies $U_n = 60 - T_n$, then

$$T_{n+1} = 0.9T_n + 0.2(60 - T_n) = 0.7T_n + 12$$

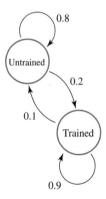

FIGURE 2.9

The solution of the linear equation $T_{n+1} = 0.7T_n + 12$ is

$$T_n = T_0(0.7)^n + 12\frac{(0.7)^n - 1}{0.7 - 1} = T_0(0.7)^n - 40((0.7)^n - 1)$$

Beginning with all untrained dolphins means that $T_0 = 0$, and so

$$T_n = 0 - 40((0.7)^n - 1) = 40(1 - (0.7)^n)$$

Since $T_6 = 40(1 - (0.7)^6) \approx 35.29$, then after 6 weeks approximately 35 dolphins will be trained, and consequently $60 - 35 = 25$ will be untrained. ∎

Sometimes in a Markov process, when the total population involved is large, rather than dealing with the actual population itself of each state, the *fraction* of the total is used instead. This means that with each iteration, the model generates the next fraction of the total shared by each state. The only practical difference here is that we use a total population size of $N = 1$ in our computations.

EXAMPLE 4

Suppose that in a certain part of the country 90% of the children of college educated parents get a college education, but 10% do not. Also, among parents who are not college educated, 40% of their children get a college education, but the other 60% do not. Assuming that the current distribution of parents in that region includes 35% who are college educated and 65% who are not, what will that distribution be three generations from now?

Solution: A transition diagram for this process is shown in Figure 2.10. Letting C_n and N_n be the respective fractions of parents who are college educated and not college educated n generations from now, we see that

$$C_{n+1} = 0.9C_n + 0.4N_n$$

Since C_n and N_n represent the fractions of the total number of parents involved, then $C_n + N_n = 1$ or $N_n = 1 - C_n$, which means

$$C_{n+1} = 0.9C_n + 0.4(1 - C_n) = 0.5C_n + 0.4$$

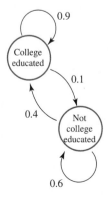

FIGURE 2.10

The equation $C_{n+1} = 0.5C_n + 0.4$ has the solution

$$C_n = C_0(0.5)^n + 0.4 \frac{(0.5)^n - 1}{0.5 - 1} = C_0(0.5)^n - 0.8((0.5)^n - 1)$$

Using $C_0 = 0.35$ this becomes

$$C_n = 0.35(0.5)^n - 0.8((0.5)^n - 1) = 0.8 - 0.45(0.5)^n$$

So three generations from now the fraction of college educated parents will be

$$C_3 = 0.8 - 0.45(0.5)^3 = 0.74375$$

or about 74%, which implies that approximately 26% will not be college educated. ■

Exercises 2.5

1. Suppose someone is saving for retirement by depositing equal amounts every quarter into a 401(k) plan estimated to earn 7% annual interest compounded quarterly. If the account initially had a zero balance: (a) How much will be available at retirement time in 20 years if each deposit is $2000? (b) How much should each deposit be to have $500,000 at retirement in 20 years?

2. If an account earns 9% annual interest compounded monthly, which will yield more at the end of the year: (*i*) A single deposit of $1000 at the beginning of the year, or (*ii*) Deposits of $100 at the end of every month of the year?

3. If $200 is deposited every 2 weeks into an account paying 6.5% annual interest compounded bi-weekly with an initial zero balance: (a) How long will it take before $10,000 is in the account? (b) During this time how much is deposited and how much comes from interest?

4. If $5000 is borrowed for 2 years at an annual interest rate of 12% paid monthly: (a) What are the monthly payments? (b) How much interest is paid during this time?

5. (a) How much can be borrowed at an annual interest rate of 6% paid quarterly for 5 years in order to have the payments equal $1000 every 3 months. (b) What is the unpaid balance on this loan after 4 years?

6. Suppose a couple with a combined annual income of $72,000 would like to purchase their first home. They have $50,000 available as a down payment and can get a mortgage for the rest at 8% annual interest paid monthly for 30 years. However, the lender will not allow their monthly mortgage payment to exceed 1/4 of their monthly income. (a) What is the maximum price home they can afford under these conditions? (b) What would they have to get their annual income up to in order to afford a $300,000 home?

7. In the city/suburbs Markov process discussed in this section: (a) Find an iterative equation for S_n alone. (b) Solve this equation for S_n and then find C_n using $C_n = 100,000 - S_n$. (c) Show that these are the same solutions found earlier.

8. In the dolphin training Markov process of Example 3: (a) Find an iterative equation for U_n alone. (b) Solve this equation for U_n and then find T_n using $T_n = 60 - U_n$. (c) Show that these are the same solutions found earlier.

9. In the college education Markov process of Example 4: (a) Find an iterative equation for N_n alone. (b) Solve this equation for N_n and then find C_n using $C_n = 1 - N_n$. (c) Show that these are the same solutions found earlier.

For each of the transition diagrams shown in Exercises 10–13: (a) Find the iterative equation for the population of either state, x_n or y_n, assuming $x_n + y_n = 1$. (b) Find formulas for the solutions x_n and y_n in terms of x_0 and y_0.

10.

12.

11.

13.

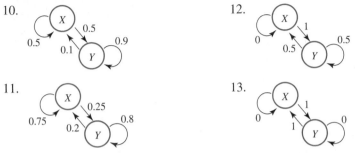

14. In the city/suburbs problem discussed in this section, suppose instead that 5% of city residents move to the suburbs and 10% of suburban residents move to the city each decade. If initially 50,000 live in the city and another 50,000 live in the suburbs, how many will live in each 5 decades later?

15. In the dolphin training problem of Example 3, suppose instead that 15% of untrained dolphins become trained each week and 5% of trained dolphins revert to untrained. If initially all are untrained, how many of the 60 will be trained and how many untrained after 6 weeks?

16. In the college education problem of Example 4, suppose instead that 80% of children of college educated parents get a college education themselves, and that 50% of children of non–college educated parents get a college education. Assuming still that presently 35% of all parents are college educated and the rest are not, how many will and how many won't be after four generations?

17. Suppose that of 1000 businesses being followed, 90% of those that make a profit in any year will also make a profit the following year and 10% will not, and 50% of those that didn't make a profit in some year will make a profit the following year and 50% will not. If presently 750 of those businesses are making a profit and 250 are not, how many will be and how many won't be making a profit in 5 years?

18. Suppose there are two products A and B competing for market share. Each month 3% of consumers switch from using A to using B, and 5% switch from B to A. The rest stay with the one they currently use. If A currently has a 70% market share and B has 30%, what percent will each have after 1 year?

19. Suppose that each year 4% of renters become homeowners, and 1% of homeowners sell their home and become renters. If curently 45% of the population are homeowners and 55% are renters, what will those percentages be a decade from now?

20. Suppose that every 5 years 8% of workers in a certain county move from below to above the national mean income, and 6% move from above to below that mean income. If currently 80% are below and 20% are above the national mean income, what will these percentages be 50 years from now?

21. Suppose that in the city/suburbs problem discussed in this section the transitions between states are not constant from one decade to the next. Instead, during decade n the fraction of the city population moving to the suburbs is $0.1 + (-0.1)^{n+1}$, and the fraction moving from the suburbs to the city is $0.15 + (0.1)^{n+1}$. Construct a non-autonomous model for this process.

22. Suppose that in the college education problem of Example 4 the transitions between states are not constant from one generation to the next. Instead, for the nth generation of parents, the fraction of their children getting a college education is $0.8 + 0.1(-1)^n$ for college educated parents and $0.4 + 0.1(-1)^n$ for non–college educated parents. Construct a non-autonomous model for this process.

■■■■■ Computer Projects 2.5

1. Create a time-series graph for the bi-weekly account balances for the first 40 weeks of the savings scenario described in Exercise 3.

2. Create a time-series graph for the unpaid balances each quarter for the loan process described in Exercise 5.

3. For the city/suburbs problem described in this section, create time-series graphs for C_n and S_n for $n = 0, \ldots, 20$. Describe your observations.

4. Repeat the previous project for C_n and N_n from the college education problem described in Example 4.

5. For the non-autonomous model constructed in Exercise 21, create time-series graphs for C_n and S_n for $n = 0, \ldots, 20$. Describe your observations.

6. Repeat the previous project for C_n and N_n from the non-autonomous college education problem described in Exercise 22.

SECTION
2.6 Dynamics of Linear Equations

Having derived formulas for the exact solutions of all autonomous linear equations, whether homogeneous or not, one might consider any further analysis of these discrete problems unnecessary. In one sense, that is essentially true. But in the following chapters, we will be investigating discrete dynamical systems of much greater complexity, for which formulas for exact solutions are either completely unobtainable, or at least quite tedious to derive.

We therefore take an altogether different approach in this section to the study of autonomous linear equations. We aim to see what sort of dynamics we can uncover about them *without* making use of their exact solutions. The techniques developed here are in fact about the only ones we will be able to carry forward with us in the investigations of the more complex problems in the chapters that follow.

Qualitative Dynamics and Asymptotic Behavior

Recall the assertion in Chapter 1 concerning the adequacy of knowing in many cases the general *qualitative* dynamics of a model, rather than always attempting to uncover its precise *quantitative* solution. While having the formula for the exact solution would of course always be preferable, in many instances knowledge of the overall dynamics will suffice, especially if that formula is either too difficult or impossible to come by. These types of modeling situations, it was argued, far outnumber those for which exact quantitative solution is essential.

Having explored in this chapter many different linear models from a variety of applied settings, one may finally see the validity of that argument. Except for the elementary financial models, which are precise descriptions of the processes they model, all others discussed here have been such crude approximations to reality that relying strictly upon their exact quantitative solutions would be rather foolish. For most of the linear models we have considered, concentrating on the dynamics that are likely to be reflected in the true underlying process being modeled would generally be a much wiser strategy.

One might also recall from Chapter 1 the point that was made there regarding the important role that asymptotic or long-term behavior plays in a model's overall qualitative dynamics. After what often turns out to be a relatively short transient period, the asymptotic behavior is what we ordinarily expect to see in a model's dynamics most of the time. Although uncovering the asymptotic behavior, as well as any other significant qualitative features that might be present, would therefore be of great benefit in virtually any modeling situation, at the present time in mathematical history, only those dynamical systems that are autonomous have lent themselves to this sort of analysis in any significant way. We therefore restrict the discussion in the remainder of this section to autonomous linear equations only.

Dynamics of Autonomous Linear Equations

Through an analysis of its asymptotic behavior, the qualitative dynamics of any discrete model of the form $x_{n+1} = ax_n + b$ can always be determined. Surprisingly, this can be accomplished equally well either with or without the use of the formula for the exact solution. In what follows we will first make use of the exact solution for x_n to determine long-term behavior, as well as other dynamics, by computing $\lim_{n \to \infty} x_n$. We will then demonstrate how the same conclusions might have been obtained from the iterative model alone. Here, we will make no distinction between the homogeneous and non-homogeneous cases, since the dynamics of the former are just a special case of those of the latter with $b = 0$.

We begin by putting the solution for x_n into the form

$$x_n = x_0 a^n + b \, \frac{a^n - 1}{a - 1} = x_0 a^n + \frac{b \, a^n}{a - 1} - \frac{b}{a - 1}$$

which can be written more conveniently as

$$x_n = \left(x_0 - \frac{b}{1 - a} \right) a^n + \frac{b}{1 - a} \tag{1}$$

Two distinct situations must now be explored separately. If $|a| > 1$ then a^n is unbounded, and so the limit

$$\lim_{n \to \infty} x_n = \lim_{n \to \infty} \left(x_0 - \frac{b}{1 - a} \right) a^n + \frac{b}{1 - a}$$

does not exist unless x_0 equals the one value that will make the coefficient of a^n exactly equal to 0. That is,

$$x_0 - \frac{b}{1-a} = 0 \quad \text{or} \quad x_0 = \frac{b}{1-a}$$

All solutions x_n therefore satisfy $|x_n| \to \infty$ as $n \to \infty$, except for the one with $x_0 = b/(1-a)$. Using (1) we see that for this one initial value $x_0 = b/(1-a)$ the solution x_n must instead satisfy

$$x_n = \left(\frac{b}{1-a} - \frac{b}{1-a} \right) a^n + \frac{b}{1-a} = 0 \cdot a^n + \frac{b}{1-a} = \frac{b}{1-a}$$

for all $n \geq 0$. Much more will be said below with regard to this special quantity $b/(1-a)$.

For the case $|a| < 1$, where a^n now approaches 0, equation (1) gives

$$\lim_{n \to \infty} x_n = \lim_{n \to \infty} \left(x_0 - \frac{b}{1-a} \right) a^n + \frac{b}{1-a} = \left(x_0 - \frac{b}{1-a} \right) \cdot 0 + \frac{b}{1-a} = \frac{b}{1-a}$$

So every solution x_n in this case approaches the special value $b/(1-a)$ as n increases. It can once again be checked using (1) that letting $x_0 = b/(1-a)$ will yield the solution $x_n = b/(1-a)$ for all $n \geq 0$.

Although the dynamics have been described identically so far for the cases $a > 0$ and $a < 0$, some significant differences between them should be pointed out. First, if $a > 0$ then the sign of

$$\left(x_0 - \frac{b}{1-a} \right) a^n \tag{2}$$

is either always positive or always negative, unless of course $x_0 = b/(1-a)$. This sign does not change with n. If $x_0 \neq b/(1-a)$, this means that the solution x_n from (1) is a monotonically increasing or decreasing sequence that satisfies one of the following:

$$\text{if } x_0 > \frac{b}{1-a} \text{ then all } x_n > \frac{b}{1-a} \quad \text{or} \quad \text{if } x_0 < \frac{b}{1-a} \text{ then all } x_n < \frac{b}{1-a}$$

But if $a < 0$ then a^n has an alternating sign. This implies that the sign of (2) must also alternate, unless of course $x_0 = b/(1-a)$. Except for that one choice of x_0, all solutions x_n from (1) must have the form

$$x_n = \left(x_0 - \frac{b}{1-a} \right) a^n + \frac{b}{1-a} = \frac{b}{1-a} \pm \left| \left(x_0 - \frac{b}{1-a} \right) a^n \right|$$

where the $\pm$ sign alternates with each successive n. Solutions x_n therefore satisfy

$$\text{either} \quad x_0 > \frac{b}{1-a}, \quad x_1 < \frac{b}{1-a}, \quad x_2 > \frac{b}{1-a}, \quad x_3 < \frac{b}{1-a}, \quad \dots$$

$$\text{or} \quad x_0 < \frac{b}{1-a}, \quad x_1 > \frac{b}{1-a}, \quad x_2 < \frac{b}{1-a}, \quad x_3 > \frac{b}{1-a}, \quad \dots$$

In either situation, this means that solutions oscillate around the value $b/(1-a)$.

EXAMPLE 1

Draw a time-series graph for the first 10 iterations of $x_{n+1} = ax_n + 6$ with $x_0 = 2$ and:
(a) $a = -1.5$; (b) $a = -0.8$.

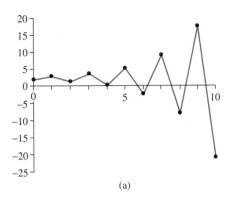

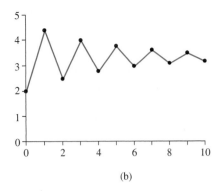

(a)   (b)

FIGURE 2.11

Solution: The graphs for parts (a) and (b) are shown in Figures 2.11(a) and 2.11(b) respectively. Note the divergence of iterates x_n from the point $b/(1-a) = 2.4$ in the former case, their convergence to the point $b/(1-a) \approx 3.33$ in the latter, and the oscillation around the respective value of $b/(1-a)$ in both. ■

Fixed Points, Stability and Oscillation

Special names are associated with the types of dynamics we have just witnessed for solutions of $x_{n+1} = ax_n + b$. The following is the first of these.

DEFINITION

For any autonomous linear equation $x_{n+1} = ax_n + b$ with $a \neq 1$, the point $p = b/(1 - a)$ is called the **fixed point** of the equation. Letting $x_0 = p$ will make $x_n = p$ for all $n \geq 0$, and so p is an **equilibrium point** of the model.

Since $x_0 = p$ makes $x_n = p$ for all $n \geq 0$, then, of course, $x_{n+1} = p$. This provides an alternate means of calculating the fixed point p without relying upon the formula $p = b/(1-a)$. We simply substitute the unknown x for both x_n and x_{n+1} in the original iterative equation $x_{n+1} = ax_n + b$, and then solve for x. That is, we solve $x = ax + b$. As long as $a \neq 1$ there will always be a unique solution for x and therefore a unique fixed point $p = x = b/(1 - a)$.

For the case $a = 1$, which rarely arises in a modeling situation, it can easily be checked that there is no fixed point in the non-homogeneous case, but every real number *acts like* a fixed point when the equation is homogeneous. That is, all real values x satisfy $x = ax + b$ when $a = 1$ and $b = 0$, but it is questionable as to whether they should all be referred to as fixed points.

EXAMPLE 2

Find the fixed point of $x_{n+1} = 16 - 1.5x_n$.

Solution: Letting both $x_{n+1} = x$ and $x_n = x$ in the equation yields $x = 16 - 1.5x$, which can easily be solved, giving $x = 32/5$. So, the fixed point is $p = 6.4$. ■

EXAMPLE 3

A species with a relatively small intrinsic growth rate of $r = 0.95$ per generation would eventually die out. But suppose immigration is also occurring at a rate of 640 per generation. In this case there is a natural positive equilibrium such that once the population reaches that level, it will remain there for all future generations. Find that equilibrium population.

Solution: The model may be written $P_{n+1} = 0.95P_n + 640$. The equilibrium population is the fixed point p of the equation, which can be found by solving $x = 0.95x + 640$. The solution is $p = 12,800$. It can be checked that letting $P_0 = 12,800$ will make $P_n = 12,800$ for all $n \geq 0$.   ∎

Another interesting dynamical feature observed above occurs when $|a| < 1$. In this case all solutions x_n converge to what we now see is the fixed point $p = b/(1-a)$. Just the opposite occurs when $|a| > 1$. That is, all solutions in this case diverge from the fixed point and continue out to $\pm\infty$, except of course for the fixed point solution in which all $x_n = p$. There are terms used to describe these properties.

DEFINITION

If all solutions of $x_{n+1} = ax_n + b$ approach the fixed point $p = b/(1-a)$ as $n \to \infty$, then p is referred to as **stable.** Otherwise, it is **unstable** and all solutions, except the fixed-point solution, diverge from p and continue out to $\pm\infty$.

Based on our previous observations, we can state in a rather simple way the criteria for stability and for instability of a fixed point.

THEOREM   *Stability*

The fixed point $p = b/(1-a)$ of the equation $x_{n+1} = ax_n + b$ is stable if $|a| < 1$ or unstable if $|a| > 1$.   ∎

We purposely did not include here the borderline cases $a = \pm 1$, since the issue of stability becomes ambiguous and is seldom pursued in these rare situations. More will be said of this in the next chapter when **bifurcations** are explored.

EXAMPLE 4

(a) Determine the stability of the equilibrium population found earlier for the population model of Example 3. (b) Discuss the significance of this.

Solution: (a) The coefficient a of the previous population model $P_{n+1} = 0.95P_n + 640$ is $a = 0.95$. Since $|a| = 0.95 < 1$, the equilibrium population, previously found to be $p = 12,800$, must be stable.

(b) This means that regardless of its initial size P_0, the population P_n will approach the equilibrium level 12,800 as n increases. Once at that level (or sufficiently close to it that one cannot distinguish the difference) P_n will remain there for all $n \geq 0$.   ∎

EXAMPLE 5

(a) Find the equilibrium point of the price model $P_{n+1} = 39 - 1.25P_n$ and determine its stability. (b) Discuss the significance of these findings.

Solution: (a) Solving $x = 39 - 1.25x$ gives the fixed point $p \approx \$17.33$. Since $|a| = |-1.25| > 1$, this equilibrium is unstable.

(b) Although in theory if P_0 were chosen to be the exact equilibrium price p, then the price P_n would remain there for all $n \geq 0$. But since it is unrealistic to expect that P_0 could ever really attain the precise value of p, we should instead expect that P_n will diverge from p and either increase to infinity or become negative, at which point the model breaks down.   ∎

One should note carefully the differing dynamics of the two previous examples. Although both had respective fixed points, the stable process of Example 4 is the only scenario in which we should expect to actually see the equilibrium level attained. Since all solutions approach a stable equilibrium in that former case, one can be quite certain that for any choice of initial condition every solution will eventually be virtually indistinguishable from that equilibrium level. When the fixed point is not stable, as in Example 5, the equilibrium has, in effect, a theoretical existence only. It is unlikely that it will ever be observed, since all solutions diverge from it. Because of this, unstable processes are inherently less predictable than stable ones, which is why determining stability is essential in most modeling situations.

The last type of behavior we observed earlier involved the issue of oscillation. The following summarizes those findings.

THEOREM   *Oscillation*

For $a > 0$ all solutions of $x_{n+1} = ax_n + b$ are monotonic, but for $a < 0$ all solutions oscillate around the fixed point $p = b/(1 - a)$.   ∎

EXAMPLE 6

(a) Determine whether prices oscillate around the equilibrium point of the price model of Example 5. (b) Discuss the significance of this.

Solution: (a) The equation involved is $P_{n+1} = 39 - 1.25P_n$, which means $a = -1.25 < 0$. Prices P_n therefore oscillate around the equilibrium price, which was previously found to be $p \approx \$17.33$. Since p was also found to be unstable, then virtually all solutions oscillate with increasing amplitude away from this equilibrium.

(b) Price oscillations may be the result of an unjustifiably high price that causes a decrease in the demand, and consequently a decrease in the price itself. The new lower price may encourage bargain-hunting, which results in an increase in demand, and therefore also eventually in the price. This latest price may again be perceived as too high by consumers, and so the demand and price fall again. This creates an oscillating price. When oscillation is combined with instability as in the present case, the price may *spiral* outward, creating highly volatile price fluctuations.   ∎

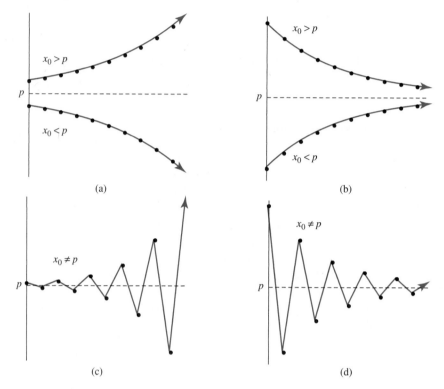

FIGURE 2.12

Dynamics and Changes of Variable

Our analysis has shown that four distinct types of dynamics are possible for any autonomous linear equation $x_{n+1} = ax_n + b$ whose fixed point is $p = b/(1 - a)$:

- If $a > 1$ solutions diverge monotonically from p (see Fig. 2.12(a)).
- If $0 < a < 1$ solutions converge monotonically to p (see Fig. 2.12(b)).
- If $a < -1$ solutions diverge from p with oscillation (see Fig. 2.12(c)).
- If $-1 < a < 0$ solutions converge to p with oscillation (see Fig. 2.12(d)).

It is interesting to note that these stability and oscillation criteria do not involve the constant b in any way. Only the magnitude and sign of a appear to effect such dynamics. Although this may at first seem a bit odd, using a change of variable may shed some light on the reason this must necessarily be so.

Recall from Section 2.4 that making the substitution

$$x_n = u_n + \frac{b}{1 - a} \quad \text{converts} \quad x_{n+1} = a\,x_n + b \quad \text{into} \quad u_{n+1} = a\,u_n$$

Since we have just learned that $p = b/(1 - a)$ is the fixed point of $x_{n+1} = ax_n + b$, this change of variable may now be viewed as follows: It is a means of transforming any non-homogeneous equation $x_{n+1} = ax_n + b$ into a homogeneous one $u_{n+1} = au_n$ with the same coefficient a, so that the fixed point p of the former turns into the fixed point 0 of the latter.

No dynamics are altered in this transformation. Rather, they are only shifted or *translated* to a different and more convenient location on the real axis. The two problems $x_{n+1} = ax_n + b$ and $u_{n+1} = au_n$ are sometimes described as being **topologically conjugate.**

For any particular coefficient a, since all equations of the form $x_{n+1} = ax_n + b$ can be transformed this way into the same unique homogeneous equation $u_{n+1} = au_n$, all should therefore share the same dynamics. Stability, instability and oscillation for $u_{n+1} = au_n$, which depend only on the magnitude and sign of a, must therefore extend to all other equations $x_{n+1} = ax_n + b$ having that same coefficient a, regardless of the value of b. This means that b must of necessity play no role with regard to these issues.

Dynamical Systems Approach

These discussions and examples should provide convincing evidence of an important point: To determine the qualitative dynamics of any autonomous linear equation, it is not necessary to solve that equation exactly. Instead, fixed points, stability and oscillation can all be determined using only the three theorems given above.

The significance of this cannot be overstated, since it sets the stage for what is to follow. The dynamics we have described in this section, plus some fascinating types of behavior we have not yet encountered, are precisely what we will be pursuing in subsequent chapters with regard to nonlinear equations and systems of equations, which generally have much greater complexity than those we have considered so far. Since formulas for exact solutions will either be entirely impossible to obtain, or at least very tedious to derive and make use of, they will rarely be at issue.

The focus instead will be on those techniques that allow a determination of the dynamics of a problem without relying upon a formula for the exact solution. This might be called the **dynamical systems approach** to investigating the problem. As we shall soon discover, attempting this approach on the nonlinear equations and systems of equations that we will be considering will often pose a much greater challenge than have any of the autonomous linear equations investigated in this chapter.

EXAMPLE 7

(a) Use the dynamical systems approach to investigate the migration model $P_{n+1} = 1.1 P_n - 150$. (b) Discuss the significance of these findings.

Solution: (a) Solving $x = 1.1x - 150$ gives the fixed point $p = 1500$. Since $a = 1.1 > 1$ then this equilibrium population is unstable, and no oscillation occurs around it. Any population P_n with $P_0 > 1500$ must therefore increase monotonically to $+\infty$. Any P_n with $P_0 < 1500$ decreases monotonically to $-\infty$, but in this case the model breaks down as soon as $P_n \leq 0$. We would say that extinction occurs at that time. Time-series graphs of this process would resemble those of Figure 2.12(a).

(b) Although this equilibrium population $p = 1500$ is unstable and therefore unlikely to be maintained, determining its value may nevertheless be important, since it represents the smallest initial population for which the species will survive and flourish. For initial populations below 1500 extinction will eventually occur. Such an equilibrium level is sometimes called a **threshold.** ∎

■■■■■■■■   **EXAMPLE 8**

Use the dynamical systems approach to investigate the city/suburbs Markov process from Section 2.5.

Solution: In the previous section it was shown that the city population after the nth decade is modeled by $C_{n+1} = 0.75C_n + 15{,}000$. Solving $x = 0.75x + 15{,}000$ yields the fixed point $p = 60{,}000$, and since $0 < 0.75 < 1$, then the situation resembles that of Figure 2.12(b). Regardless of its initial size, the city population will monotonically approach over time the stable equilibrium population 60,000. Since the total population of the region is 100,000, then the suburban population will consequently stabilize at 40,000.   ■

We remark that a Markov process in which the magnitudes of all transitions between states are non-zero, such as the one above, is called **regular**. Although the proof is left as an exercise, it can be shown that a stable equilibrium always exists in this case, and that it lies between 0 and the total population size involved. All solutions therefore approach this equilibrium, which is called the **steady-state** of the Markov process.

Exercises 2.6

In Exercises 1–8 find the fixed point and determine both its stability and whether or not oscillation occurs.

1. $x_{n+1} = 2x_n - 3/4$

2. $x_{n+1} = 0.7x_n$

3. $x_{n+1} = \frac{1}{2}x_n + 4$

4. $x_{n+1} = (x_n + 1)/2$

5. $x_{n+1} = -1.75x_n$

6. $x_{n+1} = 0.9(1 - x_n)$

7. $x_{n+1} = 4(x_n - 1)/5$

8. $x_{n+1} = (7 - 3x_n)/2$

In Exercises 9–14 find the equilibria of the models described in previous exercises. Also determine both stability and whether or not oscillation occurs.

9. Exercise 23 of Section 2.3

10. Exercise 24 of Section 2.3

11. Exercise 25 of Section 2.3

12. Exercise 21 of Section 2.4

13. Exercise 22 of Section 2.4

14. Exercise 23 of Section 2.4

15. Suppose that there are currently 25,000 unemployed workers in some state. Each month 8% of all those unemployed find jobs but another 1500 become unemployed.

   (a) How many will be unemployed 6 months from now?
   (b) At what level will the number of unemployed workers stabilize over time?

16. Suppose that each day during flu season, 15% of those who have the flu in a certain town recover from it, while another 600 people come down with the flu. If there are currently 1800 cases of the flu:

   (a) How many cases will there be two weeks from now?
   (b) At what level will the number of cases eventually stabilize?

17. Suppose that someone always carries an unpaid balance on a certain credit card. Each month the credit card company charges 1% interest on any previous unpaid balance, and the person pays off 10% of that previous balance. Also, during each month another $200 is charged on that credit card.

    (a) Write an iterative equation for the monthly unpaid balance U_n.
    (b) What level will the unpaid balance gradually approach?

18. Suppose that each year 2% of all the trees in a certain forest are destroyed naturally. Also each year 5000 mature trees are harvested for lumber, but 7500 new trees are either planted or sprout up on their own.

    (a) Write an iterative equation for the yearly number of trees T_n in that forest.
    (b) If there are currently estimated to be 100,000 trees in that forest, what is the maximum number of trees the forest will ever have?

In Exercises 19–24 find the stable steady-states of the Markov processes described in the following previous examples and exercises.

19. Example 3 of Section 2.5   22. Exercise 18 of Section 2.5

20. Example 4 of Section 2.5   23. Exercise 19 of Section 2.5

21. Exercise 17 of Section 2.5   24. Exercise 20 of Section 2.5

25. Determine whether the Markov processes with transition diagrams shown in the following previous exercises have stable steady states:

    (a) Exercise 12 of Section 2.5   (b) Exercise 13 of Section 2.5

26. If a linear equation $x_{n+1} = ax_n + b$ with $a \neq 0$ is solved for x_n and iterated *backward* instead of *forward* for some x_0, a sequence of *inverse iterates* $x_{-1}, x_{-2}, x_{-3}, \ldots$ can be obtained. Prove that a fixed point p is stable under forward iteration if and only if it is unstable under backward iteration, by showing the following: (a) If p is unstable under forward iteration then $x_{-n} \to p$ as $n \to \infty$. (b) If p is stable under forward iteration then $x_{-n} \to \pm\infty$ as $n \to \infty$ for any $x_0 \neq p$.

■ Computer Projects 2.6

For each of the stable processes given in Projects 1–4: (i) Find the fixed point p*; (ii) Use* $x_0 = p + 1$ *and a computer to find the smallest value of* n *for which* $|x_n - p| < 0.001$; *(iii) Do you see a relationship between* $|a|$ *and* n *from (ii)?*

1. $x_{n+1} = 0.9x_n + 2$ 3. $x_{n+1} = -\frac{1}{4}x_n$

2. $x_{n+1} = \frac{1}{2}x_n - 3$ 4. $x_{n+1} = 0.01x_n + 1$

For each of the unstable processes given in Projects 5–8: (i) Find the fixed point p*; (ii) Use* $x_0 = p + 1$ *and a computer to find the smallest value of* n *for which* $|x_n - p| > 1000$; *(iii) Do you see a relationship between* $|a|$ *and* n *from (ii)?*

5. $x_{n+1} = 1.1x_n - 1$ 7. $x_{n+1} = 10x_n + 18$

6. $x_{n+1} = -2x_n + 3$ 8. $x_{n+1} = 100x_n$

2.7   Empirical Models and Linear Regression

Although the general forms of many linear models have been developed in this chapter, these models would not prove very useful in practice unless their parameters can be determined in specific circumstances. In this section we take one last look at linear equations with an investigation of **linear regression,** which is the most commonly used method for constructing a linear model based upon an empirical set of data.

Finding Linear Equations

Constructing an autonomous linear model of the form $x_{n+1} = ax_n + b$ by determining values for the parameters a and b is really just a special case of a more general issue, i.e., finding the equation of a straight line $y = mx + b$ by determining the values of the slope m and the y-intercept b. Anyone who has studied algebra is likely to be familiar with that problem. When only one piece of information regarding the line is given, such as one particular point (x_1, y_1), or perhaps the slope m, it is easy to see that an infinite number of lines go through that point or have that slope. Not enough information is given in this case to provide a unique answer to the question.

When two particular points (x_1, y_1) and (x_2, y_2) are given, however, then a unique straight line going through them always exists, as does its corresponding unique linear equation. That equation is generally found by first determining the slope of the line $m = \dfrac{y_2 - y_1}{x_2 - x_1}$ and then using the point-slope formula $y - y_1 = m(x - x_1)$ to obtain the equation, provided that the line is not vertical.

But if one were asked to find the equation of the straight line that goes through three or more given points, a major obstacle is immediately encountered. To have a straight line that goes through three or more different points, these points would have to all line up. Since this is not very likely, no such line ordinarily exists.

This assumes, of course, that we want the line to actually *go through* all those points. But in most cases, when a set of data points is gathered empirically, i.e., through experimentation or observation, the measurements made are not likely to be 100% accurate. In the real world, experimental errors and roundoff errors nearly always occur in the measurements of the quantities involved. No point is so reliable that the line must be forced to actually go through it. Instead, one is usually interested in finding the best straight line that approximates the set of data in its entirety. What is meant here by *best* may be a matter of opinion, but a standard interpretation is commonly given to the term in most research settings.

Least-Squares Regression

In the natural and social sciences when analyzing empirical data that suggests a linear relationship between two quantities the most commonly used technique for deriving the corresponding linear equation, is **least-squares regression.** The regression line is considered to be the best one to use, for both theoretical and practical reasons. On one hand, this particular line minimizes (in some sense) the total amount of error that any such approximate

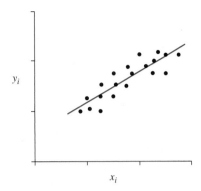

FIGURE 2.13

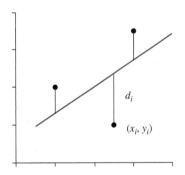

FIGURE 2.14

line would have with respect to the given set of data points. On the other hand, it can be rather quickly determined using some arithmetic and elementary algebra (although the derivation of the technique, as we shall now see, does require some multivariate calculus).

To derive the method, consider the problem of finding the line that best approximates the set of N points: $(x_1, y_1), (x_2, y_2), \ldots, (x_N, y_N)$. These data points may, for example, be the result of gathering together observations made during N repetitions of a certain experiment. Graphing a typical collection of such points in the (x, y) plane, as in Figure 2.13, shows that in general no single straight line could possibly pass through all of them. Despite this, one may still have reason to believe that the relationship between the variables x and y is nevertheless a linear one. If that is the case, then it is always possible to find the unique values of m and b that make the linear equation $y = mx + b$ minimize a certain measurement of the total error that must exist between any such straight line and the given points.

The key is to define carefully what is meant by the *total error*. To obtain a useful measurement of this error, first consider for each data point (x_i, y_i) the vertical distance d_i between the line $y = mx + b$ and that point, as shown in Figure 2.14. Each d_i, which may be viewed as the ith error, is the difference between the y value as predicted by the linear equation for $x = x_i$, and the actual observed y value corresponding to that x value, which is y_i. Since the y value predicted by the equation $y = mx + b$ for $x = x_i$ is $y = mx_i + b$, then this difference must be

$$d_i = mx_i + b - y_i$$

The total error E is then defined to be the sum of the squares of these N errors d_i. That is,

$$E = d_1^2 + d_2^2 + \cdots + d_N^2 = \sum_{i=1}^{N} d_i^2 = \sum_{i=1}^{N} (mx_i + b - y_i)^2$$

The rationale behind defining E this way, rather than, say, $\sum_{i=1}^{N} |d_i|$, is twofold. First,

$$\sqrt{E} = \sqrt{d_1^2 + d_2^2 + \cdots + d_N^2}$$

resembles the distance formula for N-dimensional space, and so in a sense we are actually dealing with the N-dimensional distance between the line and the set of points. Also, this definition of E constitutes a differentiable function of the unknowns involved, and so computing necessary derivatives, as we shall be doing next, is easily accomplished. This is not true when absolute values are involved.

Earlier, it was decided that the line $y = mx + b$ we seek is the one that minimizes the total error. This has now come to mean the line that minimizes E. Since all of the x_i and y_i are known coordinates from the original set of points that was given, the total error

$$E = \sum_{i=1}^{N} (mx_i + b - y_i)^2 \tag{1}$$

may be viewed as a differentiable function of just the two unknown variables m and b, and may therefore be written $E = E(m, b)$. Calculus can now be used to determine the values of these two variables that minimize E. However, since $E(m, b)$ is a function of two variables rather than just one, multivariate calculus is needed. For those familiar with single variable calculus only, the following is a brief review of some of the ideas involved.

Partial Derivatives

When optimizing a function of two or more variables, the technique of equating the derivative to 0 is still valid. However, now there is more than one kind of derivative. There are derivatives with respect to *each* of the variables involved. They are called **partial derivatives,** and are computed essentially the same way as for functions of only one variable. The only difference is that when computing the partial derivative with respect to any particular variable, all other variables that may appear in the equation are treated as constants.

EXAMPLE 1

Compute the partial derivatives of: (a) $f(x, y) = x^2 + 5y - 3$; (b) $f(x, y) = (2x - 3y)^2 + x^2 - 36y$.

Solution: (a) There are two partial derivatives of $f(x, y) = x^2 + 5y - 3$, one with respect to x and the other with respect to y. If we call them $f_x(x, y)$ and $f_y(x, y)$ respectively, then $f_x(x, y) = 2x$ since -3 is a constant, and $5y$ is temporarily treated like one. Also, $f_y(x, y) = 5$, since now both -3 and x^2 are constants.

(b) This time $f(x, y) = (2x - 3y)^2 + x^2 - 36y$. Using the chain rule, we have

$f_x(x, y) = 2(2x - 3y) \cdot 2 + 2x$ which simplifies to $f_x(x, y) = 10x - 12y$, and
$f_y(x, y) = 2(2x - 3y) \cdot (-3) - 36$ which simplifies to $f_y(x, y) = -12x + 18y - 36$

■

After computing all the partial derivatives of a function, to maximize or minimize that function, each of these derivatives should be set individually to 0. This will create a system of equations that must be solved simultaneously to find the values of the variables that optimize the function. For example, for a function of two variables $f(x, y)$ this would mean solving the two simultaneous equations

$$f_x(x, y) = 0 \quad \text{and} \quad f_y(x, y) = 0$$

to find the values of x and y that generate the relative maxima and minima of the function.

EXAMPLE 2

Find the values of x and y that minimize the multivariate function $f(x, y) = (2x - 3y)^2 + x^2 - 36y$ from part (b) of Example 1.

Solution: Since we previously found that

$$f_x(x, y) = 10x - 12y \quad \text{and} \quad f_y(x, y) = -12x + 18y - 36$$

then letting each of these equal 0 yields the system

$$\begin{array}{l} 10x - 12y = 0 \\ -12x + 18y - 36 = 0 \end{array} \quad \text{which is equivalent to} \quad \begin{array}{l} 5x - 6y = 0 \\ -2x + 3y = 6 \end{array}$$

The unique solution $x = 12$ and $y = 10$ is easily found. Further analysis (which we shall not go into here) would show that the point $(12, 10)$ corresponds to both a relative and absolute minimum of the function.   ■

Normal Equations

With this background, it is now possible to minimize $E = E(m, b)$ from (1), which will therefore minimize the total error between a set of data points and the regression line. Since $E(m, b)$ is a function of the two variables m and b, each of its partial derivatives $E_m(m, b)$ and $E_b(m, b)$ must first be computed, and then set equal to 0. Using the chain rule on $E(m, b) = \sum_{i=1}^{N}(mx_i + b - y_i)^2$, the first of these becomes

$$E_m(m, b) = \sum_{i=1}^{N} 2(mx_i + b - y_i)x_i = 0$$

since all of the x_i and y_i are constants, and b is being treated like one. Dividing by 2 and distributing x_i, this can be written more simply as

$$\sum_{i=1}^{N} \left(mx_i^2 + x_i b - x_i y_i \right) = 0$$

But since

$$\sum_{i=1}^{N} \left(mx_i^2 + x_i b - x_i y_i\right) = \sum_{i=1}^{N} mx_i^2 + \sum_{i=1}^{N} x_i b - \sum_{i=1}^{N} x_i y_i$$

$$= m\sum_{i=1}^{N} x_i^2 + b\sum_{i=1}^{N} x_i - \sum_{i=1}^{N} x_i y_i$$

then we must have

$$m\sum_{i=1}^{N} x_i^2 + b\sum_{i=1}^{N} x_i - \sum_{i=1}^{N} x_i y_i = 0 \quad \text{or} \quad m\sum_{i=1}^{N} x_i^2 + b\sum_{i=1}^{N} x_i = \sum_{i=1}^{N} x_i y_i$$

This last equation, which may be written in the form

$$\left(\sum_{i=1}^{N} x_i^2\right) m + \left(\sum_{i=1}^{N} x_i\right) b = \left(\sum_{i=1}^{N} x_i y_i\right) \tag{2}$$

is the first of two linear equations involving the unknowns m and b that we need. To find the other, we go back to $E(m, b) = \sum_{i=1}^{N}(mx_i + b - y_i)^2$ and compute the partial derivative

$$E_b(m, b) = \sum_{i=1}^{N} 2(mx_i + b - y_i)$$

Equating this to 0 yields

$$\sum_{i=1}^{N} 2(mx_i + b - y_i) = 0 \quad \text{or} \quad m\sum_{i=1}^{N} x_i + \sum_{i=1}^{N} b - \sum_{i=1}^{N} y_i = 0$$

Since $\sum_{i=1}^{N} b = Nb$, this may also be written

$$m\sum_{i=1}^{N} x_i + Nb - \sum_{i=1}^{N} y_i = 0 \quad \text{or} \quad m\sum_{i=1}^{N} x_i + Nb = \sum_{i=1}^{N} y_i$$

Rewriting this last equation and combining it with (2) yields the system

$$\left(\sum_{i=1}^{N} x_i^2\right) m + \left(\sum_{i=1}^{N} x_i\right) b = \left(\sum_{i=1}^{N} x_i y_i\right)$$
$$\left(\sum_{i=1}^{N} x_i\right) m + Nb = \left(\sum_{i=1}^{N} y_i\right) \tag{3}$$

which are called the **normal equations.**

Since N and all the values of x_i and y_i are known in advance, these normal equations constitute a linear system of equations for the unknowns m and b. Solving the system will generally provide the unique values of m and b that minimize the total error E from (1). The line $y = mx + b$ with these particular values of m and b, called the **least-squares regression line,** is the one that best approximates the given set of data points.

EXAMPLE 3

Find the least-squares regression line that approximates the data

$$(-1, 1), \ (0, 1), \ (0, 2), \ (1, 2), \ (2, 3)$$

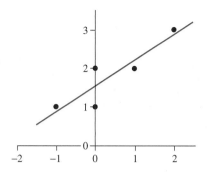

FIGURE 2.15

Solution: Since there are obviously $N = 5$ data points, in order to write the normal equations, four additional quantities are needed:

$$\sum_{i=1}^{5} x_i = 2, \quad \sum_{i=1}^{5} y_i = 9, \quad \sum_{i=1}^{5} x_i^2 = 6, \quad \sum_{i=1}^{5} x_i y_i = 7$$

The normal equations from (3) therefore become

$$\begin{aligned} 6m + 2b &= 7 \\ 2m + 5b &= 9 \end{aligned} \quad \text{whose unique solution is} \quad m = \frac{17}{26} \quad \text{and} \quad b = \frac{20}{13}$$

The regression line is therefore $y = \frac{17}{26}x + \frac{20}{13}$. This line is graphed along with the data points in Figure 2.15.   ■

We remark that when the number of data points is large, a computer is ordinarily enlisted to compute the sums involved, and perhaps even to solve the normal equations themselves. Many software packages and spreadsheet programs have built-in functions to accomplish these tasks.

Empirical Linear Models

With some slight modifications, the above discussion can be adapted to allow the construction of the best approximate linear iterative model of the form $x_{n+1} = ax_n + b$ using an empirical set of data. The following describes how the optimal values of a and b may be determined whenever a series of consecutive iterates x_i is known.

For some process, suppose that the set of $N + 1$ values $x_1, \ldots, x_{N+1}$ is observed. If we are assuming that this data represents consecutive iterates generated by an autonomous linear equation, then we are in fact assuming that for some a and b $x_{i+1} = ax_i + b$ for $i = 1, \ldots, N$. This means that in theory each of the points (x_i, x_{i+1}) should lie on the straight line $y = ax + b$. Since this is empirical data, however, graphing these N points (x_i, x_{i+1}) for $i = 1, \ldots, N$ would likely reveal that they do not really lie along a true straight line (see Fig. 2.16).

Instead, linear regression must be used to find the best a and b to use when approximating this data by the straight line $y = ax + b$. Of course, when using the normal equations (3) in this setting, m must be replaced with a, and the data points are given by

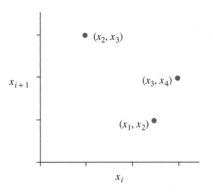

FIGURE 2.16

$(x_i, y_i) = (x_i, x_{i+1})$ for $i = 1, \dots, N$. Besides these adaptations, the remainder of the technique is essentially unchanged. Solving the normal equations will yield the best straight line that approximates these data points and therefore also the best empirical linear model of the form $x_{n+1} = ax_n + b$.

EXAMPLE 4

Suppose that the approximate population in thousands of some species for each of the first few generations for which it is observed is

$$2, \quad 3.5, \quad 2.5, \quad 3, \quad 2, \quad 4, \quad 0.5, \quad 5, \quad 1.5$$

respectively. Construct the autonomous linear model $P_{n+1} = aP_n + b$ that best approximates the population growth of that species.

Solution: Since the populations are given for $N + 1 = 9$ generations, then $N = 8$, and the actual data points $(x_i, y_i) = (P_i, P_{i+1})$ are

$$(2, 3.5), \quad (3.5, 2.5), \quad (2.5, 3), \quad (3, 2), \quad (2, 4), \quad (4, 0.5), \quad (0.5, 5), \quad (5, 1.5)$$

Computing the four sums needed in the normal equations gives

$$\sum_{i=1}^{8} x_i = 22.5, \quad \sum_{i=1}^{8} y_i = 22, \quad \sum_{i=1}^{8} x_i^2 = 76.75, \quad \sum_{i=1}^{8} x_i y_i = 49.25$$

The normal equations are therefore

$$\begin{aligned} 76.75a + 22.5b &= 49.25 \\ 22.5a + 8b &= 22 \end{aligned}$$ whose solution is $a \approx -0.937$ and $b \approx 5.386$

The iterative linear equation that best models the population growth of this species is therefore $P_{n+1} = -0.937P_n + 5.386$, where P_n represents the size of the population in thousands. This is graphed as a line along with the eight data points (P_i, P_{i+1}) in Figure 2.17.   ∎

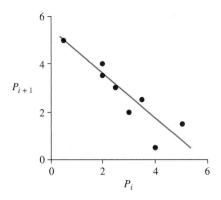

FIGURE 2.17

Exercises 2.7

In Exercises 1–4 find the partial derivatives $f_x(x, y)$ *and* $f_y(x, y)$.

1. $f(x, y) = 3x^4 + 5x - 8y^3$ 3. $f(x, y) = 4x^2 y^5$

2. $f(x, y) = (2x + 3y)^4$ 4. $f(x, y) = x/y$

In Exercises 5–8 find the partial derivatives $S_a(a, b)$ *and* $S_b(a, b)$.

5. $S(a, b) = \displaystyle\sum_{i=1}^{5} (a\,i - b)^2$ 7. $S(a, b) = \displaystyle\sum_{i=0}^{4} (a\,2^i + b)^2$

6. $S(a, b) = \displaystyle\sum_{i=1}^{5} (a + b + i)^2$ 8. $S(a, b) = \displaystyle\sum_{i=1}^{10} (a^2 - b + i)^2$

In Exercises 9–12 find the critical points, i.e., the points (x_0, y_0) *that satisfy both* $f_x(x_0, y_0) = 0$ *and* $f_y(x_0, y_0) = 0$.

9. $f(x, y) = x^2 + y^2 + 5$ 11. $f(x, y) = (x - 2)(y + 1)$

10. $f(x, y) = 3x^4 + 5x - 8y^3$ 12. $f(x, y) = \displaystyle\sum_{i=0}^{4} (2^i x^2 + y^2)^2$

A critical point (x_0, y_0) *of* $f(x, y)$ *can be determined to be a maximum, minimum or neither by comparing the value of* $f(x_0, y_0)$ *to nearby values of* $f(x, y)$. *In Exercises 13–16 use this approach to classify the critical points of the given function* $f(x, y)$.

13. $f(x, y)$ from Exercise 9   15. $f(x, y)$ from Exercise 11

14. $f(x, y)$ from Exercise 10   16. $f(x, y)$ from Exercise 12

In Exercises 17–20 first write the Normal Equations using the given data, and then solve to find the least squares line $y = mx + b$.

17. $(0, 0), (1, 2), (2, 3)$ 19. $(-2, 10), (-1, 6), (0, 5), (0, 4), (2, 0)$

18. $(0, 0), (0, 1), (1, 0), (1, 1)$ 20. $(-2, -5), (-1, -2), (0, 0), (1, 1), (2, 1)$

In Exercises 21–24 use the given data to construct the best approximate linear iterative model.

21. $x_0 = 1$, $x_1 = 3$, $x_2 = 10$, $x_3 = 20$

22. $S_0 = 2$, $S_1 = 5$, $S_2 = 10$, $S_3 = 30$

23. $P_0 = 60$, $P_1 = 25$, $P_2 = 10$, $P_3 = 5$, $P_4 = 2$, $P_5 = 1$

24. $T_0 = 50$, $T_1 = 20$, $T_2 = 12$, $T_3 = 8$, $T_4 = 5$, $T_5 = 4$

25. (a) Instead of using the Chain Rule, find the partial derivatives E_m and E_b of $E = \sum_{i=1}^{N}(mx_i + b - y_i)^2$ by first squaring each term being added and then differentiating. (b) Show that this yields the same E_m and E_b as before.

Computer Projects 2.7

In Projects 1 and 2 write the Normal Equations using the given data and a computer to perform the necessary calculations. Solve to find the least-squares line $y = mx + b$.

1. $(0, 40)$, $(10, 60)$, $(15, 70)$, $(20, 75)$, $(20, 80)$, $(25, 100)$, $(30, 110)$, $(30, 120)$

2. $(-25, 10)$, $(-20, 8)$, $(-15, 5)$, $(-10, 2)$, $(-5, 0)$, $(0, -4)$, $(5, -8)$, $(10, -10)$, $(15, -20)$, $(20, -25)$, $(25, -35)$, $(30, -40)$

In Projects 3 and 4 use the given data, and a computer to perform the necessary calculations, to construct the best approximate linear iterative model.

3. $x_0 = 2$, $x_1 = 3$, $x_2 = 15$, $x_3 = 40$, $x_4 = 90$, $x_5 = 200$, $x_6 = 350$, $x_7 = 650$, $x_8 = 900$

4. $P_0 = 10{,}000$, $P_1 = 6500$, $P_2 = 3550$, $P_3 = 3250$, $P_4 = 2800$, $P_5 = 2750$, $P_6 = 2720$, $P_7 = 2675$, $P_8 = 2650$, $P_9 = 2625$, $P_{10} = 2620$

3

Nonlinear Equations and Models

In the previous chapter we considered discrete dynamical systems based upon iteration of linear functions. An exact solution was obtained for those that are autonomous, and a complete overall theory was established for their asymptotic behavior. We also investigated a number of mathematical models of this type from the natural and social sciences.

We now begin an investigation of nonlinear dynamical systems, setting our sights on equations and models that involve iteration of nonlinear functions. As we shall see, the problem here is much more complex, as well as much more interesting, which is why these types of dynamical systems have received so much attention during the past several decades. Due to their complexity, a number of findings in this chapter had to wait until the age of computers to be first revealed to researchers. While the behavior of linear equations was established long ago, some of what we see here, especially toward the end of the chapter, has only recently been discovered. Nonlinear dynamical systems are presently at the forefront of contemporary mathematical and scientific research.

3.1   Some Nonlinear Models

Nonlinear equations have application in many of the same areas of investigation that we previously encountered in our study of linear equations. This section provides some examples from the fields of ecology, epidemiology and micro-economics. The nonlinear models considered here often provide much better explanations than do linear ones for the complex phenomena we sometimes observe in our world. They therefore allow for greater understanding and more accurate prediction of the natural and scientific processes being investigated.

Density-Dependent Population Models

Nonlinear equations frequently arise when modeling population growth. The linear population models considered in the previous chapter, it turns out, are often overly simplistic. While they may accurately reflect the way that a relatively small population grows, different growth characteristics generally apply when the population is large.

This is due to the principle of **density dependence.** That is, the *larger* the population, the *smaller* its growth rate is likely to be. At first this may sound a bit strange, but it is a commonly recognized ecological principle. As the population of a species increases, the resources it needs for survival and continued growth may become strained. For example, food, water, space, etc., may be insufficient to support such a large population. And perhaps for the human species, there may also be an increase in pollution, crime, war, etc., all of which endanger life. If the population reaches a sufficiently large size, its very survival may be threatened. A species that was once flourishing may experience a later decline due to overpopulation.

Linear models do not take density dependence into consideration. Rather, they are based upon the assumption that the same growth characteristics always apply to the population, regardless of size. As a result, linear models like $P_{n+1} = r P_n$ always predict one of two types of behavior: either exponential growth for $r > 1$, or exponential decay to 0 or extinction of the species, for $0 < r < 1$. But except for bacteria and virus populations, this is seldom the type of population growth we see in the natural world. Nonlinear equations, on the other hand, are quite successful at modeling density dependence and their effects. Consequently, the qualitative behaviors of their solutions are much more varied, and more realistically reflect true population dynamics.

One way to construct a model that embodies density dependence is to modify the linear model $P_{n+1} = r P_n$ by replacing the constant growth rate r with a growth rate that is a *decreasing function* of the population size P_n. The model then becomes

$$P_{n+1} = R(P_n) \cdot P_n \tag{1}$$

where $R(P_n)$ represents this new growth rate function. A decreasing function is appropriate here since the growth rate should become smaller as the population becomes larger, due to the corresponding increase in density. Of course, the best function to use can never really be known, and many satisfy the right criteria. Each leads to a slightly different equation with its own specific characteristics.

Perhaps the simplest type of decreasing growth rate function is a linear one such as

$$R(P_n) = r \left(1 - \frac{P_n}{C} \right)$$

where r is a constant again called the **growth rate,** and C is a constant representing the **carrying capacity,** i.e., the largest population the environment can sustain. Substituting this $R(P_n)$ into equation (1) above yields the following model:

NONLINEAR POPULATION MODEL I

The nonlinear equation

$$P_{n+1} = r \, P_n \left(1 - \frac{P_n}{C} \right)$$

where $r \geq 0$ and $C > 0$, called the **logistic equation,** is a density-dependent model for the population P_n of a species after n generations.

This model is classified as **nonlinear** since the equation being iterated is a nonlinear function of P_n. In fact, if we replace P_n with x and P_{n+1} with y, then the resulting equation $y = rx(1 - x/C)$ is quadratic.

Because of its simplicity, the logistic equation is perhaps the most commonly used model of density dependence. But in order to have it represent a realistic model, certain restrictions often need to be put on its parameters, as well as its initial value. In particular, we must have $0 \leq r \leq 4$ and $0 \leq P_0 \leq C$, otherwise the population P_n may become negative for some n. These resrictions sometimes limit the equation's use.

Another popular density-dependent model, with fewer restrictions, results from letting the growth rate function be

$$R(P_n) = r e^{-P_n/N}$$

where r is again the constant growth rate parameter and N is a constant representing the population level that produces the maximum population. This time the following model results:

NONLINEAR POPULATION MODEL II

The exponential equation

$$P_{n+1} = r \, P_n \, e^{-P_n/N}$$

where $r \geq 0$ and $N > 0$, constitutes another density-dependent population model.

In this model the only other restriction is that $P_0 \geq 0$. We shall later see that, despite some obvious differences between this equation and the logistic equation, the dynamics of the two have much in common.

Contagious-Disease Models

Another use of nonlinear equations is to model the spread of a contagious disease through a population. To avoid possible epidemics, medical researchers may try to predict the future number of people who will be infected if a certain disease is left to run its course naturally. Depending upon their findings, health officials may decide whether or not some type of public-health intervention is warranted.

A number of factors must be considered when attempting to predict how many will be infected at any given time. For example, certain illnesses are short-lived, such as a cold or the flu, and recovery is relatively quick. Immunity to re-infection may or may not be conferred after recovery. Other diseases are chronic, and so once infected, an individual remains so.

Other factors that affect the model involve the communicability of the disease. When exposed to someone infected, is it easy or difficult to catch? Are all healthy individuals susceptible? Are all infected individuals able to infect others? And does it spread only through direct person-to-person contact, or also in some other ways?

Each of these considerations can be factored into an epidemiological model, provided of course it is nonlinear. For the simple model considered here, we will assume that everyone is either infected or susceptible, all those infected are able to infect others, which occurs through direct contact only, and there is no immunity after infection and recovery, which implies that all recovered individuals return to being susceptible. We also assume that a certain fixed fraction r of those infected recover at each step. The fraction r is called the **recovery rate,** and if recovery is not possible, we simply let $r = 0$.

In a population of size N, if we let I_n represent the number of infected individuals at any given time n, then it makes sense to say that the next number infected I_{n+1} equals the number currently infected I_n, minus the number of those who have recently recovered $r I_n$ (if that is possible), plus the number of new cases. That is,

$$I_{n+1} = I_n - r I_n + New\ Cases \qquad (2)$$

where r is a constant satisfying $0 \le r \le 1$. It is the last quantity in that equation that is the trickiest to estimate. We must ask ourselves: *What determines the number of new cases?*

The answer to this question involves a generally accepted principle: the number of recent contacts between infected and susceptible individuals, and hence the number of new cases, are each directly proportional to the size of the infected population I_n, multiplied by the size of the susceptible population, in this case $N - I_n$. The reason for this comes from epidemiological studies indicating that when either population is small, there are fewer contacts, and therefore fewer new cases. Most contacts and new cases occur when the sizes of the infected and susceptible populations are roughly equal. Concluding that the number of contacts and number of new cases are each proportional to the *product* of I_n and $N - I_n$ matches both of those observations.

Based on this, the number of new cases in our partially completed model (2) above must be

$$New\ Cases = k\,I_n\,(N - I_n) = k\,N\,I_n \left(1 - \frac{I_n}{N}\right)$$

where k is a non-negative constant. For simplicity we let $s = kN$, which is called the **infection rate.** A small positive value of s indicates that the disease is difficult to get, but a large positive value of s means the illness is easily transmitted from person to person. The construction of our disease model is now complete.

NONLINEAR INFECTION MODEL I

The equation

$$I_{n+1} = I_n - r\, I_n + s\, I_n \left(1 - \frac{I_n}{N} \right)$$

where $0 \leq r \leq 1$, $s \geq 0$ and $N > 0$, models the number of individuals I_n at time n who are infected with a contagious disease.

Note that this model is nonlinear. In fact, similar to the logistic equation, it is quadratic and therefore has the same kind of restrictions on its parameters as did the logistic model. Among these restrictions are $0 \leq 1 - r + s \leq 4$ and $0 \leq I_0 \leq N$.

But this is not the only disease model possible. Others will result if we make slightly different assumptions concerning the spread of the disease. For example, the number of new cases may instead be

$$New\ Cases = k\, I_n^2\, (N - I_n)$$

which also satisfies the epidemiological principles mentioned earlier. Substituting this into (2), factoring and letting $s = kN^2$ would yield in this case:

NONLINEAR INFECTION MODEL II

The equation

$$I_{n+1} = I_n - r\, I_n + \frac{s}{N}\, I_n^2 \left(1 - \frac{I_n}{N} \right)$$

with the same restrictions on the parameters, also models the spread of a contagious disease.

In both of these simple disease models, after someone is infected and recovers, that person is once again susceptible to the disease, i.e., there is no immunity to re-infection. In a situation where such immunity does occur, a more complex model involving a *system* of nonlinear equations would be needed. These types of models will be investigated in Chapter 5.

Economic Models

The next type of model we consider comes from economic theory. As we did in Chapter 2, suppose we try to predict how the price P_n of a certain product will change over time. However, rather than making the overly simplistic assumptions that led to the Linear Price Model derived there, we now no longer assume that supply remains constant, nor do we necessarily employ all linear functions in the construction of our models.

It is commonly accepted that prices are pushed upward as the demand from consumers for that product increases, and prices are pushed downward as the supply of that product made available from vendors increases. From this, it is reasonable to conclude that the next price P_{n+1} will increase or decrease from its present price P_n according to an equation such as

$$P_{n+1} = P_n + Demand\ Force - Supply\ Force$$

Of course, the true economic forces acting on the price can never really be known, but we can arrive at a model for them by observing the following: as the current demand D_n increases, the *Demand Force* increases continuously from negative to positive; and as the current supply S_n increases, the *Supply Force* increases continuously from negative to positive.

With these principles in mind, suppose we let

$$Demand\ Force\ = a_1\ D_n - b_1 \quad and \quad Supply\ Force\ = a_2\ S_n - b_2$$

where a_1, a_2, b_1, b_2 are non-negative constants. These are about the simplest equations that satisfy the supply and demand criteria described above.

Our model for the price now becomes

$$P_{n+1} = P_n + (a_1\ D_n - b_1) - (a_2\ S_n - b_2) \tag{3}$$

but we are not quite finished. It is also well known that as prices increase, the supply that is provided by vendors increases, but the demand from consumers decreases. In other words, while the supply S_n is *directly* proportional to the price, the demand D_n is *inversely* proportional to the price. If we assume that S_n and D_n react immediately to the price P_n, this means that

$$D_n = \frac{c_1}{P_n} \quad and \quad S_n = c_2\ P_n \tag{4}$$

where c_1 and c_2 are non-negative constants.

Substituting these into (3) yields

$$P_{n+1} = P_n + \left(a_1\ \frac{c_1}{P_n} - b_1 \right) - (a_2\ c_2\ P_n - b_2) = \frac{a_1 c_1}{P_n} + (1 - a_2\ c_2)\ P_n + b_2 - b_1$$

Finally, if for simplicity we let

$$a = a_1 c_1, \quad b = 1 - a_2 c_2 \quad and \quad c = b_2 - b_1$$

then we arrive at our model.

NONLINEAR PRICE MODEL I

The nonlinear equation

$$P_{n+1} = \frac{a}{P_n} + b\ P_n + c$$

with $a \geq 0$, $b \leq 1$ and any value c, models the price P_n of a product at time n.

We remark that a number of other restrictions besides those mentioned would have to be put on the parameters a, b and c in order to keep prices P_n from becoming negative. The same is true of the price models below.

Many other versions of the model can be constructed using the same economic principles, but with different assumptions on the relationships between price, supply and demand.

If, for example, we want to prevent solutions from becoming unbounded when the price is close to 0, we could instead assume the demand in (4) has a form such as

$$D_n = \frac{c_1}{P_n + d}$$

In this case equation (3) would instead become:

NONLINEAR PRICE MODEL II

The equation

$$P_{n+1} = \frac{a}{P_n + d} + b\,P_n + c$$

where a and d are non-negative, and $b \leq 1$ is another nonlinear model of the price P_n at time n.

On the other hand, if we assume that the supply and/or demand forces above are power functions instead of linear, we might end up with a price model such as the following:

NONLINEAR PRICE MODEL III

Another nonlinear price model is

$$P_{n+1} = \frac{a}{P_n^k} + b\,P_n + c$$

where a and k are non-negative, and $b \leq 1$.

Other Models

There are some other interesting nonlinear iteration models from various disciplines that could be developed here, but the principles they are based upon and the dynamics they exhibit often parallel those already introduced. For example, to model the spread through the population of a new sociological trend, or the prevalence of a newly released product or technological tool, such as personal computers or cell phones, the equation is essentially that of a disease model. And a price equation can also be used as a learning theory model for the amount learned over time with regard to a certain topic. Some of these models will be described in the exercises.

Exercises 3.1

1. Construct the logistic model corresponding to each of the following:
   (a) $r = 2.75, C = 50{,}000$ (b) $r = 3.5, C = 10^6$
2. Identify r and C for each of the following logistic models:
   (a) $P_{n+1} = 3P_n - 0.006P_n^2$ (b) $P_{n+1} = \dfrac{9P_n}{4} - \dfrac{P_n^2}{1000}$

3. Determine the carrying capacity for a logistic model in which:

   (a) A population of 25,000 causes extinction in the next step.
   (b) A population of $P_0 = 100,000$ produces the maximum population in the next step.

4. Find the logistic model that satisfies:

   (a) $C = 20,000$, $P_0 = 5000$, $P_1 = 3000$.
   (b) $P_0 = 5000$, $P_1 = 8000$, $P_2 = 6000$.
   (c) If $P_0 = 5000$ then P_1 is twice P_0, and if $P_0 = 8000$ then P_1 is half P_0.
   (d) The maximum population $P_1 = 45,000$ occurs when $P_0 = 35,000$.

5. Construct density dependent population models that have the following growth rate functions $R(P_n)$:

   (a) $R(P_n) = r(1 - P_n/C)^2$ (c) $R(P_n) = r(1 - P_n^2/C^2)$

   (b) $R(P_n) = \dfrac{r}{1 + P_n^2/C^2}$ (d) $R(P_n) = re^{-P_n^2/N^2}$

Exercises 6–8 refer to Nonlinear Population Model II: $P_{n+1} = rP_n e^{-P_n/N}$.

6. Construct the model with: (a) $r = 4$, $N = 65,000$; (b) $r = 10$, $N = 10^7$.

7. Identify r and N for each of the following models:

   (a) $P_{n+1} = 6.5P_n e^{-0.005P_n}$ (b) $P_{n+1} = P_n e^{2 - P_n 10^{-8}}$

8. Find the model that satisfies:

   (a) $N = 1000$, $P_0 = 100$, $P_1 = 250$
   (b) $r = 5$, $P_0 = 100$, $P_1 = 250$
   (c) $P_0 = 2000$, $P_1 = 6000$, $P_2 = 4000$
   (d) The maximal population $P_1 = 7500$ occurs when $P_0 = 5000$.

Exercises 9–11 refer to Nonlinear Infection Model I: $I_{n+1} = I_n - rI_n + sI_n(1 - I_n/N)$.

9. Construct the model with:

   (a) $r = 0.3$, $s = 2$, $N = 10,000$ (b) $r = 1$, $s = 3.5$, $N = 10^6$

10. Identify r, s and N for each of the following models:

   (a) $I_{n+1} = 0.7I_n + 2.9I_n(1 - I_n/7500)$
   (b) $I_{n+1} = 3.25I_n(1 - 0.0025I_n)$

   (c) $I_{n+1} = 10I_n - \dfrac{I_n^2}{100}$ in a population of size 1000

   (d) $I_{n+1} = 5.75I_n - \dfrac{I_n^2}{1000}$ in a population of size 5000

11. Construct a model that satisfies:

   (a) Each week 80% of those ill in a population of size 10^6 recover, and if 1000 are ill one week then the following week 1500 are ill.
   (b) Each week 40% of those ill in a population of size 400,000 are still sick, and the maximum number of new cases possible in any week is 50,000.
   (c) In a population of size 10,000, recovery is not possible, and if 100 are ill, that number will double in the following week.

12. Construct a disease model assuming that at each step rI_n recover and the number of new cases is proportional to the product of:

   (a) I_n^2 and $(1 - I_n/N)^2$ (b) I_n^2 and $(1 - I_n^2/N^2)$ (c) I_n and $e^{-I_n/N}$

Exercises 13–16 refer to Nonlinear Price Model I: $P_{n+1} = \dfrac{a}{P_n} + bP_n + c.$

13. Construct the model with: (a) $a = 3, b = 0.5, c = 0$ (b) $a = 1, b = -2, c = -1$

14. Identify a, b and c for each of the following models:

    (a) $P_{n+1} = -0.1\left(P_n - \dfrac{1}{P_n}\right)$

    (b) $P_{n+1} = \dfrac{2}{P_n} + 0.1P_n - 1.75$

    (c) $P_{n+1} = \dfrac{2 - P_n - 5P_n^2}{P_n}$

15. (a) If $a = 1$, $b = 0.5$ and $c = 2$, there is one price P_0 that gives the minimum price for P_1. Find that P_0 and P_1.

    (b) If $a = 1$, $b = -1$ and $c = 2$, is there a price P_0 that gives the minimum price for P_1? Explain.

    (c) If $a = 1$, $b = 0$ and $c = 2$, is there a price P_0 that gives the minimum price for P_1? Explain.

16. Find the model that satisfies each of the following:

    (a) $c = 0$, $P_0 = 1$, $P_1 = 4$, $P_2 = 2$

    (b) $c = 2$, and $P_0 = 3.5$ gives the minimum price $P_1 = 2.5$

17. For the price model $P_{n+1} = \dfrac{9}{P_n + 1} + bP_n - 4$:

    (a) For $b = 1$ find the minimum price P_1 for any $P_0 > 0$

    (b) For $b = 0$ find the maximum price P_1 for any $P_0 \geq 0$

18. When modeling how a new technological tool spreads through a population, the following principles may apply: In a population of size N, if the number using the technology in any year is T_n, then T_{n+1} equals T_n minus a fraction r of those T_n who revert, plus the number of new users, which we assume is proportional to the product of the number of users and the number of non-users. Show that this produces the model $T_{n+1} = T_n - rT_n + sT_n(1 - T_n/N)$, where $0 \leq r \leq 1$ and $s > 0$.

19. What model results if in the previous exercise we assume the number of new users is proportional to the product of the square of the number of users and the square of the number of non-users.

20. When learning a new topic the following principles may apply: If the current amount learned is A_n, then A_{n+1} equals A_n minus the fraction r of A_n forgotten, plus the new amount learned, which we assume is inversely proportional to the amount already learned A_n. Show that this produces the model $A_{n+1} = A_n - rA_n + \dfrac{s}{A_n}$, where $0 \leq r \leq 1$ and $s > 0$.

21. What model results if in the previous exercise we assume that the new amount learned is inversely proportional to A_n^k for some $k > 0$.

22. Many nonlinear models we have seen involve quadratic functions of the form $y = ax^2 + bx$. To obtain such a model from a collection of data points $(x_1, y_1), \ldots, (x_N, y_N)$, a form of regression can be used to find the values of a and b that minimize $\sum_{i=1}^{N}(ax_i^2 + bx_i - y_i)^2$. Derive a set of *normal equations* that a and b must satisfy in this case.

23. Using the data points $(1, 3)$, $(2, 3)$, $(3, 2)$, $(4, 1)$, find a model of the form $y = ax^2 + bx$ by solving the normal equations for a and b from the previous exercise.

24. Some nonlinear models we have seen involve functions of the form $y = a/x + bx$. To obtain such a model from collection of data points $(x_1, y_1), \ldots, (x_N, y_N)$, a form of regression can again be used to find the values of a and b that minimize the sum of the squares of the errors. Derive a set of normal equations for a and b in this case.

25. Using the data points $(1, 5)$, $(2, 2)$, $(3, 2)$, $(4, 1)$, $(5, 1)$, find a model of the form $y = a/x + bx$ by solving the normal equations for a and b from the previous exercise.

▰▰▰▰ Computer Projects 3.1

1. Use Exercise 22 and a computer to find the least squares model of the form $y = ax^2 + bx$ for the data points $(50, 100)$, $(105, 160)$, $(110, 200)$, $(130, 250)$, $(150, 300)$, $(200, 400)$, $(215, 450)$, $(245, 350)$, $(250, 325)$, $(270, 300)$, $(300, 250)$, $(325, 250)$, $(350, 200)$, $(375, 100)$, $(400, 50)$.

2. Use Exercise 24 and a computer to find the least squares model of the form $y = a/x + bx$ for the data points $(1, 6)$, $(1, 5)$, $(1.25, 5)$, $(1.5, 4.5)$, $(2, 3)$, $(2.5, 3)$, $(2.5, 3.25)$, $(3, 3.5)$, $(3.5, 3.5)$, $(3.75, 4)$, $(4, 4.5)$, $(4.25, 4.5)$, $(5, 4.75)$, $(5.5, 5)$, $(6, 5.5)$.

SECTION
3.2   Autonomous Equations and Their Dynamics

Now that we have seen how rich with applications nonlinear dynamical systems are, in this section we begin an investigation of their properties, their solutions and their qualitative dynamics.

Autonomous and Non-Autonomous Equations

An important property, first seen in Chapter 2 for linear equations, is shared by the models of the previous section. Namely, the variable n does not appear in the equation except as a subscript on the discrete quantity being investigated. Whenever this occurs, just as in the linear case, the equation is called *autonomous*. However, now we are talking about autonomous nonlinear equations. In this chapter we focus mainly on this type of problem, so we define the following.

DEFINITION

The general form of an **autonomous nonlinear equation** for x_n is

$$x_{n+1} = f(x_n)$$

where $f(x)$ is a nonlinear real valued function. When an initial value x_0 is provided, a unique solution x_n exists for all $n \geq 0$.

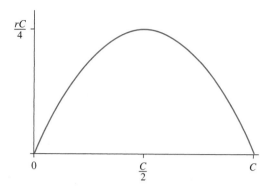

FIGURE 3.1

In addition to assuming $f(x)$ is real valued, for technical reasons we must assume its range is either contained within or exactly equals its domain. This way we know for certain a solution will always exist. We will also generally assume differentiability of $f(x)$, or at least continuity, at all points of interest in the domain. For the types of models we consider, these are reasonable assumptions.

EXAMPLE 1

Determine the function $f(x)$ corresponding to the logistic equation $P_{n+1} = rP_n(1 - P_n/C)$ and discuss its general features.

Solution: Replacing P_n by x on the right-hand side of that equation yields the function $f(x) = rx(1 - x/C)$. The general shape of its graph $y = f(x)$ for a typical choice of r and C is shown in Figure 3.1. As can be seen, it is a quadratic function that increases from the origin to a maximum and then decreases again. Its shape suggests that small populations close to 0 grow slightly larger with each successive step. But when the population is sufficiently large, closer to C, growth is inhibited and the population decreases again. This reflects density dependence. Note that this model can be used only for populations between 0 and C, and for r values between 0 and 4; otherwise, solutions become negative and the model breaks down.   ■

A function $f(x)$ that has exactly one relative maximum or minimum (but not both), such as the quadratic function above, is sometimes called **unimodal.** Many of the models we considered in the last section involve unimodal functions, including the next.

EXAMPLE 2

Determine the function $f(x)$ corresponding to Nonlinear Population Model II of Section 3.1, i.e., $P_{n+1} = rP_ne^{-P_n/N}$, and compare its advantages and disadvantages to those of the logistic equation.

Solution: In this case $f(x) = rxe^{-x/N}$. A typical graph is shown in Figure 3.2, which indicates that this function is also unimodal. Among its advantages is that, unlike the logistic equation where both the population and r must never exceed certain values, this exponential

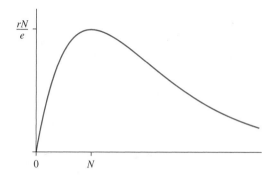

FIGURE 3.2

model makes sense for all positive population sizes and r values. A disadvantage of this model, however, is its complexity, since it is based upon an exponential rather than a quadratic function. ∎

Autonomous linear equations $x_{n+1} = ax_n + b$ treated in Chapter 2 are also of the form $x_{n+1} = f(x_n)$ where $f(x) = ax + b$, and so could be included within the investigation of this chapter. But since we have already established their dynamics, we will frequently use what we have learned about them to give some direction in our study of nonlinear equations.

Unlike those linear problems, where there is always a formula for the exact solution, nonlinear equations are far more difficult to solve. In fact, it is usually impossible to find explicit formulas for their solutions, even if autonomous. For example, there are no exact solutions known for the logistic equation or any other model described in Section 3.1 (except in some special cases). Consequently, we will seldom even attempt to find formulas for exact solutions. Instead, the focus will be mainly on either direct numerical iteration or techniques for determining the qualitative dynamics, especially the asymptotic behavior.

We will usually omit the word *autonomous* when referring to such equations, since we assume so for virtually all nonlinear problems we treat. As we shall see, the task we are faced with in dealing just with autonomous equations is substantial, even without considering the non-autonomous case.

The significance of identifying a model as being autonomous cannot be overstated. The primary implication is that the equation being iterated $x_{n+1} = f(x_n)$ does not vary with time or with each step of the iteration process. While the values of x_n are certainly likely to change with n, the model itself is *static*. The rule for determining x_1 when x_0 is known is the same as that for computing x_{n+1} knowing x_n. That general rule is given by the function $f(x)$. In short,

$$\text{Next Step} = f(\text{Present Step})$$

As a result, the analysis of either the model or its solutions can be reduced to an investigation of the function $f(x)$.

This is not true for **non-autonomous equations,** which have the form

$$x_{n+1} = f(n, x_n)$$

and involve functions $f(n, x)$ of *two* independent variables. Models such as these depend upon n and therefore *do* change with each successive step.

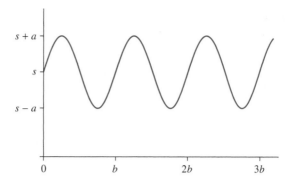

FIGURE 3.3

EXAMPLE 3

The non-autonomous equation

$$I_{n+1} = I_n - r\,I_n + (s + a\sin 2\pi n/b)\,I_n\left(1 - \frac{I_n}{N}\right)$$

where r, s, a, b, N are parameters, is another version of a contagious disease model. Identify and discuss the significance of its corresponding function $f(n, x)$.

Solution: Since n and I_n are the only variables on the right-hand side of the equation (the others are parameters), we therefore have a function of two variables:

$$f(n, x) = x - r\,x + (s + a\sin 2\pi n/b)\,x\left(1 - \frac{x}{N}\right)$$

In this model the constant s representing the infection rate parameter in Nonlinear Infection Model I of Section 3.1 has been replaced with $s + a\sin(2\pi n/b)$, which is a periodic function of n. The graph of $y = s + a\sin(2\pi n/b)$ is shown in Figure 3.3. The implication is that the communicability of the illness *oscillates periodically,* perhaps with the season or with some other fluctuation in the environment. Sometimes, this is a reasonable assumption to make for the spread of a disease, or for other natural and social processes.   ■

Some Exact Solutions

The first observation we will make with regard to the solutions of nonlinear equations is that they exhibit a wider diversity of possible behavior than do linear ones. This can be seen by first finding exact solutions for a few rare cases for which this is possible, and then iterating some others numerically.

EXAMPLE 4

Find a formula for the exact solution of $x_{n+1} = x_n^3$, and use it to determine the asymptotic behavior for (a) $x_0 = 2$ and (b) $x_0 = 1/2$.

Solution: By looking at the first few iterates

$$x_1 = x_0^3, \quad x_2 = x_1^3 = \left(x_0^3\right)^3 = x_0^{3^2}, \quad x_3 = x_2^3 = \left(x_0^{3^2}\right)^3 = x_0^{3^3}, \quad \dots$$

it is clear that the exact solution for all n is $x_n = x_0^{3^n}$. In part (a) therefore, where $x_0 = 2$, this means that $x_n = 2^{3^n}$. To investigate the asymptotic behavior, we compute

$$\lim_{n \to \infty} x_n = \lim_{n \to \infty} 2^{3^n} = 2^\infty = +\infty$$

and so the solution diverges to $+\infty$ as $n \to \infty$. In part (b), where $x_0 = 1/2$, for all n we have $x_n = (1/2)^{3^n}$, and so

$$\lim_{n \to \infty} x_n = \lim_{n \to \infty} \left(\frac{1}{2}\right)^{3^n} = \lim_{n \to \infty} \frac{1}{2^{3^n}} = \frac{1}{2^\infty} = 0$$

This means the solution converges to 0 as $n \to \infty$. ∎

It is easily checked in this example that all solutions for $-1 \le x_0 \le 1$ remain finite, while solutions for all other initial conditions diverge to $\pm\infty$. This type of behavior, where different intervals of initial conditions lead to very different dynamics, has not been observed and indeed never occurs for linear equations.

From this example one may see that there is a similar formula for the solution of *any* iterative equation of the form $x_{n+1} = x_n^a$, where a is constant.

THEOREM   *An Exact Solution*

The equation $x_{n+1} = x_n^a$ has the exact solution $x_n = x_0^{a^n}$ for all n. ∎

Of course, here we must assume that either $x_0 \ge 0$ or a is an integer, otherwise there may not exist any real solution. An exact solution can also be derived for $x_{n+1} = b x_n^a$ where a and b are constants, but that is left as an exercise.

EXAMPLE 5

Find the exact solution of $x_{n+1} = \sqrt{x_n}$ for any $x_0 \ge 0$, and use it to determine the asymptotic behavior for (a) $x_0 = 0$ and $x_0 = 1$; (b) any $x_0 > 0$.

Solution: With the problem written as $x_{n+1} = x_n^{1/2}$, we see that the exact solution for any $x_0 \ge 0$ must be $x_n = x_0^{1/2^n}$. For part (a) if $x_0 = 0$ or $x_0 = 1$ then

$$x_n = 0^{1/2^n} = 0 \quad \text{or} \quad x_n = 1^{1/2^n} = 1$$

respectively for all $n \ge 0$. For part (b) with $x_0 > 0$ we have

$$\lim_{n \to \infty} x_n = \lim_{n \to \infty} x_0^{1/2^n} = x_0^{1/\infty} = x_0^0 = 1$$

which means that all solutions x_n converge to 1 when $x_0 > 0$. This implies that 1 behaves somewhat as did a stable fixed point of a linear equation, and 0 as an unstable fixed point. Except that now both occur for the same equation. Again, such behavior never occurs for linear equations. ∎

The examples above suggest that the analysis of nonlinear equations will not be as simple as it was for linear ones. For example, if different ranges of initial conditions may lead to very different asymptotic behavior, then how many choices of initial conditions must be tried and how many iterations must be performed to make meaningful long-term predictions? And what other types of complexity may there be for nonlinear equations that we have not yet encountered?

Some Numerical Solutions

The following examples demonstrate that indeed there are some additional and rather surprising types of dynamics not seen above. Since no exact solutions are known for these equations, we must rely on other means of analysis. For now, the only alternative is numerical iteration.

EXAMPLE 6

In the logistic model assume the carrying capacity is $C = 10,000$ and the initial population is $P_0 = 2500$. Use a computer to generate a table and graph of the first 20 iterates for (a) $r = 3.2$ and (b) $r = 3.9$.

Solution: The data for parts (a) and (b) are shown in Tables 3.1(a) and 3.1(b) respectively, and the graphs are shown in Figures 3.4(a) and 3.4(b) respectively. The effects of density dependence can be observed in each solution. The initially small population $P_0 = 2500$ grows larger, but never exceeds the carrying capacity $C = 10,000$. Whenever the population approaches that maximum level, it falls sharply in the next iteration, only to rise again in the steps that follow. In each case, an oscillation is the result, but that's where the similarity ends.

TABLE 3.1

n	P_n	n	P_n	n	P_n	n	P_n
0	2500			0	2500		
1	6000	11	5134	1	7313	11	4074
2	7680	12	7994	2	7664	12	9416
3	5702	13	5131	3	6981	13	2146
4	7842	14	7995	4	8219	14	6573
5	5415	15	5131	5	5709	15	8785
6	7945	16	7995	6	9554	16	4162
7	5225	17	5130	7	1662	17	9476
8	7984	18	7995	8	5404	18	1937
9	5151	19	5130	9	9686	19	6091
10	7993	20	7995	10	1185	20	9286

(a)   (b)

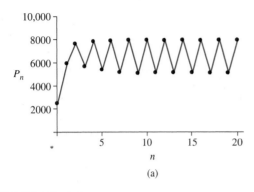
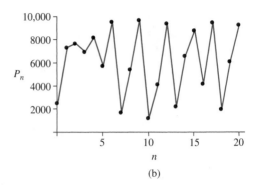

(a)   (b)

FIGURE 3.4

The solution of (a) seems to eventually settle down into an oscillation between two values, but no obvious pattern can be discerned for (b). Neither of these two types of behavior is possible for linear equations. ■

EXAMPLE 7

For $r > 0$ Nonlinear Infection Model II of Section 3.1,

$$I_{n+1} = I_n - r\, I_n + \frac{s}{N}\, I_n^2 \left(1 - \frac{I_n}{N}\right)$$

exhibits the **threshold** effect. That is, if I_n ever falls below a certain threshold level, the disease dies out regardless of the infection rate s. But for any I_n above that threshold value, the disease spreads more widely throughout the population. Verify this for $r = 1$, $s = 5.83$ and $N = 10{,}000$ by using a computer to generate and graph the first 20 iterates starting with (a) $I_0 = 2000$ and (b) $I_0 = 6000$.

Solution: The data for parts (a) and (b) are shown in Tables 3.2(a) and 3.2(b) respectively, and the graphs in Figures 3.5(a) and 3.5(b) respectively. For part (a) where $I_0 = 2000$ the solution decays to 0, which suggests that 2000 lies below the threshold. The solution for part (b) where $I_0 = 6000$ is evidently above the threshold, since the number infected sustains itself and eventually settles into a periodic oscillation. If one looks carefully, four different values may be discerned that keep repeating. Again, this type of behavior is not possible for linear equations. ■

Numerical iteration can certainly provide a solution for a specific nonlinear equation and initial condition, up to any finite value of n. Numerically iterating for several initial conditions and parameters can also indicate the types of asymptotic behaviors possible for the problem. But those alone are insufficient to provide an understanding of the underlying principles that are at play. Additional techniques are needed to help explain the dynamics we are seeing, why they occur, and perhaps also to allow us to determine in advance how a particular mathematical model will behave.

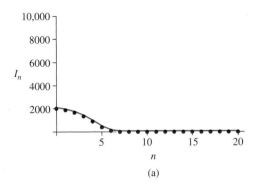

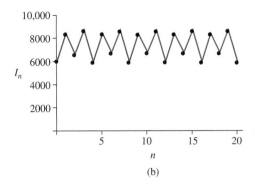

FIGURE 3.5

TABLE 3.2

n	I_n	n	I_n	n	I_n	n	I_n
0	2000			0	6000		
1	1866	11	0	1	8395	11	8635
2	1651	12	0	2	6594	12	5933
3	1326	13	0	3	8634	13	8346
4	889	14	0	4	5937	14	6717
5	420	15	0	5	8349	15	8636
6	99	16	0	6	6709	16	5932
7	6	17	0	7	8636	17	8346
8	0	18	0	8	5931	18	6718
9	0	19	0	9	8345	19	8635
10	0	20	0	10	6720	20	5932

(a)   (b)

Fixed Points of Nonlinear Equations

Recall that a fixed point of the linear equation $x_{n+1} = ax_n + b$ is a point p such that if $x_0 = p$ then all $x_n = p$. Such a p can be found by substituting x in place of both x_n and x_{n+1}, and solving the resulting equation $x = ax + b$. Provided $a \neq 1$, this always yields the fixed point $p = b/(1 - a)$, which corresponds to the equilibrium point for the application being modeled. Since fixed points play such a large role in the dynamics of linear equations, it is natural to expect that this should also be true in the nonlinear case.

A fixed point p of $x_{n+1} = f(x_n)$ should behave like one for a linear equation: if $x_0 = p$ then all $x_n = p$. Any such p can again be found by first substituting x in place of both x_n

and x_{n+1}, but this time solving $x = f(x)$, or equivalently $f(x) = x$. We therefore define the following:

DEFINITION

A point p is called a **fixed point** of $x_{n+1} = f(x_n)$, or more simply a **fixed point** of $f(x)$, if $f(p) = p$. Letting $x_0 = p$ will make $x_n = p$ for all $n \geq 0$, and so p is an **equilibrium point** of the model.

Of course, solving $f(x) = x$ to find such a p may be quite difficult, if not impossible, depending upon the complexity of $f(x)$. Complicating the situation is the fact that unlike linear equations a nonlinear equation of the form $f(x) = x$ may have multiple solutions, or perhaps no solution. This means there may not be a unique fixed point.

Despite these difficulties, fixed points of nonlinear equations can often be found, as the following examples demonstrate.

■ EXAMPLE 8

Find all fixed points, if any, of $x_{n+1} = x_n^2$.

Solution: We must solve

$$f(x) = x^2 = x \quad \text{or} \quad x^2 - x = 0$$

Factoring yields $x(x - 1) = 0$. So there are two fixed points 0 and 1.   ■

In the previous example note that either solution 0 or 1 can be verified to be truly a fixed point by substituting that solution for x_n into the original equation $x_{n+1} = x_n^2$, and seeing if the same value emerges for x_{n+1}. That is,

$$\text{for } x_n = 0: \quad x_{n+1} = x_n^2 = 0^2 = 0 \quad \text{and} \quad \text{for } x_n = 1: \quad x_{n+1} = x_n^2 = 1^2 = 1$$

This verifies both solutions. This can always be done to verify the correctness of the calculation of a fixed point.

■ EXAMPLE 9

Find all equilibrium price levels, if any, for Nonlinear Price Model I of Section 3.1 with $a = 1, b = 1$ and $c = 0$, i.e., $P_{n+1} = \dfrac{1}{P_n} + P_n$.

Solution: Here we must solve

$$f(x) = \frac{1}{x} + x = x \quad \text{or} \quad \frac{1}{x} = 0$$

Since no real solutions exist, there are no fixed points and no equilibrium prices.   ■

■ EXAMPLE 10

Find the equilibrium population levels, if any, for Nonlinear Population Model II of Section 3.1 with $r = 2$ and $N = 10,000$, i.e., $P_{n+1} = 2P_n e^{-P_n/10,000}$.

Solution: This time we must solve the equation

$$f(x) = 2xe^{-x/10,000} = x \quad \text{or} \quad x(2e^{-x/10,000} - 1) = 0$$

Obviously, $x = 0$ is one solution. The other must satisfy

$$e^{-x/10,000} = 1/2 \quad \text{or} \quad \frac{-x}{10,000} = \ln(1/2)$$

This means

$$x = -10,000 \ln(1/2) = -10,000(\ln 1 - \ln 2) = -10,000(0 - \ln 2) = 10,000 \ln 2$$

So there are two fixed points: 0 and $10,000 \ln 2$, and two equilibrium populations: 0 and $10,000 \ln 2 \approx 6931$. A population starting at either one of these should remain there (at least for a while).

Since finding the fixed point $10,000 \ln 2$ turned out to be rather complex, it is wise to verify that our calculations were correct. Substituting $10,000 \ln 2$ for P_n in the original equation gives

$$P_{n+1} = 2 \cdot (10,000 \ln 2) \cdot e^{-(10,000 \ln 2)/10,000} = 20,000 \ln 2 \cdot e^{-\ln 2}$$

$$= 20,000 \ln 2 \left(\frac{1}{2} \right) = 10,000 \ln 2$$

Since the same value $10,000 \ln 2$ is obtained for P_{n+1}, this must correctly be the fixed point.   ■

EXAMPLE 11

The equation

$$P_{n+1} = \frac{r P_n}{1 + P_n^2/C^2}$$

for positive r and C is sometimes used as a density-dependent population model. The graph of its corresponding function resembles that of Nonlinear Population Model II, but no exponentials are involved. Find all equilibrium population levels of this model, if any, for $r = 5$ and $C = 1000$.

Solution: Since

$$f(x) = \frac{5x}{1 + x^2 10^{-6}} \quad \text{we must solve} \quad \frac{5x}{1 + x^2 10^{-6}} = x$$

If we multiply by the denominator and bring all terms to one side, we obtain

$$x(1 + x^2 10^{-6}) - 5x = 0 \quad \text{which simplifies to} \quad x(x^2 - 4,000,000) = 0$$

There are three solutions $x = 0$, $x = 2000$ and $x = -2000$, but since this is a population model, we are not interested in the negative one. The equilibrium populations are therefore 0 and 2000. These can easily be verified to be correct by substituting them into the original equation.   ■

We see from these examples that a nonlinear equation $x_{n+1} = f(x_n)$ may indeed have one or many fixed points, or perhaps none. This differs from what we know of linear equations $x_{n+1} = ax_n + b$, where a unique fixed point always exists (as long as $a \neq 1$).

Stability of Nonlinear Equations

Another difference between linear and nonlinear dynamical systems involves the concept of stability. As they do for linear equations, the notions of stability and instability of fixed points exist for nonlinear ones as well. But their definitions must be modified somewhat.

DEFINITION

A fixed point p of $x_{n+1} = f(x_n)$ is **locally stable** or an **attractor** if $|x_{n+1} - p| < |x_n - p|$, and so $\lim\limits_{n \to \infty} x_n = p$, for all x_0 close to p. Otherwise, p is **unstable** or a **repeller.**

As the word *locally* implies, stability of a fixed point p of a nonlinear function generally involves an interval around p in which solutions converge to p. The largest such interval is sometimes called the **basin of attraction** of the fixed point. What happens outside that interval is irrelevant to local stability. Note that this issue does not arise for linear equations, since either all solutions converge to the fixed point p, or all diverge to $\pm\infty$ (except for p itself). That type of stability is often called **global stability,** and the corresponding basin of attraction consists of the entire real line. Since local stability is the only type of stability we consider for nonlinear equations, we often dispense with the words *local* and *locally*.

EXAMPLE 12

Find the fixed points of $x_{n+1} = \sqrt{x_n}$ and determine their stability.

Solution: Solving $\sqrt{x} = x$ gives two fixed points 0 and 1. In Example 5 it was found that all solutions for $x_0 > 0$ converge to 1 as $n \to \infty$. This means that 1 is an attractor, and in fact it is a global attractor if we exclude $x_0 = 0$. On the other hand, 0 is unstable since all solutions with $x_0 \neq 0$ diverge from 0 as $n \to \infty$.   ∎

EXAMPLE 13

Find the fixed points of $x_{n+1} = x_n^3$ and determine their stability.

Solution: Solving $f(x) = x^3 = x$ gives three fixed points 0, 1 and -1. In Example 4 the exact solution $x_n = x_0^{3^n}$ was found. It can easily be seen that if $-1 < x_0 < 1$, then

$$\lim_{n \to \infty} x_n = \lim_{n \to \infty} x_0^{3^n} = x_0^\infty = 0$$

which implies that 0 is an attractor. Also for $x_0 < -1$ or $x_0 > 1$,

$$\lim_{n \to \infty} x_n = \lim_{n \to \infty} x_0^{3^n} = x_0^\infty = \pm\infty$$

which means that the other two fixed points are repellers.   ∎

EXAMPLE 14

For Nonlinear Infection Model I of Section 3.1

$$I_{n+1} = I_n - r I_n + s I_n \left(1 - \frac{I_n}{N}\right),$$

TABLE 3.3

n	I_n	n	I_n		n	I_n	n	I_n
0	1000				0	5000		
1	550	11	3		1	1750	11	8
2	315	12	2		2	897	12	5
3	184	13	1		3	498	13	3
4	109	14	1		4	286	14	2
5	65	15	0		5	168	15	1
6	39	16	0		6	99	16	1
7	23	17	0		7	59	17	0
8	14	18	0		8	35	18	0
9	8	19	0		9	21	19	0
10	5	20	0		10	13	20	0

(a)   (b)

it is easily checked by direct substitution that 0 is always a fixed point. (a) Determine whether 0 is stable or unstable for $r = 0.9$, $s = 0.5$ and $N = 10,000$. (b) Discuss the significance of the result.

Solution: (a) In this case, a formula for the exact solution is not available. So, in order to determine stability, for now we must rely just upon numerical computations. Table 3.3(a) shows the result of 20 iterations beginning with $I_0 = 1000$, and Table 3.3(b), 20 more iterations for $I_0 = 5000$. Based on this data, 0 appears to be stable.

(b) The significance of knowing that 0 is a stable fixed point is that we can conclude the disease we are modeling will likely die out on its own without intervention. Since 0 appears to attract all solutions, we can be confident in this conclusion even though we will probably never know the exact number P_0 of those initially ill. One of the benefits of stability is that, since all nearby solutions are attracted to the fixed point, precisely measuring P_0 is unnecessary. The dynamics are identical for all points in the basin of attraction. ∎

EXAMPLE 15

For the model $P_{n+1} = 2P_n e^{-P_n/10,000}$ discussed in Example 10, determine the stability of its equilibrium populations 0 and $10,000 \ln 2$.

Solution: Again, there is no formula for the exact solution, and so we iterate numerically, this time displaying the results graphically. Figures 3.6(a) and 3.6(b) show the graphs of the first 20 iterates for $P_0 = 1000$ and $P_0 = 5000$ respectively. It appears that 0 is unstable, but $10,000 \ln 2$, or about 6931, is a stable equilibrium population. Again this means that we can be fairly confident that the population will stabilize at around 6931, regardless of the initial population P_0 (as long as it's not 0, that is). ∎

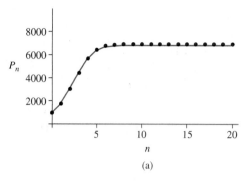

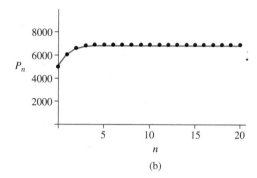

FIGURE 3.6

Since we saw earlier that different initial conditions could lead to different asymptotic behavior for a nonlinear equation, perhaps in the two previous examples there exist some initial values for which solutions do not converge to the respective fixed point. How do we know for sure that these fixed points are stable and attract *all* nearby solutions? If there are solutions that do not converge, this puts in doubt the validity of many conclusions that might be made with regard to the model.

What is needed is a more reliable method of determining the stability of such fixed points, preferably a method that does not involve direct numerical iteration of the equation, nor use of the formula for the exact solution, which is rarely available. This will be taken up in the next section.

Exercises 3.2

1. In each of the following identify the function $f(x)$ being iterated:

   (a) $x_{n+1} = 2x_n - \dfrac{7}{x_n}$

   (b) $x_{n+1} = \dfrac{5x_n}{x_n + 8}$

2. In each of the following write the equation corresponding to $x_{n+1} = f(x_n)$:

   (a) $f(x) = \dfrac{x^2}{5} + \dfrac{5}{x^2}$

   (b) $f(x) = \sqrt{x^2 + x}$

3. Determine which of the following are autonomous and which are non-autonomous. In each case identify the function $f(x)$ or $f(n, x)$ being iterated.

   (a) $x_{n+1} = \sqrt{x_n + n}$

   (c) $x_{n+1} = (x_n - 1)/(n - 1)$

   (b) $x_{n+1} = e^{x_n} \sin x_n$

   (d) $x_{n+1} = x_n \ln x_n$

4. Each of the following is a parameterized autonomous or non-autonomous equation. In each case identify the function $f(x)$ or $f(n, x)$ being iterated and the parameter(s).

   (a) $x_{n+1} = a - bx_n^2$

   (c) $x_{n+1} = \sqrt{x_n + k}$

   (b) $x_{n+1} = rx_n^2 + sn^2$

   (d) $x_{n+1} = \dfrac{x_n + t}{x_n + n}$

5. (a) Show that the maximum value of $f(x) = rx(1 - x/C)$ for $r, C > 0$ occurs at the point $(C/2, rC/4)$. (b) Show that the maximum value of the parameterized function $f(x) = rxe^{-x/N}$ for $r, N > 0$ occurs at the point $(N, rN/e)$.

In Exercises 6–9 sketch the graph of $y = f(x)$ *and find the coordinates of the maximum point.*

6. $f(x) = rx(1 - x^2 10^{-6})$ for: (a) $r = 1/2$; (b) $r = 2$.

7. $f(x) = \dfrac{rx^2}{3000}(1 - x/3000)$ for: (a) $r = 1$; (b) $r = 6$.

8. $f(x) = \dfrac{rx^2}{1000}(1 - x/1000)^2$ for: (a) $r = 1$; (b) $r = 10$.

9. $f(x) = \dfrac{rx^2}{500}e^{-x/500}$ for: (a) $r = 1$; (b) $r = 3$.

10. For the equation $x_{n+1} = bx_n^a$ where a and b are positive constants:

(a) Compute $x_1, \ldots, x_5$, in terms of x_0.

(b) Can you construct a formula for the exact solution x_n for all n?

(c) Use the geometric series to simplify your solution in part (b).

11. Compute $x_1, \ldots, x_5$ if: (a) $x_{n+1} = 1/x_n$ and $x_0 = 2$; (b) $x_{n+1} = 5/x_n$ and $x_0 = 8$. (c) From these can you predict what happens to all solutions of $x_{n+1} = c/x_n$?

12. The formula for the exact solution of $x_{n+1} = (x_n - c)^a + c$, where c is a constant, can be derived as follows:

(a) Use the substitution $u_n = x_n - c$ to convert $x_{n+1} = (x_n - c)^a + c$ into $u_{n+1} = u_n^a$.

(b) Write the solution for u_n for all n.

(c) Use the substitution $u_n = x_n - c$ again to find a formula for x_n for all n.

In Exercises 13–16 find all fixed points (if any) of each function.

13. $f(x) = 2x^2 - 3$

14. $f(x) = 4x^5$

15. $f(x) = \sin(\pi x/2)$

16. $f(x) = (\pi/4)\tan x$

In Exercises 17–20 find all fixed points (if any) of each equation.

17. $I_{n+1} = 4I_n - I_n^2/1000$

18. $P_{n+1} = 4P_n e^{-P_n}$

19. $P_{n+1} = 5 + 6/P_n$

20. $A_{n+1} = 10/(A_n + 3)$

A fixed point of $f(x)$ *may be viewed as a point where the graphs of* $y = f(x)$ *and* $y = x$ *intersect. Determine graphically whether the functions in Exercises 21–24 have fixed points and how many.*

21. $f(x) = 5x^3 - 15x^2 + 10$

22. $f(x) = e^x + 2$

23. $f(x) = 0.5\sin x$

24. $f(x) = 8e^{-x}$

In Exercises 25–28 find all fixed points and determine their stability using the formula for the exact solution $x_n = x_0^{a^n}$*. Find the basins of attraction for those that are stable.*

25. $x_{n+1} = x_n^2$

26. $x_{n+1} = x_n^{1/3}$

27. $x_{n+1} = 1/x_n^2$

28. $x_{n+1} = 1/\sqrt{x_n}$

In Exercises 29–32 find all fixed points and determine their stability using the formula for the exact solution (see Exercise 10). Find the basins of attraction for those that are stable.

29. $x_{n+1} = x_n^2/2$

30. $x_{n+1} = 4x_n^3$

31. $x_{n+1} = \sqrt{x_n}/3$

32. $x_{n+1} = 5\sqrt{x_n}$

In Exercises 33–36 find all fixed points and determine their stability using the formula for the exact solution (see Exercise 12). Find the basins of attraction for those that are stable.

33. $x_{n+1} = (x_n - 2)^3 + 2$

34. $x_{n+1} = 5 + \sqrt{x_n - 5}$

35. $x_{n+1} = x_n^2 - 2x_n + 2$

36. $x_{n+1} = x_n^2 + 6x_n + 6$

Computer Projects 3.2

In Projects 1–3 find all fixed points and determine their stability by generating at least the first 100 iterates for various choices of initial values and observing the dynamics.

1. $I_{n+1} = I_n - r I_n + s I_n (1 - I_n 10^{-6})$ for: (a) $r = 0.5$, $s = 0.25$; (b) $r = 0.5$, $s = 1.75$; (c) $r = 0.5$, $s = 2.0$; (d) $r = 0.25$, $s = 3.0$.

2. $P_{n+1} = 1/P_n + 0.75 P_n + c$ for: (a) $c = 0$; (b) $c = -1$; (c) $c = -1.25$; (d) $c = -1.38$.

3. $x_{n+1} = a x_n (1 - x_n^2)$ for: (a) $a = 0.5$; (b) $a = 1.5$; (c) $a = 2.25$; (d) $a = 2.3$.

4. The population model $P_{n+1} = r \dfrac{P_n^2}{500} e^{-P_n/500}$ has a *threshold*. That is, regardless of the growth rate r, small initial populations $P_0 > 0$ always lead to $P_n \to 0$ or extinction of the species. But initial populations P_0 just beyond the threshold value lead to an increasing population P_n. (The threshold is the upper end-point of the basin of attraction of the fixed point 0.) In each of the following cases, use *trial and error* to find the threshold population level, i.e., the smallest initial population for which the population does not become extinct. (a) $r = 3$; (b) $r = 4.5$; (c) $r = 6$. From this can you conclude what happens to the threshold as the growth rate r gets larger?

5. The price model $P_{n+1} = \dfrac{3}{P_n} + \dfrac{P_n}{4} - c$ for $0 \le c \le 1.7$ and $P_0 > 0$ has a positive stable equilibrium for all values of the parameter c close to 0. But for larger values of c this equilibrium is no longer stable. Instead, different dynamics exist that may or may not be classified as stable. (We'll be talking about those issues later on.) Use *trial and error* to estimate to at least 3 decimal places the largest value of c for which the equilibrium price is positive and stable.

6. For each of the following compute at least 100 iterates of the non-autonomous equation $P_{n+1} = (3 + \sin(2\pi n/b)) P_n (1 - P_n/1000)$ starting with $P_0 = 500$. Describe your observations for each. (a) $b = 1$ (b) $b = 4$ (c) $b = 8$.

SECTION
3.3 Cobwebbing, Derivatives and Dynamics

In the previous section it was found that fixed points often play a dominant role in the dynamics of nonlinear equations, just as they do for linear ones. Although simple algebra is all that is needed to find such a fixed point, a quick and reliable technique for determining its stability and whether or not solutions oscillate around it has not yet been demonstrated. For linear equations $x_{n+1} = a x_n + b$ we learned in Chapter 2 that the fixed point is stable if $|a| < 1$ or unstable if $|a| > 1$, and solutions oscillate around the fixed point if $a < 0$. For nonlinear functions, determining these dynamics is not so straightforward. Additional tools must be developed in that endeavor. In this section we address that issue.

Cobweb Graphs

One simple technique for investigating the behavior of $x_{n+1} = f(x_n)$, especially when $f(x)$ is nonlinear, is called **cobwebbing.** This is a quick and convenient *pencil-and-paper* approach that doesn't require any numerical or algebraic calculations, provided the graph of $y = f(x)$ is easily drawn. A **cobweb graph** is a very different kind of graph than we have been generating. Rather than a time-series graph of the points (n, x_n), the goal of a cobweb graph is to locate on *both* axes the position of each point x_n, along with lines connecting successive iterates.

If we are iterating $x_{n+1} = f(x_n)$ for any given function $f(x)$, then we begin by sketching the graphs of $y = f(x)$ and the line $y = x$ together in the same coordinate system. Then plot the initial point x_0 on the x-axis. To locate the next point x_1, draw a vertical line from x_0 on the x-axis to the graph of $y = f(x)$, and then a horizontal line from there to the y-axis. Since $x_1 = f(x_0)$ that point on the y-axis must be the location of x_1 (see Fig. 3.7(a)).

In order to continue, we must next place x_1 on the x-axis. This can be done using the line $y = x$ to *reflect* x_1 from its position on the y-axis to the same position on the x-axis. So, draw a horizontal line from x_1 on the y-axis to the line $y = x$, and then a vertical line from there to the x-axis. This is the location of x_1 on the x-axis (see Fig. 3.7(b)).

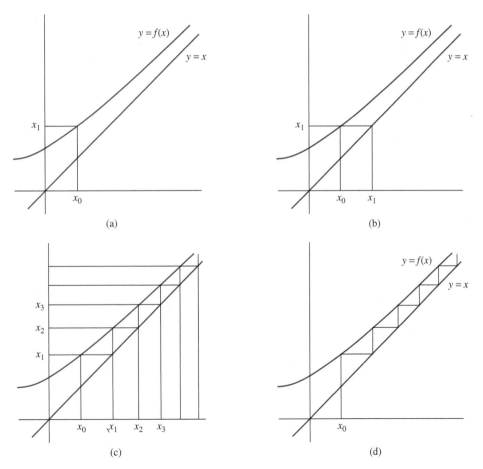

(a)

(b)

(c)

(d)

FIGURE 3.7

Like x_1, each subsequent point x_n will also have two locations created this way, one on the x-axis and one on the y-axis. To find the two locations of x_2, we draw a vertical line from x_1 on the x-axis to the graph of $y = f(x)$; a horizontal line from there to the y-axis, which the first location of x_2; then a horizontal line from this x_2 to the line $y = x$; and finally a vertical line from there to the x-axis, which is the second location of x_2. Continue this process to find x_3, x_4, etc., on the axes (see Fig. 3.7(c)).

Finally, *erase* all line segments that are *not* between the graphs of $y = f(x)$ and $y = x$, except for the first vertical line starting at x_0 on the x-axis. The result is shown in Figure 3.7(d).

The resulting graph with its horizontal and vertical line segments resembles a cobweb, and so the technique is called *cobwebbing*. In this case it also resembles a staircase, but we shall see in other examples that it may instead resemble a spiral. The entire first vertical line segment starting at x_0 on the x axis and ending at $y = f(x)$ is usually left on the cobweb to indicate the starting point. This is especially useful when plotting on the same graph cobwebs starting from several different initial points x_0, which is often the case when trying to determine such dynamics as stability, instability or more complex behavior.

A More Efficient Cobwebbing Method

Now that we have seen what a cobweb graph is, we next shorten the steps to create it. After all, the technique is supposed to be a fast and simple way to determine the behavior. Cobweb graphs are usually created in the following way.

Cobweb Graphing

Draw a vertical line from the point x_0 on the x axis to $y = f(x)$, followed by a horizontal line from that point on $y = f(x)$ directly to $y = x$. From there draw another vertical line to $y = f(x)$, followed by another horizontal line to $y = x$. Repeat the steps: draw a vertical line to $y = f(x)$ followed by a horizontal line to $y = x$, for as many iterations as desired.

If this cobwebbing method is used for the function depicted in Figure 3.7, then the cobweb in Figure 3.7(d) is again obtained. However, this time we didn't extend the lines to the axes, and so there is less drawing and no erasing. As a result, it is a much faster technique.

When generating a cobweb in this shorter way, the locations of x_1, x_2, x_3, etc. will not be specified on the x and y axes. If desired, they can be easily located on either axis by extending the horizontal or vertical line segments to that axis, perhaps using dotted lines (see Fig. 3.8). As we shall later see, specifying and labeling these locations is generally unnecessary. It is really where the cobweb leads us as n increases that we wish to see.

The following are some additional examples of cobwebbing.

EXAMPLE 1

Draw a cobweb graph for $x_{n+1} = x_n^2$ starting with $x_0 = 1.1$.

Solution: See Figure 3.9. The cobweb grows upward and to the right, resembling a staircase with increasingly larger steps each time. If continued, the cobweb would grow to ∞.

■

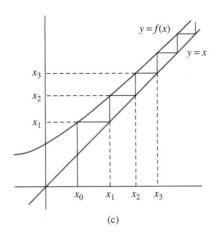

(c)

FIGURE 3.8

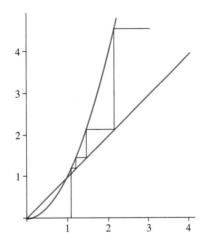

FIGURE 3.9

EXAMPLE 2

Draw a cobweb graph for $x_{n+1} = 1 - x_n^2$ starting with $x_0 = 0.5$.

Solution: See Figure 3.10. This cobweb resembles a spiral that keeps growing outward, although always bounded inside the square with $0 \le x \le 1$ and $0 \le y \le 1$. ∎

It is important to remember that even though a cobweb graph looks very different from a time-series graph of the points (n, x_n), there is a correspondence between the two.

EXAMPLE 3

Compare the cobweb and time-series graphs for the first 20 iterations of $x_{n+1} = x_n^2$ starting with $x_0 = 0.75$.

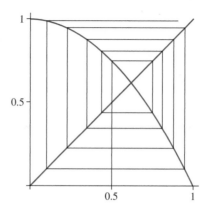

FIGURE 3.10

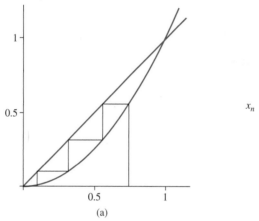

(a)

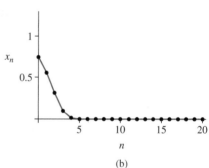

(b)

FIGURE 3.11

Solution: The cobweb graph shown in Figure 3.11(a) resembles a staircase with steps that keep shrinking. It grows to the left and approaches the origin. The points (n, x_n) of the corresponding time-series graph in Figure 3.11(b) show a pattern that resembles exponential decay.   ∎

EXAMPLE 4

Compare the cobweb and time-series graphs for 20 iterations of the price model $P_{n+1} = \dfrac{2}{P_n} + 1$ starting with $P_0 = 4$.

Solution: The cobweb graph in Figure 3.12(a) spirals inward to the point where the graphs of $y = \dfrac{2}{x} + 1$ and $y = x$ intersect. The corresponding time-series graph in Figure 3.12(b) shows an oscillation in the price that slowly decays to constant value, but not to 0.   ∎

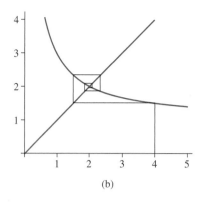

(b)

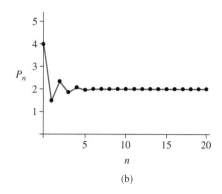

(b)

FIGURE 3.12

Cobweb Graphs and Dynamics

One may be wondering how cobwebbing helps in the investigation of the dynamics of $x_{n+1} = f(x_n)$. To see this we first consider what cobweb graphs of the general linear equation $x_{n+1} = ax_n + b$ must look like for various choices of a.

If we assume that $a > 1$, then the line $y = ax + b$ must have a slope that exceeds 1, and so must be steeper than $y = x$. The resulting line and cobwebs must therefore appear as in Figure 3.13(a). On the other hand, if $0 < a < 1$, then $y = ax + b$ must be drawn with a positive slope, but less steep than $y = x$. So the line and corresponding cobwebs must resemble those in Figure 3.13(b).

Note that in Figure 3.13(a) where $a > 1$, as $n \to \infty$, each cobweb resembles a growing staircase that is diverging from the point where the graphs of $y = ax + b$ and $y = x$ intersect, but in Figure 3.13(b), where $0 < a < 1$, the staircase shrinks as it approaches that point of intersection. To find the coordinates of that point, we solve $ax + b = x$, which gives $x = b/(1 - a)$. It is no coincidence that this is the fixed point of $x_{n+1} = ax_n + b$, which we previously determined to be unstable for $a > 1$ and stable for $0 < a < 1$. The cobwebs in these two situations give graphical depictions of instability and stability respectively.

We next consider cobweb graphs of $x_{n+1} = ax_n + b$ for $a < 0$. If $a < -1$ we must draw $y = ax + b$ as a line with a negative slope that is steeper than $y = -x$. The result is shown in Figure 3.13(c). If $-1 < a < 0$ the line $y = ax + b$ must be less steep than $y = -x$ (see Fig. 3.13(d)). In these cases too, the cobweb graphs correspond precisely to the behavior of the exact solution. For $a < -1$ there is divergence from the unstable fixed point as $n \to \infty$, while convergence to the stable fixed point occurs for $-1 < a < 0$. Also, since $a < 0$, one may notice from either the exact solution or the cobweb that solutions oscillate around the fixed point. Whenever this occurs the cobweb will resemble a spiral.

Based on this analysis, it appears that a cobweb graph can reveal important qualitative behavior concerning $x_{n+1} = ax_n + b$. In particular, the fixed point can be identified as the place where the graphs of $y = ax + b$ and $y = x$ intersect. And by choosing x_0 close to that fixed point and following the resulting cobweb, it is possible to determine whether that fixed point is an attractor or a repeller. Any oscillation around the fixed point can also be observed.

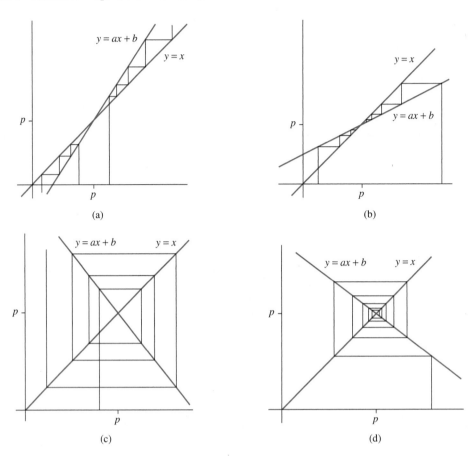

FIGURE 3.13

It is therefore reasonable to expect that cobweb graphs might allow us to see similar dynamics for nonlinear equations. The following examples show that indeed this is often the case.

EXAMPLE 5

Use cobweb graphs for $x_{n+1} = x_n^3$ to determine the stability of its fixed points 0, 1 and -1.

Solution: Figure 3.14 shows four cobwebs, one in each of the four intervals between the fixed points. These webs diverge from $(-1, -1)$ and $(1, 1)$, but those between -1 and 1 converge to $(0, 0)$. Using the formula for the exact solution, it was shown in Example 13 of Section 3.2 that 0 is stable and the other two are unstable. Cobweb graphs verify these previous findings.   ∎

EXAMPLE 6

Use cobweb graphs for $x_{n+1} = \sqrt{x_n}$ to determine the stability of its fixed points 0 and 1.

Solution: Figure 3.15 shows that a cobweb starting at any x_0 between 0 and 1 diverges from $(0, 0)$ as n increases, but converges to $(1, 1)$. A web starting at $x_0 > 1$ also converges to

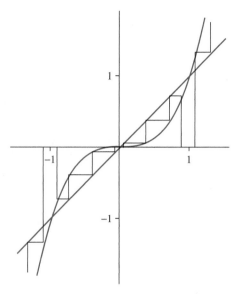

FIGURE 3.14

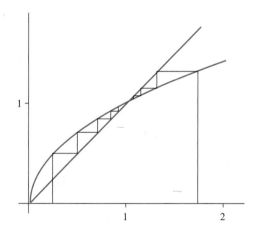

FIGURE 3.15

(1, 1) as n increases. This indicates that 0 is unstable and 1 is stable, which is consistent with the conclusions of Example 12 from Section 3.2, where the formula for the exact solution was used. ∎

EXAMPLE 7

In Example 14 of Section 3.2 it was found numerically that 0 is a stable fixed point of the disease model $I_{n+1} = I_n - 0.9I_n + 0.5I_n(1 - I_n/10{,}000)$. Verify this using a cobweb graph.

Solution: Figure 3.16 shows a cobweb starting with $P_0 = 5000$ that approaches (0, 0). This indicates that 0 is an attractor, which is consistent with those previous findings. ∎

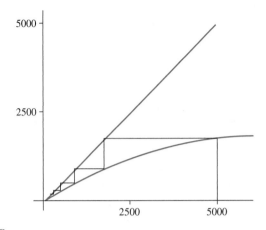

FIGURE 3.16

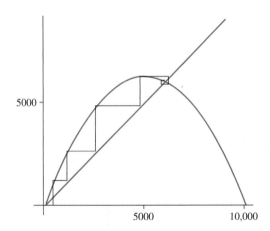

FIGURE 3.17

EXAMPLE 8

For the logistic model $P_{n+1} = 2.5P_n(1 - P_n/10{,}000)$, find the equilibrium population levels and use a cobweb graph to determine the dynamics.

Solution: In this case $f(x) = 2.5x(1 - x/10{,}000)$. The equilibria occur at the places where $y = 2.5x(1 - x/10{,}000)$ and $y = x$ intersect. These are found by solving

$$2.5x(1 - x/10{,}000) = x \quad \text{or} \quad 1.5x - \frac{x^2}{4000} = \frac{x(6000 - x)}{4000} = 0$$

which yields the fixed points 0 and 6000. Figure 3.17 shows a cobweb starting with $x_0 = 500$. It indicates that 0 is a repeller and 6000 is an attractor. Also, oscillation occurs around 6000. Cobwebs starting from all other initial points would lead to the same results, providing very convincing evidence for these conclusions.   ■

In summary, what we may correctly deduce from all these examples with regard to cobweb graphs for any problem, either linear or nonlinear, is the following.

Cobweb Dynamics

For the equation $x_{n+1} = f(x_n)$, any point p where the graphs of $y = f(x)$ and $y = x$ intersect is a fixed point. If all cobwebs starting close to p converge to (p, p), then p is stable. If a cobweb diverges from (p, p), then p is unstable. And a spiral cobweb around (p, p) indicates that solutions oscillate around p.

Derivatives and Dynamics

There is another way besides cobwebbing to determine the behavior of solutions near a fixed point when $f(x)$ is nonlinear. The method is especially useful when the dynamics are too complex to be investigated graphically, or when the precise graph is not available, such as in our later investigation of parameterized families of functions.

Recall that the derivative $f'(x)$ of a function $f(x)$ gives the slope of the line that is tangent to the graph of $y = f(x)$ at any point. Suppose we have already found a fixed point p of $x_{n+1} = f(x_n)$, and we then evaluate $f'(x)$ at p. The result $f'(p)$ is the slope of the tangent line to $y = f(x)$ at p.

To see what $f'(p)$ can tell us about the stability of p, let us first assume that $f'(p) > 1$. Since p is a fixed point of $f(x)$, the graphs of $y = f(x)$ and $y = x$ must intersect at $(x, y) = (p, p)$, and since $f'(p) > 1$ the graph of $y = f(x)$ must be *steeper* than the graph of $y = x$ at (p, p). More precisely, the slope of the tangent line to the graph of $y = f(x)$ at (p, p) must be greater than the slope of $y = x$, which equals 1. In this case suppose we pick initial points x_0 on either side of p and do some cobwebbing. The result must appear as in Figure 3.18(a). The cobwebs seem to indicate that the fixed point p is unstable.

Next, let us again assume that p is a fixed point of $f(x)$, but now $0 < f'(p) < 1$. So $y = f(x)$ must again intersect $y = x$ at (p, p), but this time at a positive slope that is less than the slope of $y = x$. If we again cobweb on either side of p, as shown in Figure 3.18(b), this time the cobwebs tell us that p is stable.

If we repeat the previous procedure two more times, once with $f'(p) < -1$ and once more with $-1 < f'(p) < 0$, we see that p is unstable in the former case and stable in the latter, as shown in the cobweb graphs of Figures 3.18(c) and 3.18(d) respectively. One may also notice that oscillation around the fixed point p occurs in both cases.

What we may learn from all this is that there is a completely non-graphical way to determine stability, instability and oscillation, which is summarized by the following rule. Notice its similarity to the corresponding criteria for linear equations.

T H E O R E M   *Stability and Oscillation*

For any fixed point p of $x_{n+1} = f(x_n)$, if $|f'(p)| < 1$, then p is locally stable, but if $|f'(p)| > 1$ then p is unstable. Also, if $f'(p) < 0$, then solutions oscillate locally around p, but if $f'(p) > 0$ they do not. ■

We remark that to use this rule, the function $f(x)$ must be differentiable, at least at the point p, i.e., the derivative must exist at that point. Also, we do not consider the borderline case in which $|f'(p)| = 1$. Similar to linear equations $x_{n+1} = ax_n + b$ with $|a| = 1$, this situation rarely occurs in practice.

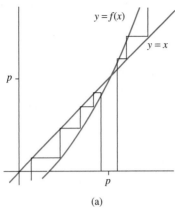

(a)

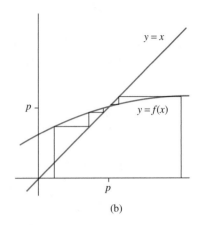

(b)

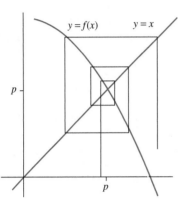

(c)

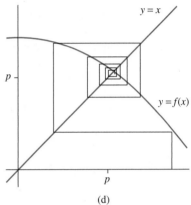

(d)

FIGURE 3.18

EXAMPLE 9

Find all fixed points for the population model discussed in Example 11 of Section 3.2 using $r = 0.5$ and $C = 1000$, i.e., $P_{n+1} = \dfrac{0.5 P_n}{1 + P_n^2\, 10^{-6}}$, and determine their stability using the previous theorem.

Solution: We must first solve $f(x) = \dfrac{0.5x}{1 + x^2 10^{-6}} = x$, which simplifies to

$$x(x^2 + 500{,}000) = 0$$

The only solution is $x = 0$. To determine stability, the quotient rule gives

$$f'(x) = \frac{(1 + x^2 10^{-6}) \cdot 0.5 - 0.5x \cdot 2x\, 10^{-6}}{(1 + x^2 10^{-6})^2} = \frac{0.5 - 0.5 \cdot x^2 10^{-6}}{(1 + x^2 10^{-6})^2}$$

Letting $x = 0$ there gives $f'(x) = 0.5$. So according to the theorem, 0 is stable. ∎

EXAMPLE 10

Previously, 0 and 10,000 ln 2 were found to be equilibria of the model $P_{n+1} = 2P_n e^{-0.0001 P_n}$, and numerical data indicated that the former is a repeller and the latter an attractor. Prove these numerical observations are correct.

Solution: Since $f(x) = 2xe^{-0.0001x}$, we must use the product rule to compute

$$f'(x) = 2xe^{-0.0001x}(-0.0001) + 2e^{-0.0001x} = 2(1 - 0.0001x)e^{-0.0001x}$$

Substituting $x = 0$ gives $f'(0) = 2$ and so 0 is unstable. With $x = 10,000 \ln 2$

$$f'(10,000 \ln 2) = 2(1 - \ln 2)e^{-\ln 2} = 2(1 - \ln 2)\frac{1}{2} = 1 - \ln 2$$

Since $1 - \ln 2 \approx 0.307$ then 10,000 ln 2, or about 6931, is a stable equilibrium population for the model.   ∎

The use of the word *locally* in the above theorem once again means that the stability or instability of p described in the theorem, as well as any oscillation around it, applies only to iterates x_n that are sufficiently close to p. The derivative can tell only what will happen as long as solutions stay within a certain interval around p. How large that interval or basin of attraction is and what occurs outside that interval cannot be determined from $f'(p)$ alone. Other means are again needed, such as cobwebbing. The following example demonstrates the limitations of the derivative.

EXAMPLE 11

Find the fixed points of $x_{n+1} = -x_n^3$, determine their stability using the derivative and draw cobweb graphs for several initial points.

Solution: Solving $f(x) = -x^3 = x$, it can be verified easily that 0 is the only fixed point. Computing the derivative gives $f'(x) = -3x^2$, and so $f'(0) = 0$. So according to the above theorem, we should expect to see stability at 0 and oscillation around it. We next draw some cobweb graphs, starting first with $x_0 = 0.95$, and then with $x_0 = 1.05$. The results are shown in Figure 3.19. The cobwebs show that oscillation around 0 occurs in both cases. For $x_0 = 0.95$ iterates x_n converge to 0 as $n \to \infty$, which corresponds to stability. But for $x_0 = 1.05$ the solution diverges to $\pm\infty$. Cobweb graphs are able to determine that latter fact, while the derivative test is not.   ∎

Still, despite its limitations, the derivative is an important tool for determining the dynamics of a problem $x_{n+1} = f(x_n)$ near a fixed point, even if it is just the local dynamics.

Proof of the Stability and Oscillation Theorem

Although the geometric arguments presented earlier may offer convincing evidence that stability, instability and oscillation with regard to a fixed point p of $x_{n+1} = f(x_n)$ are determined by the magnitude and sign of $f'(p)$, that does not really constitute a mathematical proof of the theorem above. We now provide that proof.

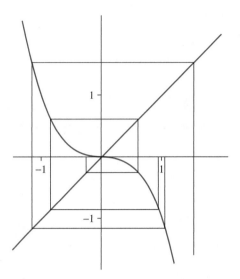

FIGURE 3.19

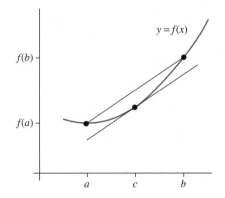

FIGURE 3.20

We begin by recalling the Mean Value Theorem from calculus, which describes a certain property that any differentiable function $f(x)$ must satisfy: Given any two points a and b there must exist a point c between them where the slope of the tangent line $f'(c)$ equals the slope of the secant line connecting the two points $(a, f(a))$ and $(b, f(b))$ (see Fig. 3.20). More concisely,

$$\frac{f(b) - f(a)}{b - a} = f'(c) \quad \text{or, equivalently,} \quad f(b) - f(a) = f'(c)(b - a)$$

If we restate this theorem with $a = p$ and $b = x_0$, then there must exist a point c between them that satisfies

$$f(x_0) - f(p) = f'(c)(x_0 - p)$$

But if p is a fixed point of $f(x)$ then $f(p) = p$, and if we call $x_1 = f(x_0)$ as we usually do, this means that

$$x_1 - p = f'(c)(x_0 - p) \tag{1}$$

Let us now suppose that $|f'(p)| < 1$. What does this really imply? With a little thought we see that since $f'(x)$ is continuous, then for all x in some small interval I centered at p we must have $|f'(x)| < d$, where d is some number between 0 and 1. If x_0 is any point in this interval I then the point c that was provided above by the Mean Value Theorem, and lies between x_0 and p, must also be in I. This implies $|f'(c)| < d$. Combining this with (1) gives

$$|x_1 - p| = |f'(c)| \cdot |x_0 - p| < d\,|x_0 - p| \quad \text{or} \quad |x_1 - p| < d\,|x_0 - p| \tag{2}$$

Since $0 < d < 1$, not only does this inequality say that x_1 is closer to p than x_0 is, but it also implies that we can repeat the same argument all over again, since we now know that x_1 is also in I. If we go back to the Mean Value Theorem, this time starting with the points $a = p$ and $b = x_1$, we will arrive at an inequality similar to equation (2) above

$$|x_2 - p| < d\,|x_1 - p|$$

and if this is combined with that previous one then

$$|x_2 - p| < d\,|x_1 - p| < d \cdot d\,|x_0 - p| = d^2\,|x_0 - p| \quad \text{or} \quad |x_2 - p| < d^2\,|x_0 - p|$$

If these arguments are repeated, then after the nth step we will arrive at

$$|x_n - p| < d^n\,|x_0 - p|$$

Finally, suppose we take the limit as $n \to \infty$ on both sides of this last inequality. Since $0 < d < 1$ then $d^n \to 0$, and so

$$\lim_{n \to \infty} |x_n - p| < \lim_{n \to \infty} d^n\,|x_0 - p| = 0 \cdot |x_0 - p| = 0$$

This proves that $|x_n - p| \to 0$ as $n \to \infty$, or in other words x_n converges to p.

We have therefore shown that for any initial point x_0 in I the solution x_n will converge to p. In this case, the interval I constitutes at least part of the basin of attraction of p. A similar argument would show that if $|f'(p)| > 1$, then an interval exists around p in which solutions diverge from p, but that is left as an exercise.

With regard to oscillation, if $f'(p) < 0$ then $f'(x) < 0$ for all x in an interval J centered at p. If x_0 is chosen from J then the Mean Value Theorem implies the existence of a point c in J that satisfies (1). Since $f'(c) < 0$ then $x_0 - p > 0$ (or equivalently $x_0 > p$) in (1) implies that $x_1 - p < 0$ (or $x_1 < p$). Similarly, if $x_0 - p < 0$ (or $x_0 < p$) in (1), then $x_1 - p > 0$ (or $x_1 > p$). In other words, x_0 and x_1 lie on opposite sides of p. A similar argument would show that x_1 and x_2 lie on opposite sides of p, and so too do x_2 and x_3, etc., as long as all these points remain in J. This means that all solutions x_n in J oscillate around p. A similar argument would show that if $f'(p) > 0$ then all points of a solution x_n in some interval J centered at p lie on the same side of p, which implies that solutions do not oscillate around p. This proves the Theorem.

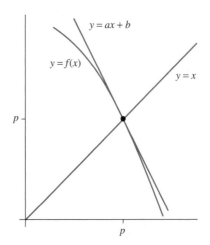

FIGURE 3.21

Linearization

There is another way to see the association between the derivative and stability. Recall from calculus that the tangent line to a curve $y = f(x)$ at some point can be used to approximate the function near that point. Since it is generally easier to work with linear rather than nonlinear functions, this often greatly simplifies the analysis of complex mathematical problems, and so it is a commonly used approach.

The idea is to approximate $y = f(x)$ by the tangent line at the point $(p, f(p))$, which can be found by using the *point-slope formula* with slope $f'(p)$. This yields

$$y - f(p) = f'(p)(x - p)$$

But if p is a fixed point, then $f(p) = p$, and so

$$y - p = f'(p)(x - p) \quad \text{or} \quad y = f'(p)x + p - pf'(p)$$

which is of the form

$$y = ax + b \quad \text{where} \quad a = f'(p) \quad \text{and} \quad b = p - pf'(p) \tag{3}$$

This is called the **linearization** of $f(x)$ at p (see Fig. 3.21).

With this a and b the linear equation $x_{n+1} = ax_n + b$ can now be used to approximate $x_{n+1} = f(x_n)$. For any initial point sufficiently close to p, the behavior of iterates should be roughly the same for both equations, as long as they remain close to p. For example, p is a fixed point for each, and the criteria for stability or instability of p under the former, $|a| < 1$ or $|a| > 1$ respectively, is consistent with the criteria for its stability or instability under the latter, $|f'(p)| < 1$ or $|f'(p)| > 1$ respectively, since $a = f'(p)$. The condition $a = f'(p) < 0$ for solutions to oscillate around p is also consistent.

EXAMPLE   12

Linearize the price model $P_{n+1} = \dfrac{2}{P_n} - 1$ at the fixed point $p = 1$ and determine its local dynamics near that point.

Solution: Since $f(x) = \dfrac{2}{x} - 1$ then $f'(x) = -\dfrac{2}{x^2}$, and so (3) becomes

$$a = f'(p) = f'(1) = -2 \quad \text{and} \quad b = p - pf'(p) = 1 - 1 \cdot (-2) = 3$$

The linearized equation is therefore $x_{n+1} = -2x_n + 3$. Since $a = -2 < -1$, this indicates that $p = 1$ is unstable and solutions oscillate around it.   ■

Since linearization provides the same stability criteria that we already had, this technique has little benefit here. However the idea of linearization can be of great importance in other settings, such as the investigation of *systems* of nonlinear equations in Chapter 5.

Exercises 3.3

In Exercises 1 and 2 perform the first few steps of cobwebbing starting at x_0.

1.
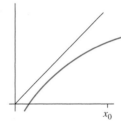

2.
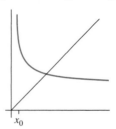

For each of the linear functions given in Exercises 3–6, draw the first few steps of a cobweb graph starting at the given value of x_0.

3. $f(x) = 2x,\ x_0 = 1$

4. $f(x) = 2x/3,\ x_0 = 9$

5. $f(x) = -5x,\ x_0 = 9$

6. $f(x) = 8 - x/4,\ x_0 = 16$

In Exercises 7 and 8 use cobwebbing to determine the stability of the fixed points A, B, C, *and any oscillation around them.*

7.
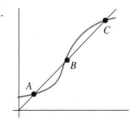

8.
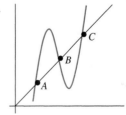

In Exercises 9–12 find all fixed points and use cobwebbing to determine their stability and any oscillation around them.

9. $x_{n+1} = x_n^2$

10. $I_{n+1} = 2I_n - I_n^2$

11. $P_{n+1} = 1.5 + 1/P_n$

12. $x_{n+1} = 0.25 \sin x_n$

In Exercises 13–16 use the derivative to determine the stability of and any oscillation around the given fixed point p.

13. $f(x) = 4x^2 + 3x$, $p = 0$ 15. $f(x) = 1/x^2$, $p = 1$

14. $f(x) = (x^2 - 3x + 5)/3$, $p = 1$ 16. $f(x) = \sin(x/4)$, $p = 0$

In Exercises 17–20 find all fixed points and use the derivative to determine their stability and any oscillation around them.

17. $x_{n+1} = x_n^3 + 2x_n$ 19. $x_{n+1} = 2/(x_n + 1)$

18. $x_{n+1} = 3x_n e^{-x_n}$ 20. $x_{n+1} = x_n^{1/4}$, $x_0 > 0$

In Exercises 21–24 find all non-negative fixed points of the given nonlinear models and use the derivative to determine their stability.

21. $P_{n+1} = r P_n (1 - P_n/1000)$ for: (a) $r = 0.5$; (b) $r = 2$.

22. $P_{n+1} = r P_n e^{-P_n/1000}$ for: (a) $r = 0.5$; (b) $r = 5$.

23. $P_{n+1} = \dfrac{r P_n^2}{1000}(1 - P_n/1000)$ for: (a) $r = 1$; (b) $r = 5$.

24. $P_{n+1} = \dfrac{2}{P_n} + c$ for: (a) $c = 1$; (b) $c = -1$.

In Exercises 25 and 26 the fixed point p *of the given function* f(x) *satisfies* f'(p) = ±1, *and so the derivative test for stability is inconclusive. Determine the stability of these fixed points in some other way.*

25. $f(x) = x^3 - x$, $p = 0$ 26. $f(x) = 2 - \dfrac{1}{x}$, $p = 1$

In Exercises 27 and 28 linearize f(x) *at each of its fixed points.*

27. $f(x) = 4x(1 - x)$ 28. $f(x) = e^{1-x}$

29. A fixed point p of a function $f(x)$ is called *superstable* if $f'(p) = 0$.
    (a) Show that $p = 1/2$ is a superstable fixed point of $f(x) = 2x(1 - x)$.
    (b) Draw several cobweb graphs for x_0 close to $1/2$.
    (c) Linearize $f(x)$ at $p = 1/2$.
    (d) Can you explain why the term *superstable* is appropriate?

30. Repeat the previous exercise for $f(x) = -x^3$ and $p = 0$.

31. The point $p = 1$ is a fixed point of both $f(x) = x^2$ and its *inverse* $f^{-1}(x) = \sqrt{x}$ (for $x \geq 0$). (a) Use cobweb graphs to show that p is stable under one of these functions and unstable under the other. (b) Confirm these observations using the derivative.

32. (a) Repeat the previous exercise for the same functions but for $p = 0$. (b) Repeat the previous exercise for $f(x) = e^{x/2} - 1$, $f^{-1}(x) = 2\ln(x + 1)$ and $p = 0$.

Computer Projects 3.3

Each of the models in Projects 1 and 2 has two unstable equilibrium points. First verify this using algebra and calculus. Then, for several random choices of P_0, create a table and graph of $P_0, \ldots, P_{100}$. Describe the dynamics. Does there appear to be some other form of stability?

1. $P_{n+1} = r P_n (1 - P_n/1000)$ for: (a) $r = 3.4$; (b) $r = 3.5$; (c) $r = 4$.
2. $P_{n+1} = r P_n e^{-P_n/10,000}$ for: (a) $r = 10$; (b) $r = 14.5$; (c) $r = 20$.

In Projects 3–6 each function f(x) *has a stable fixed point* p *that cannot be found exactly using algebra. In each case perform at least 100 iterations to find an approximation to* p *and verify its stability by computing* |f′(p)|.

3. $f(x) = 1 - 0.5x^3$

4. $f(x) = 0.25e^x$

5. $f(x) = \cos x$

6. $f(x) = 2 + \ln x$

7. For $0 < r < 1$ the fixed point $p = 0$ of $x_{n+1} = r x_n (1 - x_n)$ is stable. (a) For each of the values $r = 0.1, \ 0.5, \ 0.75, \ 0.99$ compute $|f'(p)|$. (b) For each of those r values, start with $x_0 = 1/2$ and perform enough iterations to determine the smallest n needed to have $|x_n| < 0.001$. (c) Compare the results of (a) and (b). Do you see a connection between the two?

SECTION
3.4   Some Mathematical Applications

We temporarily pause our investigation of the general behavior of nonlinear dynamical systems to present several well known applications involving some of the ideas developed so far. In addition to the applicability of nonlinear iteration, fixed points and stability to the analysis of density-dependent population models and other natural processes, there are also purely mathematical applications of that theory. Here we describe two of the most important: the **Newton Root-Finding Method** and the **Euler Method** for approximating the solution of a differential equation.

Newton Root-Finding Method

Finding Roots of Equations

Suppose $g(x)$ is a nonlinear function. One of the most basic algebra questions one can ask is: *What are the solutions of* g(x) = 0, *or what are the roots of* g(x)*?* The question may look deceptively simple, but actually finding the exact solution or solutions of that equation may be quite difficult, depending upon the complexity of $g(x)$. In fact, it is usually impossible. For example, try solving

$$x^5 - 26x^3 + 7 = 0 \quad \text{or} \quad 6x - \sin 5x = 0 \quad \text{or} \quad 3xe^{2x} - 4 = 0$$

Many who study algebra may not be aware that no methods exist for finding the exact solutions of equations such as these, since the only types of equations they have ever been faced with are either linear or quadratic. Occasionally, they may also solve higher-degree polynomial equations, but usually these have a simple form that can be easily factored into linear or quadratic terms. Sometimes, if the numbers are just right, certain trigonometric, exponential or logarithmic equations can also be solved exactly. But in the *real world* those types of simple equations are rarely the case.

Since most nonlinear equations cannot be solved exactly, we therefore try an alternate approach. That is, we find an approximate solution that is accurate enough for the purpose

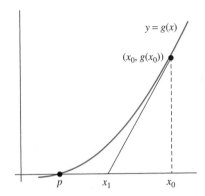

FIGURE 3.22

at hand. The Newton Method is among the best and most popular of these approximation techniques. It works by turning the problem into an iterative one, where each successive step brings us closer to the unknown exact solution.

To derive the method, first assume $g(x)$ is a differentiable function that has an unknown root p. This means $g(p) = 0$ and so the graph of $y = g(x)$ meets the x-axis at $x = p$. Of course, we don't know the value of p, but often by looking at the graph of $y = g(x)$ or just by some intelligent guessing, it is possible to come up with an initial estimate for p. Call that initial estimate x_0.

We will next find an estimate x_1 for p that is better than x_0. Consider the graph of the line tangent to $y = g(x)$ at the point $(x_0, g(x_0))$. If that line is extended toward the x-axis, it will eventually hit that axis. Let x_1 be the point on the x-axis where that intersection occurs (see Fig. 3.22). It appears that x_1 must be closer to p than x_0 is. So it is worth finding the value of x_1, since it is a better approximation to p than we had before.

To find x_1, notice in Figure 3.22 that a right triangle has been created whose hypotenuse is the tangent line to $y = g(x)$ at $(x_0, g(x_0))$. This allows us to write the slope of that tangent line in two ways. From calculus the slope of that tangent line is $g'(x_0)$. But, dividing the height of the triangle $g(x_0)$, by the length of the base $x_0 - x_1$, also gives the slope of that tangent line. Equating the two yields

$$g'(x_0) = \frac{g(x_0)}{x_0 - x_1}$$

and solving for x_1 gives

$$x_1 = x_0 - \frac{g(x_0)}{g'(x_0)}$$

which means we now have a better approximation to p.

Of course, it is unlikely that x_1 will be close enough to p that we can stop here. Rather, we will probably need to generate even better approximations x_2, x_3, x_4, etc., until we are satisfied that we have achieved the accuracy we desire. This can easily be done by repeating the above procedure, this time starting with x_1 and arriving at a better estimate x_2, and then using x_2 to find x_3, and so on (see Fig. 3.23). In the formula for x_1 above, if x_0 is replaced with x_n and x_1 with x_{n+1}, we have the equation for each successive iterate, which gives us the following.

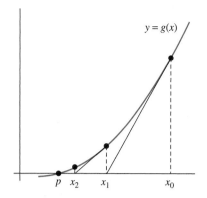

FIGURE 3.23

NEWTON ROOT-FINDING METHOD

To find a root p of $g(x) = 0$, if x_0 is any initial guess sufficiently close to p, then

$$x_{n+1} = x_n - \frac{g(x_n)}{g'(x_n)} \tag{1}$$

will generate increasingly better approximations, and $\lim_{n \to \infty} x_n = p$.

EXAMPLE 1

Using the Newton Method, determine how many iterations are needed to get a 6 decimal place approximation to the root $p = 1$ of $x^2 - 1 = 0$, starting with (a) $x_0 = 2$ and (b) $x_0 = 10$.

Solution: Here $g(x) = x^2 - 1$ and $g'(x) = 2x$. So equation (1) becomes

$$x_{n+1} = x_n - \frac{g(x_n)}{g'(x_n)} = x_n - \frac{x_n^2 - 1}{2x_n}$$

Tables 3.4(a) and 3.4(b) show the first 10 iterations beginning with $x_0 = 2$ and $x_0 = 10$ respectively. So, it takes about 4 iterations for part (a) and 7 iterations for part (b) to get 6 decimal places correct. Even though $x_0 = 10$ is quite far from the real root $p = 1$, relatively few steps were needed to arrive at this approximation. This fast rate of convergence is typical (though not always the case) for the Newton Method, which is why it is so popular.   ■

EXAMPLE 2

Use the Newton Method to find an approximation to a root of $x = \cos x$ that is accurate to 6 decimal places.

Solution: Graphing $y = x$ and $y = \cos x$ together shows that they intersect between $x = 0$ and $x = \pi/2 \approx 1.57$ (see Fig. 3.24). So we will let $x_0 = 1$. To use the Newton

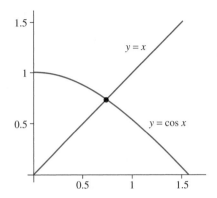

FIGURE 3.24

TABLE 3.4

n	x_n		n	x_n
0	2.000000		0	10.000000
1	1.250000		1	5.050000
2	1.025000		2	2.624010
3	1.000305		3	1.502553
4	1.000000		4	1.084043
5	1.000000		5	1.003258
6	1.000000		6	1.000005
7	1.000000		7	1.000000
8	1.000000		8	1.000000
9	1.000000		9	1.000000
10	1.000000		10	1.000000

        (a)   (b)

Method, we must first write the problem as $x - \cos x = 0$. Since $g(x) = x - \cos x$ and $g'(x) = 1 + \sin x$, then from (1) we must iterate

$$x_{n+1} = x_n - \frac{g(x_n)}{g'(x_n)} = x_n - \frac{x_n - \cos x_n}{1 + \sin x_n}$$

The results are shown in Table 3.5. Notice that the first 6 digits stop changing after about the third step. When this happens, it is a general indication that no further iterations are necessary, since the results are unlikely to start changing again. Although the exact solution is not known, we can be fairly confident that the approximation 0.739085 is accurate to 6 places. ■

TABLE 3.5

n	x_n
0	1.000000
1	0.750364
2	0.739113
3	0.739085
4	0.739085
5	0.739085
6	0.739085
7	0.739085
8	0.739085
9	0.739085
10	0.739085

It is worth mentioning that the method does not always work. If x_0 is chosen too far from a root, the method may not be able to find it. The following example demonstrates this.

EXAMPLE 3

The point $p = 0$ is the only root of $g(x) = \dfrac{x}{x^2 + 1} = 0$. Try to find it using the Newton Method starting with (a) $x_0 = 0.5$ and (b) $x_0 = 2$.

Solution: Since $g'(x) = \dfrac{1 - x^2}{(x^2 + 1)^2}$, then we must iterate

$$x_{n+1} = x_n - \frac{x_n/(x_n^2 + 1)}{(1 - x_n^2)/(x_n^2 + 1)^2} = x_n - \frac{x_n(x_n^2 + 1)}{1 - x_n^2}$$

Table 3.6 (a) shows 10 iterations for part (a) using $x_0 = 0.5$. In this case convergence is once again quite rapid. Table 3.6 (b) shows what happens for (b) when $x_0 = 2$. This time iterates go in the wrong direction and diverge to ∞. The reason for this can be seen from the graph of $y = g(x)$, shown in Figure 3.25. Since $x_0 = 2$ exceeds the point $x = 1$ where a maximum occurs for the function, tangent line projections to the x-axis point in the direction opposite to where the root $p = 0$ lies. This is why the rule says that x_0 must be *sufficiently close* to p. ∎

Proof of Convergence

The above graphical derivation of the Newton Method may provide a convincing argument to most that the method generates better approximations x_n with each step, assuming the initial guess x_0 is sufficiently close to a root p. But that is not really a proof that

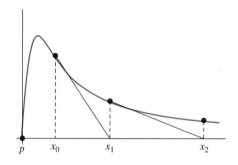

FIGURE 3.25

TABLE 3.6

n	x_n	n	x_n
0	0.500000	0	2
1	−0.333333	1	5
2	0.083333	2	11
3	−0.001166	3	22
4	0.000000	4	45
5	0.000000	5	89
6	0.000000	6	179
7	0.000000	7	358
8	0.000000	8	715
9	0.000000	9	1431
10	0.000000	10	2861

(a)   (b)

$\lim_{n \to \infty} x_n = p$. Using tools developed in previous sections, however, this can actually be proven by showing that the unknown root is a stable fixed point of the nonlinear function being iterated.

To see this, first note that equation (1) above is of the form $x_{n+1} = f(x_n)$ where

$$f(x) = x - \frac{g(x)}{g'(x)}$$

If p is a simple root of multiplicity 1 of $g(x)$, then $g(p) = 0$ and $g'(p) \neq 0$. This makes

$$f(p) = p - \frac{g(p)}{g'(p)} = p - \frac{0}{g'(p)} = p$$

Since $f(p) = p$, this means p is a fixed point of $f(x)$.

To prove local stability of that fixed point, we must compute $f'(p)$. The quotient rule gives

$$f'(x) = 1 - \frac{(g'(x))^2 - g(x)g''(x)}{(g'(x))^2}$$

Substituting p for x, and using the fact that $g(p) = 0$ and $g'(p) \neq 0$ yields

$$f'(p) = 1 - \frac{(g'(p))^2 - g(p)g''(p)}{(g'(p))^2} = 1 - \frac{(g'(p))^2 - 0 \cdot g''(p)}{(g'(p))^2} = 1 - \frac{(g'(p))^2}{(g'(p))^2} = 0$$

Since $f'(p) = 0$ lies between -1 and $+1$, then p must be stable. This proves that if p is a simple root of $g(x)$ then convergence to p will occur for any sufficiently close x_0. The proof that convergence will also occur when p is a root of multiplicity greater than 1 is left as an exercise.

Euler Method

Differential Equations Models

Another use of nonlinear iteration in solving a mathematical problem stems from its association with differential equations. As described in Chapter 1, differential equations of the form

$$\frac{dy}{dt} = g(t, y) \tag{2}$$

are used to model continuous-time processes. Here, $g(t, y)$ represents a function of two independent variables t and y. Differential equations arise most frequently when modeling in the physical sciences and engineering, although they also appear in the biological and social sciences as well.

Similar to discrete-time models, a unique solution of $y' = g(t, y)$ will exist if $g(t, y)$ is a differentiable function (of both variables), and an initial value of $y(0)$ is given. But unlike discrete-time iterates, the solution of a differential equation is a continuous function $y(t)$ defined on an entire interval of the t-axis. For example, the differential equation

$$\frac{dv}{dt} = -g$$

where $-g$ is the acceleration of gravity (about -32 ft/sec^2 or -9.8 m/sec^2) models the velocity $v(t)$ of an object that is dropped and falls with no air resistance. The unique solution $v(t) = -gt$ satisfying $v(0) = 0$ is a continuous linear function for all time $t \geq 0$ (until it hits the ground, that is).

Unlike this example, where simple integration was used to solve it, coming up with a solution $y(t)$ of a general differential equation can often be a difficult task. Learning techniques for solving such equations exactly generally constitutes an entire course for students who have studied at least differential and integral calculus. Even then, relatively few differential equations can actually be solved exactly.

For example, try solving the differential equation for the velocity of a falling object subject to air resistance

$$m\frac{dv}{dt} = -kv - mg$$

or the continuous version of the logistic equation

$$\frac{dP}{dt} = rP(1 - P/C)$$

which once again models density-dependent population growth. These can be solved by *separating variables* and then integrating, but it takes a bit of effort. Many other differential equations, however, cannot be solved at all (as far as anyone presently knows). For example, even though in theory differential equations such as

$$\frac{dy}{dt} = ty^3 + t^4y \quad \text{or} \quad \frac{dy}{dt} = e^{y+t}\sin(y - t) \quad \text{or} \quad \frac{dy}{dt} = \frac{y + t}{y^2 + t^2}$$

have solutions, they very likely cannot be solved by any known techniques.

Approximate Solutions

There are some simple ways, however, to find an approximate solution of *any* differential equation. These methods usually involve converting the continuous-time problem into a discrete-time equation and then iterating. This will generate a sequence of iterates y_n that are approximately equal to the solution $y(t)$ of the original continuous-time problem at discrete time steps. The size of these time steps is called the *step size* and affects the accuracy of the approximation: the smaller the step size, the closer the estimates y_n will be to the true solution.

Many ways can be used to convert the continuous problem into a discrete one, each with its own advantages and disadvantages. Here, we introduce the Euler Method, which is perhaps the simplest of these techniques.

To derive the method, we begin by calling $y_0 = y(0)$ and plotting the point $(0, y_0)$ in the (t, y)-plane. Since the value of $y(0)$ is given, we know the location of that point. As in the Newton Root-Finding Method, the Euler Method involves extending a tangent line, this time the line that is tangent to the graph of the unknown solution $y(t)$ at the point $(0, y_0)$. Extend that tangent line to the right, or *forward in time,* until the t-coordinate equals h. If h is small, then the right end point of that tangent line will now be a point (h, y_1) that is close to $(h, y(h))$ (see Fig. 3.26(a)).

The value of y_1 will be used as an approximation to $y(h)$. To compute that value, observe from Figure 3.26(b) that the slope of the tangent line can be computed in two ways. First, it equals the derivative $y'(0)$. Another way is to look at the right triangle created by

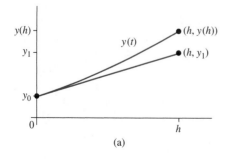
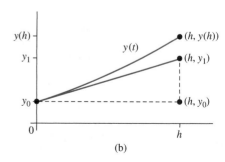

FIGURE 3.26

the three points: $(0, y_0)$, (h, y_0) and (h, y_1). The slope of the tangent line equals the height of that triangle $y_1 - y_0$, divided by the length of its base h. Since the two computations of the slope must be equal, then

$$\frac{y_1 - y_0}{h} = y'(0)$$

and solving for y_1 gives

$$y_1 = y_0 + hy'(0) \tag{3}$$

Finally, although we don't know $y(t)$, we can compute $y'(0)$ from (2), which can be written $y'(t) = g(t, y(t))$. Letting $t = 0$ then gives

$$y'(0) = g(0, y(0)) = g(0, y_0)$$

and so equation (3) becomes

$$y_1 = y_0 + h\, g(0, y_0)$$

We now have an estimate y_1 for the value of $y(h)$.

If the previous argument is repeated, this time using (h, y_1) as the starting point, it is possible to find a point $(2h, y_2)$ that is close to $(2h, y(2h))$, so that y_2 can be used to approximate $y(2h)$. Following the steps above, we compute

$$y_2 = y_1 + h\, g(h, y_1)$$

In this way it is possible to find an estimate y_n for each $y(nh)$ successively for all $n > 0$. The iterative equation

$$y_{n+1} = y_n + h\, g(nh, y_n)$$

generates the values of the y_n. A computer can now be used to numerically iterate that equation up to any value of n.

We summarize our findings:

EULER METHOD

A solution $y(t)$ of $\dfrac{dy}{dt} = g(t, y)$ can be approximated by a solution y_n of

$$y_{n+1} = y_n + h\, g(nh, y_n) \tag{4}$$

for any small step size $h > 0$. If the value of $y_0 = y(0)$ is given, then each y_n is the approximate value of $y(nh)$.

EXAMPLE 4

For the differential equation $\dfrac{dy}{dt} = t^2 + y^2$ with initial condition $y(0) = 2$, use the Euler Method with a step size of $h = 0.1$ to find the approximate values of the solution $y(t)$ at $t = 0.1, 0.2$ and 0.3.

Solution: Since $g(t, y) = t^2 + y^2$ and $h = 0.1$, then (4) becomes

$$y_{n+1} = y_n + 0.1\left((0.1n)^2 + y_n^2\right) = y_n + 0.1\left(0.01n^2 + y_n^2\right)$$

Letting $n = 0, 1, 2$ in turn gives the estimates

$$y_1 = y_0 + 0.1(0.01 \cdot 0^2 + y_0^2) = 2 + 0.1(0 + 2^2) = 2.4$$

$$y_2 = y_1 + 0.1(0.01 \cdot 1^2 + y_1^2) = 2.4 + 0.1(0.01 + 2.4^2) = 2.977$$

$$y_3 = y_2 + 0.1(0.01 \cdot 2^2 + y_2^2) = 2.977 + 0.1(0.04 + (2.977)^2) = 3.8672529$$

So, the approximations are $y(0.1) \approx 2.4$, $y(0.2) \approx 2.977$ and $y(0.3) \approx 3.8672529$.   ∎

Although in this example we computed the estimates ourselves, it is of course much more likely that in practice a computer would be utilized for this effort.

Autonomous Differential Equations

Using the Euler approximation, it is possible to investigate to some extent the dynamics of continuous nonlinear dynamical systems using some of the same tools developed earlier in this chapter for the discrete case. For example, the notions of fixed point and stability also exist for autonomous differential equations. Similar to the problem we have been investigating, $x_{n+1} = f(x_n)$, which does not depend explicitly on n, an autonomous differential equation is one that does not explicitly involve time t, and so can be written more concisely as

$$\frac{dy}{dt} = g(y) \tag{5}$$

since the function $g(y)$ no longer depends on the variable t. A fixed point may exist for such a problem, but this time it should be viewed as a constant p such that $y(t) = p$ is a solution for all $t \geq 0$. This is an equilibrium solution of the continuous model.

It makes sense that fixed points of (5) should also be fixed points of its Euler approximation, which now becomes

$$y_{n+1} = y_n + h\,g(y_n) \tag{6}$$

Since this is now of the form $y_{n+1} = f(y_n)$, where $f(y) = y + hg(y)$, then, as usual, to find those fixed points we solve

$$f(y) = y + h\,g(y) = y$$

Since for $h \neq 0$ this is equivalent to $g(y) = 0$, the fixed points of $f(y) = y + h\,g(y)$ are therefore the points p that satisfy $g(p) = 0$. This also means that $y(t) = p$ is a fixed point solution of (5) if p satisfies $g(p) = 0$. This is true because if $y(t) = p$ then $y'(t) = 0$ (since p is a constant), and so both

$$\frac{dy}{dt} = 0 \quad \text{and} \quad g(y(t)) = g(p) = 0$$

making both sides of the differential equation (5) equal for all $t > 0$.

To determine local stability of p, we check to see if $-1 < f'(p) < +1$. Since

$$f(y) = y + h\,g(y) \quad \text{then} \quad f'(p) = 1 + h\,g'(p)$$

The condition $-1 < f'(p) < +1$ therefore becomes

$$-1 < 1 + h\,g'(p) < +1 \quad \text{or} \quad -2/h < g'(p) < 0$$

Since this must be true for any $h > 0$ and

$$\lim_{h \to 0} -2/h = -\infty$$

then stability exists if $-\infty < g'(p) < 0$, or more simply $g'(p) < 0$. A similar argument would show that p is unstable for $g'(p) > 0$.

We have therefore proven:

THEOREM   *Fixed Points and Stability*

The autonomous differential equation $\dfrac{dy}{dt} = g(y)$ has a fixed point solution $y(t) = p$ for any p satisfying $g(p) = 0$. This solution is locally stable if $g'(p) < 0$ or unstable if $g'(p) > 0$.   ∎

We remark that local stability of a fixed point p of a differential equation means that for all initial points $y(0)$ sufficiently close to p, the solution $y(t)$ satisfies $\lim_{t \to \infty} y(t) = p$.

EXAMPLE 5

For the continuous logistic model of density-dependent population growth $\dfrac{dP}{dt} = rP(1 - P/C)$, find any equilibrium solutions and determine their stability for $r = 1.05$ and $C = 10{,}000$.

Solution: Since $g(y) = 1.05y(1 - y/10{,}000)$ then $g'(y) = 1.05(1 - y/5000)$. Solving

$$g(y) = 1.05y(1 - y/10{,}000) = 0$$

gives two fixed points 0 and 10,000. Also,

$$g'(0) = 1.05 > 0 \quad \text{and} \quad g'(10{,}000) = -1.05 < 0$$

So, $P(t) = 0$ is a fixed but unstable population level, and $P(t) = 10{,}000$ is a stable equilibrium population.   ∎

Accuracy of the Euler Method

One may wonder how accurate the Euler Method is at approximating solutions of differential equations. One factor affecting the accuracy is of course the step size h. For small values of h, the approximation (4) closely resembles (2). In fact, in the limit as $h \to 0$ these equations are the same. So, the smaller the h value, the smaller the error in the approximate solution will be.

A less obvious factor that affects accuracy is the stability or instability of fixed-point solutions. Figure 3.27(a) indicates that if we find an approximate solution of an autonomous differential equation of the form (5) in the vicinity of a stable fixed point p, then the **absolute error** will approach 0 as $t \to \infty$. The absolute error is the magnitude of the difference between the true solution and its approximation at any time t. On the other hand, in the vicinity of an unstable fixed point the absolute error will grow with t, as shown in Figure 3.27(b).

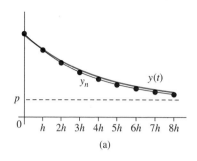 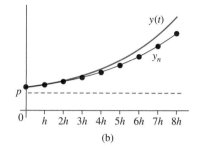

(a) (b)

FIGURE 3.27

The **relative error** is usually also smaller when there is stability. The relative error is ratio of the absolute error divided by the magnitude of the true solution. This smaller relative error means that we can be confident that the behavior of the approximate solution more accurately reflects the dynamics of the real one, when there is stability.

EXAMPLE 6

If r is any constant then the true solution of $\dfrac{dy}{dt} = r(y - 4)$ with initial condition $y(0) = 5$ is $y(t) = 4 + e^{rt}$. Use the Euler Method with $h = 0.25$ to approximate $y(1)$ if (a) $r = 1$ and (b) $r = -1$. Compare each with the value of the true solution at $t = 1$.

Solution: Since $h = 0.25$ and we are trying to estimate $y(1) = y(4 \cdot 0.25)$ then $n = 4$, and so we must compute y_4. Also, since $g(y) = r(y - 4)$, then (6) becomes

$$y_{n+1} = y_n + 0.25r(y_n - 4) = (1 + 0.25r)y_n - r$$

In part (a) where $r = 1$ this means we must iterate $y_{n+1} = 1.25y_n - 1$. Starting with $y_0 = 5$ we find

$$y_1 = 5.25, \quad y_2 = 5.5625, \quad y_3 = 5.953125, \quad \text{and} \quad y_4 = 6.4414062$$

Since the true solution for $r = 1$ is $y(t) = 4 + e^t$, then $y(1) = 4 + e \approx 6.718282$. So,

$$\text{absolute error} = |y_4 - y(1)| \approx |6.441406 - 6.718282| = 0.276876$$

$$\text{and relative error} = \left|\frac{y_4 - y(1)}{y(1)}\right| \approx \left|\frac{0.276876}{6.718282}\right| \approx 0.0412$$

In part (b) where $r = -1$ we must iterate $y_{n+1} = 0.75y_n + 1$. With $y_0 = 5$ we get

$$y_1 = 4.75, \quad y_2 = 4.5625, \quad y_3 = 4.421875, \quad \text{and} \quad y_4 = 4.3164062$$

Comparing the true solution $y(1) = 4 + e^{-1} \approx 4.367879$ for $r = 1$ with y_4 gives

$$\text{absolute error} \approx 0.051473 \quad \text{and} \quad \text{relative error} \approx 0.0118$$

The absolute error in part (a) is at least five times as great as in part (b), and the relative error in (a) almost four times as great as in (b). This can be understood by observing that $y(t) = 4$ is an unstable fixed point solution in (a) and a stable fixed point solution in (b). ∎

Exercises 3.4

In Exercises 1–6, given g(x) and x_0, find x_1, x_2, x_3 using equation (1).

1. $g(x) = 2x^2 - 8$, $x_0 = 1$

2. $g(x) = 7 - x^2$, $x_0 = 3$

3. $g(x) = 2x^3 - 3x^2 + 2x - 3$, $x_0 = 1$

4. $g(x) = 2 - e^x$, $x_0 = 0$

5. $g(x) = x - \sqrt{x} - 1$, $x_0 = 1$

6. $g(x) = 1 - x + \sin x$, $x_0 = \pi$

7. The function $g(x) = x^2 + 3x - 18$ has a root at $p = 3$. Determine graphically which of the following values of x_0 will generate a sequence x_n that converges to p using the Newton method. (a) $x_0 = 2$; (b) $x_0 = -1$; (c) $x_0 = -2$.

8. Repeat the previous exercise for $g(x) = xe^{-x}$, $p = 0$ and

    (a) $x_0 = 2$;   (b) $x_0 = 1$;   (c) $x_0 = -1$.

9. The function $g(x) = x^2 - 2x + 4$ has no real roots. Show graphically what will happen in this case if the Newton method is used beginning with $x_0 = 0$.

We have proven that a root of multiplicity 1 of g(x) is a stable fixed point of (1), and so the Newton method converges in this case. In Exercises 10–13 show that the given root p of g(x) is a stable fixed point of (1) even though p has multiplicity 2 or higher.

10. $g(x) = (x - 1)^2$, $p = 1$

11. $g(x) = (x - 2)^3$, $p = 2$

12. $g(x) = \sin^2 x$, $p = 0$

13. $g(x) = x^3 e^x$, $p = 0$

14. Equation (1) can also be derived in another way:

    (a) If x_0 is an initial estimate for a root p of $g(x)$, find the equation of the straight line tangent to $y = g(x)$ at $x = x_0$.

    (b) Find the point x_1 where this straight line crosses the x-axis.

    (c) Replacing x_0 and x_1 with x_n and x_{n+1} respectively yields (1).

15. Determine which of the following differential equations are autonomous and which are non-autonomous. In each case identify the function $g(y)$ or $g(t, y)$.

    (a) $y' = 2y^2 - y^3 + t$

    (b) $y' = 8y^2 \sin y$

    (c) $3y' - 6y = t$

    (d) $2yy' - 3y = 4$

For each of the autonomous differential equations given in Exercises 16–19 use the Euler method with $h = 0.1$ to estimate y(0.1), y(0.2), y(0.3).

16. $y' = 2y$, $y(0) = 2$

17. $y' = 3 - y$, $y(0) = 0$

18. $y' = \sqrt{1 - y^2}$, $y(0) = 0$

19. $y' = 2y/(y^2 + 1)$, $y(0) = 1$

For each of the non-autonomous differential equations given in Exercises 20–23 use the Euler method with $h = 0.1$ to estimate y(0.1), y(0.2), y(0.3).

20. $y' = 5t - 4y + 3$, $y(0) = 0$

21. $y' = (t + 1)y^2$, $y(0) = 1$

22. $y' = (y + t)/(y - t)$, $y(0) = 1$

23. $y' = -\sqrt{t^2 + y^2}$, $y(0) = 1$

For each of the autonomous differential equations given in Exercises 24 and 25 the Euler method becomes a discrete linear equation of the form $y_{n+1} = ay_n + b$. Find that linear equation and its exact solution for $h = 0.05$ and the given $y_0 = y(0)$.

24. $y' = 5y$, $y(0) = 1$

25. $y' = 8 - 2y$, $y(0) = 0$

In Exercises 26–29 find the equilibrium solutions of the given autonomous differential equation and determine their stability.

26. $y' = 2y - 7$ 28. $y' = y^3 - 3y^2 + y$

27. $y' = 5y - 3y^2$ 29. $y' = y^4 - y^2 - 6$

30. The differential equation for the velocity (in ft/sec) of a falling object subject to air resistance is $mv'(t) = -kv(t) - mg$. If $g = 32$, $m = 5$ and $k = 2/3$, compute the *terminal velocity* of the object, i.e., compute $\lim_{t \to \infty} v(t)$.

31. The differential equation for the (Fahrenheit) temperature $T(t)$ of a hot object placed in a cooler room is $T'(t) = -k(T(t) - R)$. At what temperature will the object eventually level off, if (a) $k = 2/3$ and $R = 68.5°$F; (b) $k = 0.5$, $T(0) = 120°$F and $T'(0) = -20$.

■■■■■ Computer Projects 3.4

1. Using the Newton method, determine how many iterations are needed to obtain a 6 decimal place approximation to the root $p = 2$ of $g(x) = 8 - x^3$ if

     (a) $x_0 = 3$;   (b) $x_0 = 10$;   (c) $x_0 = 100$.

2. Repeat the previous problem for $g(x) = e^x - 10$ and $p = \ln 10$.

3. Find a 6 decimal place approximation to the solution of $e^x = 1/x$.

4. Find 6 decimal place approximations to all solutions of $2x^3 - 7x^2 - 10x + 35 = 0$.

If p is a root of multiplicity m *of* g(x) *then iterating* $x_{n+1} = x_n - m\dfrac{g(x_n)}{g'(x_n)}$ *converges to* p *faster than (1) does. Demonstrate this in Projects 5 and 6 by computing* $x_1, \ldots, x_{20}$ *each way, starting with the same* x_0.

5. $g(x) = (x - 1)^2$, $p = 1$, $m = 2$ 6. $g(x) = \sin^3 x$, $p = 0$, $m = 3$

7. Through numerical experimentation, find the basin of attraction of the root $p = 0$ of $g(x) = x^3 - 4x$ using Newton's method.

8. The point $p = 3$ is a stable fixed point of $f(x) = \sqrt{2x + 3}$, and so iteration of $x_{n+1} = f(x_n)$ will converge to p. The point p is also a root of $f(x) - x = 0$, and so Newton's method with $g(x) = f(x) - x$ will also converge to p. Determine which type of iteration converges faster to p for the same x_0.

In Projects 9 and 10 find approximate solutions of the given differential equations for $0 \le t \le 1$ *using the Euler method with* h = 0.1, 0.05, 0.01. *Compare the three estimates of* y(1) *with the given exact solution. Find the absolute and relative errors in each case.*

9. $y' = -2ty^2$, $y(0) = 1$; exact solution $y(t) = 1/(t^2 + 1)$.

10. $y' = 4t + 2ty$, $y(0) = 0$; exact solution $y(t) = -2 + 2e^{t^2}$.

In Projects 11 and 12 find approximate solutions of the given autonomous differential equations for $0 \le t \le 10$ *using the Euler method with* h = 0.1. *Graph the given exact solution* y(t) *and its approximation together for* $0 \le t \le 10$. *Compute the absolute and relative errors in the approximation of* y(10).

11. $y' = 2y - 6$, $y(0) = 4$; exact solution $y(t) = 3 + e^{2t}$.

12. $y' = 6 - 2y$, $y(0) = 4$; exact solution $y(t) = 3 + e^{-2t}$.

The differential equations given in Projects 13 and 14 have no known exact solutions. Use the Euler method to find an approximate solution of each for $0 \le t \le 10$. Repeat for different values of h until you are confident your approximation is correct to 3 decimal places.

13. $y' = y^2 + 4t \sin(y + t)$, $y(0) = 1$.   14. $y' = t^2/\sqrt{4y + 3t}$, $y(0) = 2$.

SECTION
3.5 Periodic Points and Cycles

We now return to our investigation of the dynamics of $x_{n+1} = f(x_n)$. So far, the qualitative behavior we have been concentrating on has dealt primarily with fixed points, especially those that are stable. This is due to the fact that stable fixed points determine much of the overall behavior of a nonlinear dynamical system, since they attract all nearby solutions. But as we shall see, fixed points are not the only kind of attractors and repellers that may exist for a nonlinear equation. In this section we consider **periodic** dynamics, beginning with the simplest case, period 2, followed by the general case of period m for any positive integer m.

2-Cycles

Recall that a fixed point p satisfies $f(p) = p$. Suppose now that there is a pair of points a and b that instead satisfy $f(a) = b$ and $f(b) = a$. For the problem $x_{n+1} = f(x_n)$ with $x_0 = a$ this means that

$$x_1 = f(x_0) = f(a) = b, \quad x_2 = f(x_1) = f(b) = a, \quad x_3 = f(x_1) = f(a) = b, \quad \dots$$

$$\text{or} \quad x_0 = a, \quad x_1 = b, \quad x_2 = a, \quad x_3 = b, \quad x_4 = a, \quad x_5 = b, \quad \dots.$$

The x_n therefore alternate back and forth between two values a and b for all n. The following terminology is commonly used to describe this.

DEFINITION

A pair of distinct points a and b satisfying $f(a) = b$ and $f(b) = a$ is called a **2-cycle** of $x_{n+1} = f(x_n)$, and each point is called a **point of period 2** for $f(x)$.

Such points may exist for a nonlinear function $f(x)$, as the following example shows.

EXAMPLE 1

Find a 2-cycle of $x_{n+1} = 1 - x_n^2$.

Solution: Here $f(x) = 1 - x^2$. If we let $a = 0$ and $b = 1$ then

$$f(a) = f(0) = 1 - 0^2 = 1 = b \quad \text{and} \quad f(b) = f(1) = 1 - (1)^2 = 0 = a$$

So, 0 and 1 constitute a 2-cycle. ∎

Although the previous example demonstrates that 2-cycles may exist, it does not indicate how to find them. Guesswork was used there, but that is not a reliable method. A more systematic approach involves making the following observation.

Since $f(a) = b$ then substituting $f(a)$ for b in the equation $f(b) = a$ gives $f(f(a)) = a$. If we call $g(x) = f(f(x))$ then $g(a) = a$. This means that a is a fixed point of $g(x)$. A similar argument shows $g(b) = b$. So, any point of period 2 for $f(x)$ is actually a fixed point of the composition $f(f(x))$. This provides a method of finding these points, namely, to find the fixed points of $f(f(x))$ by solving

$$f(f(x)) = x \quad \text{or} \quad f(f(x)) - x = 0 \tag{1}$$

However, we need to be a bit more careful than that. Suppose p is a fixed point of $f(x)$. Then of course $f(p) = p$, but this implies

$$f(f(p)) = f(p) = p$$

So p is also a fixed point of $f(f(x))$. That is, by solving $f(f(x)) = x$, we will get not only points of period 2, but also fixed points as well.

EXAMPLE 2

For $f(x) = -x^3$: (a) Find the fixed points of $f(f(x))$. (b) Determine which are points of period 2.

Solution: (a) Since $f(f(x)) = -(-x^3)^3 = x^9$, we must solve

$$x^9 = x \quad \text{or} \quad x^9 - x = 0$$

Factoring yields

$$x^9 - x = x(x^8 - 1) = x(x^4 - 1)(x^4 + 1) = x(x^2 - 1)(x^2 + 1)(x^4 + 1)$$
$$= x(x - 1)(x + 1)(x^2 + 1)(x^4 + 1)$$

Since no further factoring can be done, we see that the solutions of

$$x(x - 1)(x + 1)(x^2 + 1)(x^4 + 1) = 0$$

are 0, 1 and -1. These are the fixed points of $f(f(x))$.
(b) Since $f(0) = -0^3 = 0$, this makes 0 a fixed point of $f(x) = -x^3$. But since

$$f(1) = -1^3 = -1 \quad \text{and} \quad f(-1) = -(-1)^3 = 1$$

then 1 and -1 constitute a 2-cycle. ■

As the example indicates, actually solving (1) to find points of period 2 can be quite difficult, much more so than solving $f(x) = x$ to find fixed points. But if $f(x)$ is a polynomial, or more generally, if equation (1) can be turned into a polynomial equation, then the following helpful trick can sometimes simplify things.

In the investigation of the dynamics of $x_{n+1} = f(x_n)$ it is likely that the fixed points of $f(x)$ have already been determined before a search for 2-cycles is undertaken. Since every fixed point p must be a solution of (1), then $x - p$ must be one of the factors of $f(f(x)) - x$.

This gives at least one and perhaps several of the factors of $f(f(x)) - x$, depending upon how many fixed points there are. We therefore have the following:

RULE

To find a 2-cycle of a polynomial $f(x)$, first divide $f(f(x)) - x$ by $x - p$ for each fixed point p. This will reduce the degree of the polynomial involved and simplify solving $f(f(x)) - x = 0$. The same may work when $f(x)$ is not a polynomial but if $f(f(x)) - x = 0$ can be converted into a polynomial equation.

The following example demonstrates this idea.

EXAMPLE 3

Find a 2-cycle of $x_{n+1} = x_n^2 + x_n - 4$.

Solution: We first find the fixed points by solving $f(x) = x^2 + x - 4 = x$. This is equivalent to

$$x^2 - 4 = 0 \quad \text{or} \quad (x - 2)(x + 2) = 0$$

So the fixed points are $p = 2$ and $q = -2$.

We must next compute

$$f(f(x)) = (x^2 + x - 4)^2 + (x^2 + x - 4) - 4 = x^4 + 2x^3 - 6x^2 - 7x + 8$$

and then solve

$$x^4 + 2x^3 - 6x^2 - 7x + 8 = x \quad \text{or}$$
$$x^4 + 2x^3 - 6x^2 - 8x + 8 = 0$$

To help solve this, we can divide both sides by

$$(x - 2)(x + 2) = x^2 - 4$$

since we already know that $p = 2$ and $q = -2$ must be solutions. This yields

$$x^2 + 2x - 2 = 0$$

which is much easier to deal with. In fact, using the quadratic formula we find the solutions are $x = -1 + \sqrt{3}$ and $x = -1 - \sqrt{3}$. Since we already found that $p = 2$ and $q = -2$ are the only fixed points, then $a = -1 + \sqrt{3}$ and $b = -1 - \sqrt{3}$ must be a 2-cycle. This can be verified by substituting each point for x in $f(x)$ and seeing that the other is obtained. ∎

Stability of 2-Cycles

To discuss the issue of stability of 2-cycles we must begin by asking ourselves what stability must mean in this case. We can gain some insight into this question from two sources. One is the logistic model $x_{n+1} = 3.2x_n(1 - x_n/10{,}000)$, previously considered in Example 6 of Section 3.2, which has a stable or attracting 2-cycle. Table 3.7 shows that the numerically computed solution approaches both points of the 2-cycle, alternating between the two

TABLE 3.7

n	x_n	n	x_n
0	2500		
1	6000	11	5134
2	7680	12	7994
3	5702	13	5131
4	7842	14	7995
5	5415	15	5131
6	7945	16	7995
7	5225	17	5130
8	7984	18	7995
9	5151	19	5130
10	7993	20	7995

values $a = 7995$ and $b = 5130$ as $n \to \infty$. One set of iterates, those with even-numbered subscripts, converges to a, while the rest, the odd-numbered ones, simultaneously converge to b.

It is easy to see that this must always occur for any stable 2-cycle. It is impossible for the even-numbered iterates to converge to one point of a 2-cycle without the odd-numbered iterates also converging to the other. Continuity of the function $f(x)$ guarantees this. As a result, any notion of stability must involve convergence to both points simultaneously.

We can also elicit some insight into stability of 2-cycles by once again viewing each point of the cycle as a fixed point of $f(f(x))$. In this way we can make use of the definition of stability of fixed points.

If these two ideas are combined, we arrive at a definition.

DEFINITION

A 2-cycle of $x_{n+1} = f(x_n)$ is **locally stable** if each point of the cycle is a stable fixed point of $f(f(x))$. Otherwise, the 2-cycle is **unstable.** Either both points are stable or both are unstable.

It is worth considering the implications of an attracting 2-cycle for the modeling process. Rather than anticipating stability to be synonymous with the presence of an attracting equilibrium point, we see now that stable dynamics could also come in the form of a solution that varies periodically. This never happens with linear equations, but it is common in nature.

For example, there are many stable periodic processes in the physical world: the motion of the earth, which gives on a regular basis day and night as well as the change of seasons.

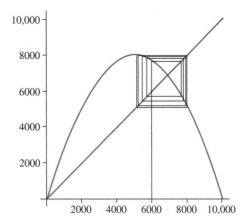

FIGURE 3.28

Wave phenomena, such as sound, light, radio waves, microwaves, etc., are also different forms of periodic energy oscillation. The yearly return of certain illnesses may be viewed as a form of periodic behavior. Business cycles in different sectors of the economy also oscillate, although one must admit that these appear neither stable nor predictable!

Since determining the stability or instability of 2-cycles, if they exist, is no less important than doing so for fixed points, we would next like to establish some simple criteria from which we can easily determine when such a cycle is indeed stable, preferably without extensive calculation. One way is to once again make use of cobweb graphs.

EXAMPLE   4

For the logistic equation $P_{n+1} = 3.2 P_n (1 - P_n/10{,}000)$, draw a cobweb graph for $P_0 = 6000$.

Solution: As Figure 3.28 shows, the cobweb oscillates as it grows, eventually stabilizing in the shape of a square with corners having coordinates equal to the values of the 2-cycle. This always occurs when there is a stable 2-cycle.   ∎

EXAMPLE   5

For $f(x) = -x^3$ use cobweb graphs to determine the stability of its 2-cycle.

Solution: In Example 2 the 2-cycle was found to consist of 1 and -1. If we draw a cobweb starting with $x_0 = 1$, a square is once again obtained, as shown in Figure 3.29. However, starting with any x_0 on either side of 1 or -1, we see that solutions oscillate away from that square, indicating that the 2-cycle is unstable.   ∎

Assuming $f(x)$ is differentiable, there is of course another way to determine whether a 2-cycle attracts or repels. We can use the derivative to determine the stability of a or b as a fixed point of $f(f(x))$. In this case, we must evaluate $\dfrac{d}{dx} f(f(x))$ at $x = a$ and compare

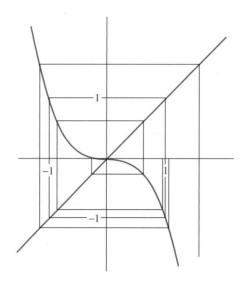

FIGURE 3.29

its absolute value with 1. Using the chain rule,

$$\frac{d}{dx} f(f(x)) = f'(f(x))f'(x)$$

and so the derivative at $x = a$ equals

$$f'(f(a))f'(a) = f'(b)f'(a) = f'(a)f'(b)$$

since $f(a) = b$. Taking the absolute value and of this expression and comparing with 1 gives the condition

$$|f'(a)f'(b)| < 1 \quad \text{for stability} \quad \text{and} \quad |f'(a)f'(b)| > 1 \quad \text{for instability}.$$

A similar calculation would show the exact same conditions for b.

We have proven the following:

THEOREM   *Stability of 2-Cycles*

If a and b are the two points of a 2-cycle of $x_{n+1} = f(x_n)$, then the cycle is locally stable if $|f'(a)f'(b)| < 1$ or unstable if $|f'(a)f'(b)| > 1$.   ■

Note what this says. Stability or instability of a 2-cycle depends upon the product of the derivatives at *both* points of the cycle. In particular, if the absolute value of that product is less than 1 the cycle attracts, or if it exceeds 1 the cycle repels. The theorem confirms mathematically what we said earlier: either both points simultaneously attract or both simultaneously repel.

EXAMPLE 6

Determine the stability of the 2-cycle of $x_{n+1} = 1 - x_n^2$.

Solution: In Example 1 the 2-cycle was found to consist of the points 0 and 1. Now, since $f(x) = 1 - x^2$ then $f'(x) = -2x$. So

$$f'(0) \cdot f'(1) = 0 \cdot (-2) = 0$$

Since this is between -1 and 1 the cycle is stable. ∎

EXAMPLE 7

Determine the stability of the 2-cycle of $x_{n+1} = x_n^2 + x_n - 4$ found in Example 3.

Solution: In Example 3 the 2-cycle was found to consist of the points $a = -1 + \sqrt{3}$ and $b = -1 - \sqrt{3}$. Since $f(x) = x^2 + x - 4$, then $f'(x) = 2x + 1$, which means

$$f'(a) = 2(-1 + \sqrt{3}) + 1 = -1 + 2\sqrt{3} \quad \text{and} \quad f'(b) = 2(-1 - \sqrt{3}) + 1 = -1 - 2\sqrt{3}$$

This makes

$$f'(a) \cdot f'(b) = (-1 + 2\sqrt{3}) \cdot (-1 - 2\sqrt{3}) = 1 - 4 \cdot 3 = -11$$

and so the cycle is unstable. ∎

Period *M* and *M*-Cycles

Fixed points and 2-cycles constitute the simplest cases of a more general idea: periodic points and cycles of any period. Since fixed points are solutions of $f(x) = x$ and each point of a 2-cycle is a solution of $f(f(x)) = x$, to extend these concepts we must discuss compositions of $f(x)$, and therefore need a more convenient notation for them.

NOTATION

For any $f(x)$ and any $n \geq 2$ the composition of $f(x)$ with itself $n - 1$ times is denoted by $f^n(x)$. That is,

$$f^2(x) = f(f(x)), \quad f^3(x) = f(f(f(x))), \quad f^4(x) = f \circ f \circ f \circ f(x), \quad \ldots$$

We remark that there should be no confusion distinguishing compositions from powers, since the latter is of little use here. To avoid any ambiguity one should use $[f(x)]^n$ to denote $f(x)$ raised to the power n.

With this new notation we see that

$$x_1 = f(x_0), \quad x_2 = f(x_1) = f(f(x_0)) = f^2(x_0)$$
$$x_3 = f(x_2) = f(f^2(x_0)) = f^3(x_0), \quad \ldots$$

and so the type of discrete dynamical system we have been studying,

$$x_{n+1} = f(x_n) \quad \text{may also be written} \quad x_n = f^n(x_0)$$

It is now possible to generalize the ideas developed above for points of period 2 and 2-cycles to points and cycles of higher periods. The following is the analogous definition of these concepts that may apply for any $m \geq 1$.

DEFINITION

A set of m distinct points $p_1, p_2, \ldots, p_m$ satisfying

$$f(p_1) = p_2, \quad f(p_2) = p_3, \quad \ldots, \quad f(p_{m-1}) = p_m, \quad f(p_m) = p_1$$

is called an **m-cycle** of $x_{n+1} = f(x_n)$. Each point p_i of an m-cycle is called a **point of period m** for $f(x)$.

It can easily be checked that for $m = 2$ this definition is consistent with the previous definition of 2-cycles and points of period 2. Under this definition even fixed points can be called 1-cycles or points of period 1, but it is for larger values of m that we most often use these terms. For example, a 4-cycle would consist of 4 distinct points p_1, p_2, p_3, p_4 that repeat after every four iterations of $f(x)$:

$$f(p_1) = p_2, \quad f(p_2) = p_3, \quad f(p_3) = p_4 \quad f(p_4) = p_1$$

Starting with any one of these points, say p_3, we can generate all the others by taking compositions:

$$f(p_3) = p_4, \quad f^2(p_3) = f(f(p_3)) = f(p_4) = p_1, \quad f^3(p_3) = f(f^2(p_3)) = f(p_1) = p_2$$
$$\text{and} \quad f^4(p_3) = f(f^3(p_3)) = f(p_2) = p_3$$

From this we see in general that, given any point p of period m, its m-cycle consists of the set of points $p, f(p), f^2(p), \ldots, f^{m-1}(p)$, with $f^m(p) = p$. Although this implies that p must be a fixed point of the composite function $f^m(x)$, the computations involved in solving the equation $f^m(x) = x$ to find such a point would likely prove a daunting task. However, if a periodic point is somehow found, the other points of its cycle can be easily generated by taking repeated compositions.

EXAMPLE 8

Given that 7/9 is a point of period 3 for $x_{n+1} = 1 - 2|x_n|$, find the other points in its 3-cycle.

Solution: Since $f(x) = 1 - 2|x|$ and $m = 3$, then according to the theorem, the other points of the cycle can be found by computing

$$f(7/9) = 1 - 2|7/9| = -5/9 \quad \text{and} \quad f^2(7/9) = f(-5/9) = 1 - 2|-5/9| = -1/9$$

So the 3-cycle consists of the points 7/9, $-5/9$ and $-1/9$. Note that

$$f^3(7/9) = f(-1/9) = 1 - 2|-1/9| = 7/9$$

which verifies that this actually is a 3-cycle.   ∎

Stability of M-Cycles

Stability and instability can also be analogously defined for higher periods. As with 2-cycles, in order for an m-cycle to be an attractor or repeller, nearby solutions must either converge to or diverge from all points of the m-cycle simultaneously. So we define:

TABLE 3.8

n	P_n	n	P_n
0	4000		
1	8400	11	8750
2	4704	12	3829
3	8719	13	8270
4	3908	14	5008
5	8333	15	8750
6	4862	16	3828
7	8743	17	8269
8	3846	18	5009
9	8284	19	8750
10	4976	20	3828

DEFINITION

An m-cycle of $x_{n+1} = f(x_n)$ is **locally stable** if each point of the cycle is a locally stable fixed point of $f^m(x)$. Otherwise it is **unstable.** Either all points of an m-cycle are stable or all are unstable.

EXAMPLE 9

Determine whether the logistic model $P_{n+1} = rP_n(1 - P_n/C)$ has a stable periodic attractor for $r = 3.5$ and $C = 10,000$ by numerically computing the first 20 iterates for $P_0 = 4000$.

Solution: The results are shown in Table 3.8. After the first dozen or so iterations, the solution settles down into a sequence of four values that keep repeating. This indicates a stable 4-cycle. ∎

As with fixed points and 2-cycles, we would of course like a practical method of determining whether or not a stable m-cycle exists, a method that does not involve extensive numerical calculation. Sometimes, if the cycle does not have too high a period, cobwebbing may be used.

EXAMPLE 10

Draw a cobweb graph for the logistic model with $r = 3.5$ and $C = 10,000$. Use several initial points P_0.

Solution: Figure 3.30(a) shows the web for a good choice of initial condition, $P_0 = 5000$, which almost immediately hits a stable point of period 4. Notice the double cycling between

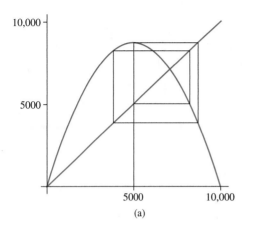
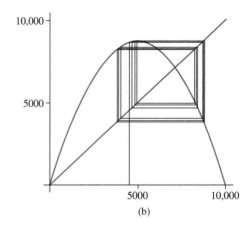

(a)   (b)

FIGURE 3.30

the 4 points of the cycle. Figure 3.30(b) shows what happens if instead $P_0 = 4500$. After several iterations, the solution arrives at the same double cycle.   ■

While cobwebbing may be useful for cycles having small periods, when m is large, often little can be discerned from a cobweb graph. The extensive oscillation between the points of the cycle usually makes the graph difficult to read. And the slightest drawing inaccuracies combined with an initial point too far from a periodic point can lead to problems detecting a stable cycle, if one should exist.

EXAMPLE 11

Use a cobweb graph to determine whether the logistic model has any stable cycles for $r = 3.55$ and $C = 10,000$

Solution: Figure 3.31(a) shows that an 8-cycle exists, and it is in fact stable. But that cobweb was created already knowing a good initial point to use, namely one of the 8 points of the cycle. A random choice for P_0 results in Figure 3.31(b). It is unlikely that a stable 8-cycle could be discerned from this cobweb graph, especially if it were drawn using pencil and paper instead of a computer graphics system.   ■

Since the precision needed to detect a cycle of high period is seldom attainable when drawing by hand, a cobweb graph is not a practical means of analysis in these cases. However, there is the derivative method, which can still be of great use, provided we are able to compute the points of the cycle and $f'(x)$.

The idea is to compute the derivative of $f^m(x)$ at some point p of the m-cycle, and then compare its absolute value with 1. In doing so, the chain rule gives

$$\frac{d}{dx} f^m(x) = \frac{d}{dx} f(f^{m-1}(x)) = f'(f^{m-1}(x)) \cdot \frac{d}{dx} f^{m-1}(x) \tag{2}$$

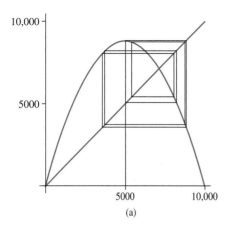

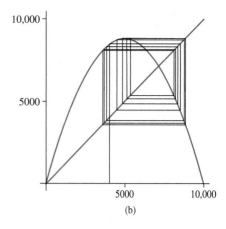

(a)   (b)

FIGURE 3.31

So we must use the chain rule again to compute

$$\frac{d}{dx} f^{m-1}(x) = \frac{d}{dx} f(f^{m-2}(x)) = f'(f^{m-2}(x)) \cdot \frac{d}{dx} f^{m-2}(x) \qquad (3)$$

Although we are not yet finished computing derivatives, combining equations (2) and (3) together gives

$$\frac{d}{dx} f^{m}(x) = f'(f^{m-1}(x)) \cdot f'(f^{m-2}(x)) \cdot \frac{d}{dx} f^{m-2}(x)$$

From this the general pattern may be seen emerging:

$$\frac{d}{dx} f^{m}(x) = f'(f^{m-1}(x)) \cdot f'(f^{m-2}(x)) \cdot f'(f^{m-3}(x)) \cdot \ \cdots \ \cdot f'(f(x)) \cdot f'(x)$$

This last equation makes $\frac{d}{dx} f^{m}(x)$ at $x = p$ equal to

$$f'(f^{m-1}(p)) \cdot f'(f^{m-2}(p)) \cdot f'(f^{m-3}(p)) \cdot \ \cdots \ \cdot f'(f(p)) \cdot f'(p)$$

But from what we saw earlier, the points $p, f(p), f^{2}(p), \ldots, f^{m-1}(p)$ are precisely the points of the m-cycle. So the value of $\frac{d}{dx} f^{m}(x)$ at any point p of an m-cycle equals the product of the derivatives $f'(x)$ evaluated at all points of the cycle. And to determine stability we compare the absolute value of that product with 1. Reversing, for convenience, the order of multiplication in that product, we have proven the following:

THEOREM   *Stability of M-Cycles*

An m-cycle $p_1, p_2, \ldots, p_m$ of $x_{n+1} = f(x_n)$ is locally stable if

$$|f'(p_1) \cdot f'(p_2) \cdot f'(p_3) \cdot \ \cdots \ \cdot f'(p_m)| < 1$$

or unstable if

$$|f'(p_1) \cdot f'(p_2) \cdot f'(p_3) \cdot \ \cdots \ \cdot f'(p_m)| > 1 \qquad \blacksquare$$

EXAMPLE 12

Determine the stability of the 3-cycle of $x_{n+1} = 1 - 2|x_n|$ computed in Example 8.

Solution: The 3-cycle was determined to consist of the points $7/9$, $-5/9$ and $-1/9$. Now, since $f(x) = 1 - 2|x|$, then

$$f'(x) = 2 \quad \text{if} \quad x < 0 \quad \text{or} \quad f'(x) = -2 \quad \text{if} \quad x > 0$$

So, $f'(7/9) = -2$, $f'(-5/9) = 2$ and $f'(-1/9) = 2$. The product of these derivatives is -8, which means the 3-cycle is unstable. ∎

We remark that in this example the derivative can be used to determine stability or instability even though $f(x)$ is not differentiable at $x = 0$. All that is required is that $f(x)$ be differentiable at each point of the cycle.

EXAMPLE 13

Determine whether the price model $P_{n+1} = \dfrac{8}{P_n} + 0.5P_n - 3$ has a stable periodic cycle.

Solution: The easiest way to find a stable cycle is to iterate numerically for several initial points, and see where the solution leads. After 1000 or so steps this results in what appears to be a stable 4-cycle:

$$p_1 = 1.018771, \quad p_2 = 5.361984, \quad p_3 = 1.172977, \quad p_4 = 4.406742$$

Although we know this cycle must be stable since iteration brought us there, it would be comforting to verify this using the product of the derivatives. Computing $f'(x) = -\dfrac{8}{x^2} + 0.5$ at each of the 4 points of the cycle gives

$$f'(p_1) = -7.207912, \quad f'(p_2) = 0.221748, \quad f'(p_3) = -5.314482, \quad f'(p_4) = 0.088040$$

The product of these is 0.747843, which lies between -1 and $+1$. This corresponds to the stability criteria stated in the theorem. ∎

Exercises 3.5

In Exercises 1–4 use guesswork to find a 2-cycle.

1. $x_{n+1} = 4 - \dfrac{x_n^2}{4}$

2. $x_{n+1} = (x_n - 2)^2 + 1$

3. $x_{n+1} = \cos(\pi x_n / 2)$

4. $x_{n+1} = \dfrac{x_n^2}{2} - 2x_n + 2$

In Exercises 5 and 6 compute and simplify $f^2(x)$ *and* $f^3(x)$.

5. $f(x) = x^2 + 1$

6. $f(x) = 1 + 1/x$

In Exercises 7 and 8 find all solutions of $f(f(x)) = x$ *and determine which are fixed points of* $f(x)$ *and which are 2-cycles.*

7. $f(x) = -4x^3$ 　　　　　　　　　8. $f(x) = -x^3/2$

In Exercises 9–12 first find all fixed points of f(x) *and then use them to help find all points of period 2.*

9. $f(x) = x^2 - 2x$ 　　　　　　　11. $f(x) = \dfrac{2}{x} - x$

10. $f(x) = -x^2 + 2x + 2$ 　　　　12. $f(x) = x + \dfrac{2}{x} - 4$

In Exercises 13–16 use both cobweb graphs and the derivative to determine the stability of the given 2-cycle.

13. The 2-cycle found in Exercise 1.

14. The 2-cycle found in Exercise 3.

15. The 2-cycle found in Exercise 7.

16. The 2-cycle found in Exercise 9.

In Exercises 17–20 given that a $= 1$ *is one point of a 2-cycle, find the other point* b *and determine the stability of the cycle.*

17. $f(x) = \dfrac{2}{x} + 3x - \dfrac{9}{2}$ 　　　19. $f(x) = \dfrac{6}{x^2 + 2}$

18. $f(x) = 2 - 2^x$ 　　　　　　　20. $f(x) = \sqrt{1 - x}$

21. The equation $P_{n+1} = r P_n(1 - P_n/C) - k$ may be viewed as a density-dependent population model with constant migration or harvesting. For $r = 4$, $C = 800$ and $k = 200$ this model has two equilibrium points and a 2-cycle.

  (a) Find the two equilibrium points and determine their stability.
  (b) Use the results from part (a) to help find the points of the 2-cycle, and then determine the stability of that cycle.

22. Repeat the previous exercise for the price model $P_{n+1} = \dfrac{1}{P_n} + \dfrac{P_n}{2} - 1$.

23. Find a 2-cycle of $x_{n+1} = 1 - 2|x_n|$ and determine its stability. *Hint:* Assume the points a, b of the 2-cycle satisfy $a < 0 < b$ and solve $f(f(x)) = x$.

24. The Intermediate Value Theorem states that if a continuous function $g(x)$ satisfies $g(a) > 0$ and $g(b) < 0$ then there must exist a point p between a and b with $g(p) = 0$. Use this theorem to prove that if a continuous function $f(x)$ has a 2-cycle a, b with $a < b$, then $f(x)$ must have a fixed point p between a and b. *Hint:* Let $g(x) = f(x) - x$.

25. We have previously seen that $1, -1$ is an unstable 2-cycle of $f(x) = -x^3$. Show that $1, -1$ is a stable 2-cycle of the inverse function $f^{-1}(x) = -x^{1/3}$.

26. Suppose a, b is a 2-cycle of a differentiable function $f(x)$, and the inverse $f^{-1}(x)$ exists.

  (a) Show that a, b is a 2-cycle of $f^{-1}(x)$.
  (b) Show that this 2-cycle is stable under $f(x)$ if and only if it is unstable under $f^{-1}(x)$.

In Exercises 27 and 28 if a, b *and* c *are constants, find a general formula for* f^n(x).

27. $f(x) = ax + b$ 　　　　　　　28. $f(x) = (x + c)^a - c$

29. If $f(x) = \dfrac{x}{x+1}$: (a) Compute and simplify $f^2(x)$ and $f^3(x)$; (b) Find a general formula for $f^n(x)$; (c) Show that $f(x)$ has no cycles of period $m > 1$.

30. (a) Suppose $f^4(p) = p$ for some point p and some function $f(x)$. What are the possible periods that p might have?
    (b) Suppose $f^{12}(p) = p$ for some point p and some function $f(x)$. What are the possible periods that p might have?

31. (a) Suppose $f(5385) = 6922$, $f(6922) = 8350$, $f(8350) = 5385$ and $f'(5385) = 1.5$, $f'(6922) = 0.6$, $f'(8350) = -1.2$. Determine the stability of that 3-cycle.
    (b) Suppose a 4-cycle p_1, p_2, p_3, p_4 of $f(x)$ satisfies $f'(p_1) = -7/2$, $f'(p_2) = 2/5$, $f'(p_3) = -3/4$, $f'(p_4) = 6/7$. Determine the stability of that cycle.

In Exercises 32 and 33 a function f(x) *and one point* p₁ *of an* m-*cycle is given. Find all other points of that cycle, and determine its period and stability.*

32. $f(x) = 1 - 4|x|$, $p_1 = -11/65$.   33. $f(x) = 2|x| - 3$, $p_1 = 1/11$.

34. (a) Show that a 4-cycle of $f(x)$ consists of two different 2-cycles of $f^2(x)$.
    (b) Show that a 6-cycle of $f(x)$ consists of two different 3-cycles of $f^2(x)$ and three different 2-cycles of $f^3(x)$.
    (c) Show that an m-cycle of $f(x)$ consists of d different cycles of $f^d(x)$, each of period m/d, for every integer d that divides m.

▬▬▬▬ Computer Projects 3.5

In Projects 1–4 determine numerically whether a stable cycle exists for the given parameter values, and if so, its period. Perform at least 200 iterations each time and if a cycle is found (approximately), use the product of derivatives to verify its stability.

1. $P_{n+1} = r\,P_n(1 - P_n/5000)$ for: (a) $r = 3.4$; (b) $r = 3.5$; (c) $r = 3.566$; (d) $r = 3.569$; (e) $r = 3.845$.

2. $P_{n+1} = r\,P_n e^{-P_n/1000}$ for: (a) $r = 5$; (b) $r = 10$; (c) $r = 14$; (d) $r = 14.5$; (e) $r = 14.75$.

3. $I_{n+1} = I_n - r\,I_n + s\,I_n(1 - I_n/1000)$ for: (a) $r = 0.25$, $s = 2.5$; (b) $r = 0.5$, $s = 2.75$; (c) $r = 0.5$, $s = 3$; (d) $r = 0.45$, $s = 3$; (e) $r = 0.17$, $s = 3$.

4. $P_{n+1} = \dfrac{2}{P_n} + 0.75 P_n + c$ for: (a) $c = -1.5$; (b) $c = -1.75$; (c) $c = -1.9$; (d) $c = -1.95$; (e) $c = -1.96$.

5. By numerical experimentation, determine whether there exists a value of $r > 0$ for which $T_{n+1} = \dfrac{r\,T_n}{T_n^2 + 1}$ has a stable 2-cycle.

6. The function $f(x) = 10xe^{-x}$ has a 2-cycle, but it is unstable and so iteration of $f(x)$ will not converge to it. Use the Newton Root-Finding Method to find the points of the 2-cycle by finding the roots of $f(f(x)) - x = 0$.

7. The equation $x_{n+1} = 2.2x_n - x_n^3$ has two different locally stable 2-cycles. Find them and their respective basins of attraction.

8. The function $f(x) = 3.5x(1 - x)$ has a nonzero unstable fixed point, an unstable 2-cycle and a stable 4-cycle. (a) Through direct numerical iteration, find the stable

4-cycle. (b) Using algebra, find the nonzero fixed point. (c) Using either algebra or the Newton Root-Finding Method on $f(f(x)) - x = 0$, find the 2-cycle. (d) Plot all these points on the x-axis. Do you see a pattern?

9. Repeat the previous project using the function $f(x) = 13xe^{-x}$.

S E C T I O N
3.6   Parameterized Families

Thus far in our analysis of nonlinear equations, we have generally treated each equation separately and independently from all others, even though many shared a similar form. For example, the logistic model

$$P_{n+1} = rP_n \left(1 - \frac{P_n}{C} \right) \tag{1}$$

actually encompasses a collection of functions that are related in the sense that all share the same basic properties, and their graphs all have the same general shape. They differ only by the particular values of the parameters r and C.

In this section we investigate entire sets of equations that are related this way, by being members of the same **parameterized family.** As we shall see, much can be learned from this approach, with regard to both the dynamics of the family of equations and the implications for the underlying applications they model.

One-Parameter Families

The logistic model, as well as the other population models we've studied, depends upon two parameters r and C. It would be beneficial to simplify the analysis of those equations by eliminating one of the two, if of course that's possible without losing any of the dynamics we wish to study.

The approach is similar to what was previously done for the non-homogeneous linear equation $x_{n+1} = ax_n + b$, which has two parameters a and b. In Chapter 2 it was shown that any problem of this form could be turned into one of the form $u_{n+1} = au_n$ using a substitution. One of the benefits of this change of variable was to see that the dynamics are the same for both equations, and depend just on a, i.e., stability for $|a| < 1$, instability for $|a| > 1$ and oscillation for $a < 0$. The parameter b is essentially inconsequential in the investigation of the asymptotic behavior of linear equations.

We would now like to do the same for nonlinear equations, i.e., to simplify by eliminating unnecessary parameters. To demonstrate how this can be accomplished for the logistic model (1), suppose we make the substitution $P_n = Cx_n$ for all $n \geq 0$. Since P_{n+1} must equal Cx_{n+1}, then (1) becomes

$$Cx_{n+1} = rCx_n \left(1 - \frac{Cx_n}{C} \right)$$

which simplifies to

$$x_{n+1} = rx_n(1 - x_n) \tag{2}$$

This is now a **one-parameter family,** which we call the **logistic family.** Since the corresponding function is now parameterized by r only, we write that function as

$$f_r(x) = rx(1 - x)$$

in order to emphasize its dependence on r. In other cases, when a different parameter is used, the subscript r should be replaced with that parameter, for example $f_a(x)$ or $f_c(x)$.

After the analysis of the logistic family (2) and its solutions x_n is complete, if desired, the solution of the original population model (1) can be recovered by again using $P_n = Cx_n$. Usually, however, we are more interested in the qualitative behavior of such a model rather than specific numerical values. So, it's enough to know that the dynamics of both are essentially the same. They differ only by the **scaling factor** C. In other words, the existence of a fixed point p of (2) implies the existence of the fixed point Cp of (1) for the same value of r. Similarly, for any parameter r stability or instability of the fixed point p under (2) implies the same for the fixed point Cp under (1). There is also a correspondence between the oscillation of respective solutions around the two fixed points. Thus the dynamics of any logistic model (1) can be determined by instead investigating the simpler equation (2) for the same parameter value r. No dynamics are lost in this change of variable.

Several other models we have considered can also be converted into one-parameter families without altering any important dynamics. For example, the substitution $P_n = Nx_n$ converts

$$P_{n+1} = rP_ne^{-P_n/N} \quad \text{into} \quad x_{n+1} = rx_ne^{-x_n} \quad \text{with} \quad f_r(x) = rxe^{-x}$$

Sometimes even a model with *three* parameters can be converted into a one-parameter family with the right substitution. While this cannot really be done for the price models we have been considering, it can for some of the contagious-disease models, as the following example demonstrates.

EXAMPLE 1

Convert the disease model

$$I_{n+1} = I_n - rI_n + sI_n\left(1 - \frac{I_n}{N}\right)$$

into a one-parameter family.

Solution: This can be accomplished by first writing the equation as

$$I_{n+1} = I_n\left(1 - r + s - \frac{sI_n}{N}\right) \tag{3}$$

and then making the change of variable

$$I_n = \frac{N(1 - r + s)x_n}{s}, \quad \text{which also implies} \quad I_{n+1} = \frac{N(1 - r + s)x_{n+1}}{s}$$

Making these substitutions in equation (3) gives

$$\frac{N(1 - r + s)x_{n+1}}{s} = \frac{N(1 - r + s)x_n}{s}\left(1 - r + s - \frac{sN(1 - r + s)x_n}{sN}\right)$$

Canceling sN from the fraction inside the parentheses, and multiplying both sides of the equation by $\dfrac{s}{N(1-r+s)}$, reduces this to

$$x_{n+1} = x_n(1-r+s-(1-r+s)x_n) = (1-r+s)x_n(1-x_n)$$

If we call $a = 1 - r + s$ then the model becomes $x_{n+1} = ax_n(1-x_n)$, which is a one-parameter family. In fact it is the same family that the logistic model was converted into, with parameter a instead of r. ∎

Intervals of Existence and Stability

Based on the previous discussion, we will now devote our attention exclusively to one-parameter families of nonlinear equations and functions. Rather than considering such issues as whether or not a fixed point or m-cycle exists, and if so, whether or not it's stable, instead we now take a much wider view of the problem by asking: *What are all the parameter values for which these occur?* The following examples show how this may be answered.

E X A M P L E   2

Find the fixed points of $x_{n+1} = ax_n^2$ and determine their stability for all parameter values $a \neq 0$.

Solution: To find fixed points we solve

$$f_a(x) = ax^2 = x \quad \text{or} \quad ax^2 - x = x(ax-1) = 0$$

For $a \neq 0$ this always has two distinct solutions $x = 0$ and $x = 1/a$. Therefore two fixed points 0 and $1/a$ exist for all $a \neq 0$. Note that the value of the latter fixed point depends upon a.

To determine their stability, cobwebbing is not practical since the shape of the function involved $f_a(x) = ax^2$ changes with a. We therefore use the derivative test. Sometimes this too can be tricky for parameterized functions, but for this $f_a(x)$ it is straightforward. We first compute $f_a'(x) = 2ax$, and then substitute the fixed points, which gives

$$f_a'(0) = 0 \quad \text{and} \quad f_a'(1/a) = 2a(1/a) = 2$$

So, 0 is stable and $1/a$ is unstable for all $a \neq 0$. ∎

This example shows that the location of the fixed point $1/a$ is itself a *function* of a. When this occurs and the parameter is a, as it is here, then the fixed point is often written as $p(a)$ to emphasize its dependence on a. In this example we have $p(a) = 1/a$, but since 0 remains in a fixed location for all a, another name for it is unnecessary. However, $p(a)$ takes on different values as a changes. For example,

$$p(1) = 1, \quad p(2) = 1/2, \quad p(5/7) = 7/5 \quad \text{and} \quad p(2.5) = 2/5$$

Often, $p(a)$ is abbreviated as p_a. In the following example the nonzero fixed point will be called $p(r)$ or p_r since the parameter is r.

■■■■■■■■ **E X A M P L E 3**

(a) For the logistic family (2) find the non-negative fixed points and determine their stability for all $r > 0$. (We say *non-negative* because of the population or disease model interpretation.) (b) Discuss the implications of these results for the population model (1).

Solution: (a) Letting $f_r(x) = rx(1 - x) = x$ we find $rx - rx^2 - x = 0$, or

$$-rx\left(x - \frac{r-1}{r}\right) = 0$$

The point 0 is a solution for all $r > 0$, and so 0 is always a fixed point. A distinct positive solution

$$p_r = \frac{r-1}{r}$$

exists for $r > 1$ only, and so p_r is a positive fixed point for all $r > 1$. Since $f_r(x) = rx - rx^2$, then to determine stability we compute

$$f_r'(x) = r - 2rx \quad \text{which makes} \quad f_r'(0) = r - 2r \cdot 0 = r$$

So, having $0 < r < 1$ makes $0 < f_r'(0) < 1$. This means 0 is an attractor for $0 < r < 1$ and a repeller for $r > 1$. With regard to the other fixed point p_r,

$$f_r'(p_r) = r - 2r \cdot \frac{r-1}{r} = 2 - r$$

which means p_r is stable when

$$-1 < 2 - r < 1 \quad \text{or} \quad 1 < r < 3$$

This implies $p_r = (r - 1)/r$ is an attractor for $1 < r < 3$, and a similar argument shows it is a repeller for $r > 3$.

(b) These observations imply that for all relatively small growth rates r that satisfy $0 < r < 1$, the fixed point 0 of both (1) and (2) is stable, and so any initially small population satisfying (1) will approach 0 over time and eventually become extinct. In fact, it can be checked numerically that all populations P_n approach 0 regardless of the initial population size P_0.

But for slightly larger growth rates satisfying $1 < r < 3$, the logistic family (2) has a stable fixed point $p(r) > 0$. Since we previously made the substitution $P_n = Cx_n$, then $p(r)$ corresponds to the equilibrium point $Cp(r)$ of (1). Looking at a graph of the function $p(r) = (r-1)/r$ for $r > 1$ as shown in Figure 3.32, we see that $p(r)$ increases as r increases (although it is bounded above by 1). This means that the larger the growth rate r, the larger the fixed point $p(r)$ of (2) becomes, and also the larger the equilibrium population $Cp(r)$ of (1) becomes.

The conclusion is that a larger growth rate $r > 1$ for the population will lead to a larger equilibrium population size, and this equilibrium will be stable (again for all initial population sizes) until r reaches 3. ■

This example shows that the existence and stability of fixed points may occur over ranges of parameter values, usually entire intervals. In particular, 0 was found to be a stable attractor of (2) for $0 < r < 1$ only. A positive fixed point p_r exists for all $r > 1$, and is

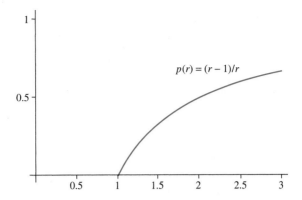

FIGURE 3.32

stable only inside the interval $1 < r < 3$. Intervals like these are often of great importance and so are given special names.

DEFINITION

The interval of parameter values for which a fixed or periodic point exists for a parameterized family is called the **interval of existence** for the point. The interval for which such a point is stable is called its **interval of stability.**

The sizes of the intervals of existence and stability have important consequences for the modeling process. If, for example, a population model has a positive stable fixed point for a relatively large interval of growth rates, then one can be fairly confident that the population being modeled will likely achieve a positive equilibrium level. In this case it is unnecessary to be that precise in measuring the true growth rate or initial population, since a large interval of stability indicates that those dynamics are prevalent. In a sense, the model itself is stable.

On the other hand, a small interval of existence or stability indicates that such dynamics are not likely to occur for the true population. Since in this case a small change in the growth rate significantly changes the dynamics, the model is essentially too sensitive to make meaningful predictions.

EXAMPLE 4

The one-parameter family $x_{n+1} = x_n(a - x_n^2)$ may be considered to correspond to a model of either density-dependent population growth or the spread of a contagious disease. For $a > 0$ find its non-negative equilibrium points and determine their intervals of existence and stability.

Solution: Letting $f_a(x) = x(a - x^2) = ax - x^3$, and solving

$$ax - x^3 = x \quad \text{or} \quad x(a - x^2 - 1) = 0$$

we find that 0 is a fixed point for all $a > 0$. Two other solutions $\pm\sqrt{a-1}$ exist when a is in the infinite interval $a > 1$, but we are interested only in the non-negative solution $p_a = \sqrt{a-1}$.

Since $f_a'(x) = a - 3x^2$, then $f_a'(0) = a$, and so the fixed point 0 exists for all $a > 0$, is stable for $0 < a < 1$ and is unstable for all $a > 1$. Also, the derivative at the positive fixed point $p_a = \sqrt{a-1}$ satisfies

$$f_a'(p_a) = a - 3(\sqrt{a-1})^2 = 3 - 2a$$

Solving $-1 < 3 - 2a < 1$ gives $1 < a < 2$. So, the fixed point p_a exists for all $a > 1$, is stable for $1 < a < 2$ and is unstable for all $a > 2$. ■

The definition above suggests that intervals of existence and stability may exist for periodic points and cycles as well as for fixed points. The following shows that this is indeed the case.

EXAMPLE 5

(a) Find the 2-cycle of the logistic family (2) for $r > 0$. (b) Determine its intervals of existence and stability.

Solution: (a) The difficulty here is trying to get the points of the 2-cycle as functions of the parameter r, as we did in Example 3 for its fixed points. Now, we must solve

$$f_r(f_r(x)) = x \quad \text{or} \quad f_r(f_r(x)) - x = 0 \tag{4}$$

where $f_r(x) = rx(1-x)$. Despite its complexity, the equation can actually be solved with the aid of the trick of dividing by certain factors based on previously computed fixed points.

Recall from Section 3.5 that any fixed point p of $f(x)$ is also a fixed point of $f(f(x))$, and if $f(x)$ is a polynomial then $x - p$ is a factor of $f(f(x)) - x = 0$. In Example 3 the fixed points of $f_r(x) = rx(1-x)$ were found to be 0 and $p(r) = (r-1)/r$. This means that

$$x - 0 = x \quad \text{and} \quad x - \frac{r-1}{r}$$

are factors of (4), which can be written as

$$r^3x^4 - 2r^3x^3 + (r^3 + r^2)x^2 + (1 - r^2)x = 0 \quad \text{or}$$

$$x^4 - 2x^3 + \frac{(r+1)}{r}x^2 + \frac{(1-r^2)}{r^3}x = 0 \tag{5}$$

Following that rule from Section 3.5, since we know x and $x - \dfrac{r-1}{r}$ are factors, then dividing by x reduces equation (5) to

$$x^3 - 2x^2 + \frac{(r+1)}{r}x + \frac{(1-r^2)}{r^3} = 0$$

and dividing by $x - \dfrac{r-1}{r}$ reduces it further to

$$x^2 - \frac{(r+1)}{r}x + \frac{(r+1)}{r^2} = 0$$

Since this is now just a quadratic equation, it can be solved using the quadratic formula, which yields the two solutions

$$p_1(r) = \frac{r+1 - \sqrt{(r+1)(r-3)}}{2r} \quad \text{and} \quad p_2(r) = \frac{r+1 + \sqrt{(r+1)(r-3)}}{2r}$$

It can be verified that

$$f_r(p_1(r)) = p_2(r) \quad \text{and} \quad f_r(p_2(r)) = p_1(r)$$

and so these are the two points of the 2-cycle as functions of r. The interval of existence of this 2-cycle can be seen by observing that $p_1(r)$ and $p_2(r)$ have real values only if $r \geq 3$, and they are distinct only for $r > 3$. At $r = 3$ they equal each other, and so do not qualify as a 2-cycle. The interval of existence of the 2-cycle is therefore $r > 3$.

(b) To determine the interval of stability, we first evaluate $f_r'(x) = r - 2rx$ at both points of the cycle:

$$f_r'(p_1(r)) = r - 2rp_1(r) = r - 2r \cdot \frac{r+1 - \sqrt{(r+1)(r-3)}}{2r} = -1 + \sqrt{(r+1)(r-3)}$$

$$f_r'(p_2(r)) = r - 2rp_2(r) = r - 2r \cdot \frac{r+1 + \sqrt{(r+1)(r-3)}}{2r} = -1 - \sqrt{(r+1)(r-3)}$$

and then multiply the results:

$$f_r'(p_1(r)) \cdot f_r'(p_2(r)) = (-1 + \sqrt{(r+1)(r-3)}) \cdot (-1 - \sqrt{(r+1)(r-3)})$$
$$= 1 - (r+1)(r-3) = -r^2 + 2r + 4$$

The interval of stability therefore corresponds to

$$-1 < -r^2 + 2r + 4 < 1 \quad \text{or} \quad 0 < r^2 - 2r - 3 < 2$$

We must find solutions r of this last inequality that are in the interval of existence of the 2-cycle, i.e., $r > 3$. It can be checked that $r^2 - 2r - 3 > 0$ is satisfied for all $r > 3$. Also, $r^2 - 2r - 3 < 2$, which is equivalent to $r^2 - 2r - 5 < 0$, is satisfied for

$$1 - \sqrt{6} < r < 1 + \sqrt{6}$$

The intersection of the two intervals

$$r > 3 \quad \text{and} \quad 1 - \sqrt{6} < r < 1 + \sqrt{6}$$

is what we are looking for. The interval of stability of the 2-cycle is therefore $3 < r < 1 + \sqrt{6} \approx 3.45$.   ∎

For some parameterized families, intervals of existence and stability may involve negative values of the parameter.

▬▬▬▬▬   **E X A M P L E   6**

For the price model $P_{n+1} = \dfrac{1}{P_n} + P_n + c$: (a) Find the positive equilibrium point and its intervals of existence and stability; (b) Find the positive 2-cycle and its intervals of existence and stability.

Solution: (a) The unique fixed point of $f_c(x) = \dfrac{1}{x} + x + c$ can easily be seen to be $p(c) = -1/c$. This means that a positive equilibrium exists for all $c < 0$. And since $f_c'(x) = 1 - \dfrac{1}{x^2}$ then

$$f_c'(p(c)) = f_c'(-1/c) = 1 - \frac{1}{(-1/c)^2} = 1 - c^2$$

For stability we must have $-1 < 1 - c^2 < 1$ or $0 < c^2 < 2$, and since $c < 0$ then $-\sqrt{2} < c < 0$. So, the positive equilibrium $p(c) = -1/c$ is stable for $-\sqrt{2} < c < 0$.

(b) To find the 2-cycle, we must solve

$$f_c(f_c(x)) = \frac{1}{\left(\frac{1}{x} + x + c\right)} + \left(\frac{1}{x} + x + c\right) + c = x$$

which, after some algebra, can be written more simply as

$$2cx^3 + 2(c^2 + 1)x^2 + 3cx + 1 = 0$$

Since $p(c) = -1/c$ is a fixed point, then $x - p(c) = x + 1/c$ must be a factor of this cubic equation. Dividing by $x + 1/c$ reduces that equation to

$$2cx^2 + 2c^2 x + c = 0 \quad \text{or} \quad 2x^2 + 2cx + 1 = 0$$

whose solutions are quickly found to be

$$p_1(c) = \frac{-c - \sqrt{c^2 - 2}}{2} \quad \text{and} \quad p_2(c) = \frac{-c + \sqrt{c^2 - 2}}{2}$$

Although these two points are real and distinct for all $c^2 > 2$, they are positive only if $c < -\sqrt{2}$. The interval of existence of the positive 2-cycle $p_1(c)$, $p_2(c)$ is therefore $c < -\sqrt{2}$.

To determine the interval of stability of this 2-cycle, we must find the values of $c < -\sqrt{2}$ that also satisfy $-1 < f_c'(p_1(c)) \cdot f_c'(p_2(c)) < 1$ where $f_c'(x) = 1 - \dfrac{1}{x^2}$. After some algebra and calculus (that the reader may wish to perform), this inequality can be dramatically simplified to

$$2 < c^2 < 2.5, \quad \text{but since } c < -\sqrt{2} \text{ then} \quad -\sqrt{2.5} < c < -\sqrt{2}$$

So the interval of stability of the 2-cycle is $-\sqrt{2.5} < c < -\sqrt{2}$.   ■

The types of computations performed in these examples to find the fixed points and 2-cycles as functions of the parameter could be generalized to find any m-cycles of a parameterized nonlinear equation. But this would seldom be attempted in practice. Except in very simple cases, the complexity of the resulting calculations is likely to prohibit higher order periodic points from ever being found this way. In the end, determining intervals of existence and stability of m-cycles is most often accomplished through direct numerical iteration, or some other computer approximation technique.

Exercises 3.6

In Exercises 1–4 convert each equation into a one-parameter family.

1. $P_{n+1} = \dfrac{r P_n^2}{C}(1 - P_n/C)$ 3. $P_{n+1} = \dfrac{r P_n}{1 + P_n^2/C^2}$

2. $P_{n+1} = r P_n(1 - P_n^2/C^2)$ 4. $I_{n+1} = I_n - r I_n + s I_n(1 - I_n^2/N^2)$

5. Show that $P_{n+1} = P_n(r - P_n/C)$ and $x_{n+1} = r x_n(1 - x_n)$ for all r, $C > 0$ comprise the same parameterized family.

6. Show that $P_{n+1} = P_n e^{s - P_n/N}$ for all real s, $N > 0$ and $x_{n+1} = r x_n e^{-x_n}$ for all $r > 1$ comprise the same parameterized family.

In Exercises 7–10 find all positive fixed points and their intervals of existence for c > 0.

7. $f_c(x) = c - x^2$ 9. $f_c(x) = x \ln(x + c)$

8. $f_c(x) = \dfrac{cx^2}{x^2 + 1}$ 10. $f_c(x) = 2x + \dfrac{2}{x} - c$

In Exercises 11–14 find the interval of stability of the fixed point 0 for r > 0.

11. $f_r(x) = r x e^{-x}$ 13. $f_r(x) = r(\tfrac{1}{6} \sin 5x - \tfrac{1}{8} \tan 7x)$

12. $f_r(x) = r x^2(1 - x)$ 14. $f_r(x) = r(2 - \sqrt{3x + 4})$

15. A fixed point p of a function $f(x)$ is called *superstable* if $f'(p) = 0$. Find the parameter value for which the positive fixed point of $f_r(x) = r x(1 - x)$ is superstable.

16. Repeat the previous exercise for $f_a(x) = x(a - x^2)$.

In Exercises 17–20 find all positive fixed points and their intervals of existence and stability for a > 0.

17. $f_a(x) = ax - x^4$ 19. $f_a(x) = ax(2 - x^2)$

18. $f_a(x) = x\sqrt{a - x}$ 20. $f_a(x) = x a^{1-x}$

In Exercises 21–24 find the 2-cycles and their intervals of existence and stability for a > 0.

21. $f_a(x) = a - x^2$ 23. $f_a(x) = \dfrac{-ax}{x^2 + 1}$

22. $f_a(x) = -ax^3$ 24. $f_a(x) = \dfrac{2}{x} + \dfrac{x}{2} - a$

25. The equation $P_{n+1} = r P_n(1 - P_n/C) - k$ may be considered a density-dependent population model with constant harvesting.

   (a) If the growth rate is $r = 3$ and the carrying capacity is $C = 6000$, what is the largest number k that could be harvested each generation so that the population has a stable positive equilibrium, and so may not become extinct?

   (b) What about if the growth rate is only $r = 2$?

26. For the disease model $I_{n+1} = I_n - rI_n + sI_n(1 - I_n/10^6)^2$, determine the largest infection rate s for which the disease will eventually die out on its own for any I_0 between 0 and 10^6, if (a) $r = 0.5$; (b) $r = 0.25$.

27. For the price model $P_{n+1} = \dfrac{a}{P_n} + P_n + c$ determine the smallest (negative) value of c for which the price will approach a stable positive equilibrium for any $P_0 > 0$, if (a) $a = 2$; (b) $a = 3$.

28. For the threshold population model $P_{n+1} = \dfrac{rP_n^2}{C}(1 - P_n/C)$, when r is small the population will become extinct regardless of the initial population size P_0. When r is larger, although small initial populations still lead to extinction (because of the threshold), there is a positive equilibrium population.

    (a) Find the smallest growth rate r for which a positive equilibrium population exists.
    (b) Find the interval of r-values for which there exists a stable positive equilibrium population.

▬▬▬ Computer Projects 3.6

1. We have already seen that $f_r(x) = rx(1 - x)$ has a positive stable fixed point for $1 < r < 3$ and a stable 2-cycle for $3 < r < 1 + \sqrt{6}$. Through numerical experimentation find the approximate intervals of stability of the (a) 4-cycle; (b) 8-cycle; (c) 16-cycle; (d) 32-cycle.

In Projects 2–5 find through numerical experimentation the approximate intervals of stability of the (a) 2-cycle; (b) 4-cycle; (c) 8-cycle; (d) 16-cycle; (e) 32-cycle.

2. $f_r(x) = rxe^{-x}$

3. $f_c(x) = \dfrac{2}{x} + 0.75x - c$

4. $f_r(x) = rx^2(1 - x)$

5. $f_a(x) = x(a - x^2)$

6. The parameter value $r = 3.83$ lies within the interval of stability of the 3-cycle of $f_r(x) = rx(1 - x)$. Through numerical experimentation find (approximately) (a) the rest of this interval of stability; (b) the interval of stability of the 6-cycle; (c) the interval of stability of the 12-cycle.

SECTION
3.7   Bifurcation and Period-Doubling

The use of intervals of existence and stability has given us a powerful tool for investigating the dynamics of parameterized families by allowing the analysis of an *infinite number* of functions at once. Instead of narrowly viewing each equation as an isolated problem, we now see that they can often be grouped together with other members of the same parameterized family that share these same intervals and therefore the same dynamics.

But the story does not end there. It is possible to take an even wider view by investigating the relationship between adjacent intervals of existence and stability for the same parameterized family, and the transitions in the dynamics of the problem as the parameter moves from one interval to the next.

Two Types of Transitions

To motivate the idea, we first turn to the linear equation $x_{n+1} = ax_n + b$. Using the terminology introduced in the last section, we could now say that $-1 < a < 1$ is the interval of stability of the fixed point $p = b/(1 - a)$ for this equation. If a were to move from smaller to larger values along the a axis, the equation would undergo two dramatic changes in its dynamics. At $a = -1$ there would be a transition from an unstable to a stable fixed point, and at $a = 1$ a transition from stable to unstable again. Such transition points can occur for nonlinear equations as well, and when they do, the resulting dynamics are often much more interesting than in the linear case.

Look, for example, at what was discovered in Example 3 of Section 3.6 concerning the logistic family $f_r(x) = rx(1 - x)$. The fixed point 0 exists for all $r > 0$ but is stable only for $0 < r < 1$. On the other hand, the positive fixed point $p_r = p(r) = (r - 1)/r$ exists for all $r > 1$ and is stable for $1 < r < 3$. Look now at one additional feature: for r values close to 1, the point $p(r) = (r - 1)/r$ is close to 0, and in fact $p(1) = 0$. In other words, the two fixed points become one at $r = 1$.

What does this imply? If we take the point of view that we did above for linear equations, this time letting r move from smaller to larger values along the r-axis, then a surprising thing happens at $r = 1$. Not only is there a transition at $r = 1$ from stability to instability of 0, but this occurs simultaneously with $p(r)$ coming into existence at precisely the same location, since $p(1) = 0$. The newly created fixed point $p(r)$ emerges from 0 and moves continuously away from it, as r leaves the interval of stability $0 < r < 1$ of 0 and enters the interval of stability $1 < r < 3$ of $p(r)$. It is as if the previously stable fixed point 0 becomes unstable at $r = 1$, and *gives birth* to a new stable fixed point $p(r)$.

But this is still not the end of the story. One might wonder what happens to the fixed point $p(r)$ as r increases. The answer to this can be found in Example 5 of Section 3.6. Recall that for $r > 3$ there exists a 2-cycle

$$p_1(r) = \frac{r + 1 - \sqrt{(r + 1)(r - 3)}}{2r} \quad \text{and} \quad p_2(r) = \frac{r + 1 + \sqrt{(r + 1)(r - 3)}}{2r} \quad (1)$$

of $f_r(x) = rx(1 - x)$ that is stable for $3 < r < 1 + \sqrt{6}$. Notice that the parameter value $r = 3$ marks precisely the end-point of the interval of stability of $p(r)$ and the beginning of the interval of existence and stability of the 2-cycle. Also, at $r = 3$ the points of the 2-cycle are equal to each other, and both equal $p(r) = (r - 1)/r$, i.e.,

$$p_1(3) = \frac{3 + 1 - \sqrt{(3 + 1)(3 - 3)}}{2 \cdot 3} = \frac{4 - 0}{6} = 2/3$$

$$p_2(3) = \frac{3 + 1 + \sqrt{(3 + 1)(3 - 3)}}{2 \cdot 3} = \frac{4 + 0}{6} = 2/3 \quad \text{and}$$

$$p(3) = \frac{3 - 1}{3} = 2/3$$

In other words, as r increases another kind of transition occurs, this time at $r = 3$, in which $p(r)$ becomes unstable and gives birth to the stable 2-cycle $p_1(r)$, $p_2(r)$. At $r = 3$ these two points emerge and move continuously away from $p(r)$ as r increases.

We have witnessed here two types of transitions for $x_{n+1} = rx_n(1 - x_n)$ as r increases. In both cases a stable fixed point becomes unstable, but in the first case this produces another stable fixed point, and in the second case a stable 2-cycle. Why the difference? This can be answered by looking at $f_r'(x)$ evaluated at the fixed points 0 and $p(r)$ for the two respective r values where the transitions occurred: $r = 1$ and $r = 3$. Since $f_r'(x) = r - 2rx$ then at $r = 1$

$$f_r'(0) = 1 - 2 \cdot 1 \cdot 0 = +1$$

and since $p(3) = 2/3$ then at $r = 3$

$$f_r'(p(r)) = 3 - 2 \cdot 3 \cdot \frac{2}{3} = -1$$

In other words, when the derivative at the fixed point passes through $+1$, another stable fixed point often comes into existence, but when it passes through -1 a stable 2-cycle usually results.

Bifurcation of Fixed Points

The two types of transitions we have described here are so common for fixed points p_r of general parameterized nonlinear functions $f_r(x)$ that they are given a name.

DEFINITION

A fixed point p_r of a parameterized family $f_r(x)$ is said to undergo a **bifurcation** at $r = r_0$ if it changes from stable to unstable, and either another stable fixed point or a stable 2-cycle emerges from it at $r = r_0$. The former corresponds to $f_r'(p_r) = +1$ and the latter to $f_r'(p_r) = -1$ at $r = r_0$.

Fixed points of many nonlinear parameterized families **bifurcate** in a manner similar to what we just observed for $f_r(x) = rx(1 - x)$, including all the models introduced in Section 3.1. The following are examples of some besides the logistic family.

EXAMPLE 1

(a) For the disease model $I_{n+1} = 0.75I_n + sI_n(1 - I_n 10^{-6})$ determine the value of the infection rate $s > 0$ at which the fixed point 0 bifurcates into a positive stable equilibrium. (b) Discuss the significance of this to the model.

Solution: Solving

$$f_s(x) = 0.75x + sx(1 - x10^{-6}) = x \quad \text{or} \quad x(-0.25 + s - sx10^{-6}) = 0$$

always gives the solution 0. For $s > 0.25$ there is also a positive solution

$$p_s = \frac{10^6(s - 0.25)}{s}$$

Computing

$$f_s'(x) = 0.75 + s - 2sx10^{-6}$$

and evaluating $f'_s(0) = 0.75 + s$ means that 0 is stable for $0 < s < 0.25$ and unstable for $s > 0.25$. The value $s = 0.25$ therefore marks the transition from stability to instability of 0, which occurs simultaneously with p_s coming into existence at the same location, since $p_s = 0$ when $s = 0.25$. Also,

$$f'_s(p_s) = 0.75 + s - 2s \cdot \frac{10^6(s - 0.25)}{s} \cdot 10^{-6} = 1.25 - s$$

So for all s-values slightly larger than 0.25, p_s is stable. This means 0 bifurcates at $s = 0.25$ into the positive stable equilibrium point p_s.

(b) The significance of this bifurcation is that an infection rate of $s = 0.25$ marks the transition between two important and very different types of dynamics. If the disease is transmitted through the population at a rate lower than this, then it will gradually die out on its own. In this case the disease does not pose a significant future threat to the overall population. On the other hand, if the infection rate exceeds this value, then the disease will never be eliminated from the population. Instead, the number of infected will stabilize so that a certain positive fraction of the population will always be infected. ∎

EXAMPLE 2

Determine whether the fixed point 0 of $f_a(x) = x(a - x^2)$ undergoes a bifurcation for some $a > 0$.

Solution: Recall from Example 4 of Section 3.6 that 0 is a stable attractor for $0 < a < 1$ and a repeller for $a > 1$. Also, the fixed point $p(a) = p_a = \sqrt{a - 1}$ exists for all $a > 1$ and is stable for $1 < a < 2$. So $a = 1$ marks the end of stability for 0 and the beginning of existence and stability of $p(a)$. In addition, at $a = 1$ the points coincide, i.e., $p(1) = \sqrt{1 - 1} = 0$. In this case the fixed point 0 bifurcates into another stable fixed point. Also recall from that previous example that $f'_a(0) = a$. At $a = 1$ this equals $+1$, which is why a new stable fixed point, rather than a stable 2-cycle, bifurcates from 0. ∎

EXAMPLE 3

Determine whether the positive fixed point $p_a = \sqrt{a - 1}$ of $f_a(x) = x(a - x^2)$ bifurcates for some $a > 1$.

Solution: In Example 4 of Section 3.6 it was found that $f'_a(p_a) = 3 - 2a$. Since $3 - 2a < +1$ for all $a > 1$, the only bifurcation possible must correspond to $f'_a(p_a) = -1$. This makes $3 - 2a = -1$ or $a = 2$. So at the upper end-point of its interval of stability, p_a may possibly undergo a bifurcation into a stable 2-cycle.

Algebraically determining whether such a 2-cycle emerges from p_a at $a = 2$ is not practical in this case, since the calculations are prohibitive. We therefore fall back on direct numerical iteration, and compute the first 20 iterates for several values of a. Tables 3.9(a) and 3.9(b) show the results for $a = 1.9$ and $a = 2.1$ respectively. In both cases, the initial point $x_0 = 1$ was used.

The numerical results show that indeed a stable 2-cycle does appear to come into existence at $a = 2$. For $a = 1.9$ the fixed point is still stable and so iterates converge to it. But for $a = 2.1$, which is slightly beyond its interval of stability, iterates converge to a

TABLE 3.9

n	x_n	n	x_n	n	x_n	n	x_n
0	1.000			0	1.000		
1	0.900	11	0.944	1	1.100	11	1.169
2	0.981	12	0.952	2	0.979	12	0.857
3	0.920	13	0.946	3	1.118	13	1.170
4	0.969	14	0.951	4	0.951	14	0.855
5	0.931	15	0.947	5	1.137	15	1.171
6	0.962	16	0.950	6	0.918	16	0.854
7	0.938	17	0.948	7	1.154	17	1.171
8	0.957	18	0.950	8	0.886	18	0.854
9	0.942	19	0.948	9	1.165	19	1.171
10	0.954	20	0.949	10	0.865	20	0.854

(a)   (b)

2-cycle. As would be expected if they emerged from p_a, the points of this 2-cycle are not far from the value of p_a at $a = 2.1$, i.e.,

$$\sqrt{2.1 - 1} = \sqrt{1.1} \approx 1.05$$

Further numerical investigation would show that the locations of the points of the stable 2-cycle move away from p_a as a increases. ∎

For some parameterized families a bifurcation occurs for a *decreasing* rather than an increasing parameter.

EXAMPLE 4

Show that, as the parameter c decreases, the positive equilibrium point of the price model $P_{n+1} = \dfrac{1}{P_n} + P_n + c$ undergoes a bifurcation.

Solution: In Example 6 of Section 3.6 the positive equilibrium point $p(c) = -1/c$, which exists for all $c < 0$, was found to be stable only if $-\sqrt{2} < c < 0$. This means that $p(c)$ changes from an attractor to a repeller as c *decreases* from above to below $-\sqrt{2}$.

Also found in that previous example was the positive 2-cycle

$$p_1(c) = \frac{-c - \sqrt{c^2 - 2}}{2} \quad \text{and} \quad p_2(c) = \frac{-c + \sqrt{c^2 - 2}}{2}$$

which exists for all $c < -\sqrt{2}$ but is stable only for $-\sqrt{2.5} < c < -\sqrt{2}$. Although the equilibrium point $p(c)$ and the points of the 2-cycle $p_1(c)$, $p_2(c)$ constitute three distinct points for all $c < -\sqrt{2}$, it is easy to check that

$$p_1(-\sqrt{2}) = \frac{1}{\sqrt{2}}, \quad p_2(-\sqrt{2}) = \frac{1}{\sqrt{2}}, \quad \text{and} \quad p(-\sqrt{2}) = \frac{1}{\sqrt{2}}$$

That is, all three points are identical for that one special parameter value $c = -\sqrt{2}$. Note also that $f_c'(p(c)) = -1$ when $c = -\sqrt{2}$. All of this taken together means that $p(c)$ becomes unstable and bifurcates into the stable 2-cycle $p_1(c)$, $p_2(c)$ as c decreases through $-\sqrt{2}$. ∎

Period-Doubling Bifurcations

Our investigation of parameterized families and bifurcation has given us a comprehensive perspective on nonlinear equations by allowing, first, the grouping together of all family members sharing the same intervals of existence and stability, and second, the identification of the type of bifurcation that binds adjacent intervals together. We next take the most panoramic view yet in our study of such equations by establishing a recently discovered but now famous **period-doubling** bifurcation scenario that is commonly played out by a large class of nonlinear parameterized families.

The identification of this scenario will allow for a powerful overall theory: a *complete* description of the *entire* range of dynamics that can occur for a large class of parameterized unimodal families, including *all* the density-dependent population models and contagious-disease models we have discussed, as well as many price models. In other words, armed with this knowledge one would be able to predict the dynamics of $x_{n+1} = f_r(x_n)$ in advance for any unimodal $f_r(x)$ just by looking at its graph and how its shape changes with r.

To establish this theory, the nature of the period-doubling mechanism must be investigated in a little more detail. For this analysis we once again use the logistic family as the paradigm. It turns out that the dynamics we observe here also apply to all the other parameterized unimodal models we have studied.

For the logistic family $f_r(x) = rx(1-x)$, so far we have seen that the non-negative fixed point $p(r) = (r-1)/r$ exists for all $r \geq 1$, and is stable for $1 < r < 3$. At $r = 3$ the 2-cycle $p_1(r)$, $p_2(r)$ given by (1) bifurcates from $p(r)$, and remains stable for $3 < r < 1 + \sqrt{6}$. As noted earlier, these three points $p(r)$, $p_1(r)$, $p_2(r)$ are actually functions of the parameter r, and each has a certain domain: $p(r)$ is defined for $r \geq 1$, and $p_1(r)$, $p_2(r)$ for $r \geq 3$.

As we saw earlier these three functions satisfy

$$p(3) = 2/3, \quad p_1(3) = 2/3, \quad \text{and} \quad p_2(3) = 2/3$$

With some algebra it can be shown that for $r > 3$ they additionally satisfy

$$p_1(r) < p(r) < p_2(r)$$

That is, for all $r > 3$ the fixed point lies between the two points of the 2-cycle that bifurcates from it at $r = 3$.

The full picture of what occurs in the bifurcation is now emerging. To help visualize the process, suppose we graph the three functions of r for their respective domains of

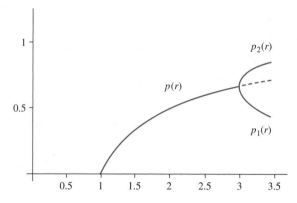

FIGURE 3.33

definition, using the horizontal axis for r and the vertical for the functions. The result must look something like Figure 3.33. The graph of the fixed point function $p(r) = (r - 1)/r$ is shown for $r \geq 1$. Part of it, the piece satisfying $1 \leq r \leq 3$, is drawn using a *solid* curve, and the remaining piece using *dashes*. The solid curve begins at the point $(1, 0)$ since $p(1) = 0$ and ends at the point $(3, 2/3)$ at which it bifurcates. This corresponds to the interval of stability of $p(r)$. The dashed section represents the continued existence of that fixed point but as a repeller.

The graphs of $p_1(r)$ and $p_2(r)$ begin at that same point $(3, 2/3)$ where $p(r)$ becomes unstable, and continue for $r > 3$. They are drawn as solid curves since they are stable.

The graph of Figure 3.33 is drawn only for $1 < r < 1 + \sqrt{6}$, since we were previously able to establish the dynamics of the logistic family $f_r(x) = rx(1 - x)$ for these r-values. But suppose we now ask: *What happens for $r \geq 1 + \sqrt{6}$?* It can be easily checked that $f'_r(p_1(r)) \cdot f'_r(p_2(r))$ passes from greater than to less than -1 at $r = 1 + \sqrt{6}$, but other than that, algebraically determining the dynamics beyond this r value is too complex.

For the remainder of this analysis we must therefore rely upon direct numerical iteration. For any r value between $1 + \sqrt{6}$ and approximately 3.54409 the following can be observed. A stable 4-cycle $q_1(r)$, $q_2(r)$, $q_3(r)$, $q_4(r)$ exists for all such r-values. Once again these are continuous functions of r, but this time it is not possible to come up with exact equations for them. Numerical data indicates, however, that for any r slightly beyond $1 + \sqrt{6}$ both $q_1(r)$ and $q_3(r)$ are close to $p_1(r)$, and both $q_2(r)$ and $q_4(r)$ are close to $p_2(r)$. In fact, at $r = 1 + \sqrt{6}$

$$q_1(r) = q_3(r) = p_1(r) \quad \text{and} \quad q_2(r) = q_4(r) = p_2(r)$$

and for all $r > 1 + \sqrt{6}$

$$q_1(r) < p_1(r) < q_3(r) \quad \text{and} \quad q_2(r) < p_2(r) < q_4(r)$$

In other words, it appears that at $r = 1 + \sqrt{6}$ another kind of bifurcation occurs, where this time a stable 2-cycle bifurcates into a stable 4-cycle.

As we shall see below, this type of bifurcation is quite important, and so we introduce the following terminology.

DEFINITION

An m-cycle $p_1(r)$, $p_2(r)$, ..., $p_m(r)$ of a parameterized family $f_r(x)$ is said to undergo a
period-doubling bifurcation at $r = r_0$, if it changes from stable to unstable, and a stable
$2m$-cycle emerges from its points at $r = r_0$. This corresponds to

$$f_r'(p_1(r)) \cdot f_r'(p_2(r)) \cdot f_r'(p_3(r)) \cdot \ \cdots \ \cdot f_r'(p_m(r)) = -1 \quad \text{at } r = r_0$$

Note that this generalizes the type of bifurcation that occurs when a fixed point bifurcates into a stable 2-cycle. In fact, the mechanism behind a period-doubling bifurcation of
an m-cycle can perhaps best be understood by instead visualizing it as a period-doubling
bifurcation of a fixed point of the composition $f_r^m(x)$. For example, for the logistic family
$f_r(x)$, each of the fixed points $p_1(x)$ and $p_2(x)$ of $f_r^2(x)$ discussed above bifurcates under
$f_r^2(x)$ at $r = 1 + \sqrt{6}$ into a stable 2-cycle. The point $p_1(x)$ bifurcates into $q_1(r)$ and $q_3(r)$,
and simultaneously the point $p_2(r)$ bifurcates into $q_2(r)$ and $q_4(r)$. This creates two different
locally stable 2-cycles of $f_r^2(x)$, although of course in reality these points constitute a single
stable 4-cycle of $f_r(x)$.

One may wonder why we didn't generalize the other type of bifurcation that can exist
for fixed points, in which a new fixed point rather than a 2-cycle emerges. It turns out that
this other type of bifurcation is quite rare for periodic points owing to the rather unusual
properties that $f_r(x)$ would need to have. Period-doubling of m-cycles, however, is quite
common for even the simplest of unimodal families.

EXAMPLE 5

Determine whether the 4-cycle of $x_{n+1} = rx_n(1 - x_n)$ undergoes a period-doubling bifurcation for some $r > 0$.

Solution: This can only be determined numerically. Table 3.10 shows a stable 8-cycle for
$r = 3.55$. Further investigation would show that this cycle does indeed bifurcate from the
4-cycle at $r \approx 3.54409$. ■

Period-Doubling Cascade

At this point, it may be worth summarizing what we have witnessed so far with regard to
the dynamics of the logistic family:

The fixed point 0 becomes unstable at $r = 1$ and bifurcates into a positive fixed point
that is stable for $1 < r < 3$; at $r = 3$ this fixed point becomes unstable and bifurcates into a
2-cycle that is stable for $3 < r < 1 + \sqrt{6}$; at $r = 1 + \sqrt{6}$ this 2-cycle becomes unstable and
bifurcates into a 4-cycle that is stable for r between $1 + \sqrt{6}$ and approximately 3.54409;
this 4-cycle becomes unstable at $r \approx 3.54409$ and bifurcates into an 8-cycle that is stable
until ....

The beginning of a pattern may now be recognized: at each step there successively
emerges through bifurcation a stable 2^m-cycle, with its own interval of stability, whose
upper end-point marks the bifurcation into a stable 2^{m+1}-cycle. As this process continues
we in turn get cycles of period: $1, 2, 4, 8, 16, 32, \ldots, 2^m, \ldots$, for every positive integer m,
each of which remains stable until it bifurcates into the next.

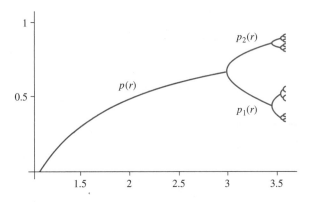

FIGURE 3.34

TABLE 3.10

n	x_n	n	x_n
0	0.541		
1	0.882	11	0.828
2	0.370	12	0.506
3	0.828	13	0.887
4	0.506	14	0.355
5	0.887	15	0.813
6	0.355	16	0.541
7	0.813	17	0.882
8	0.541	18	0.370
9	0.882	19	0.828
10	0.370	20	0.506

This *cascade* of period-doubling bifurcations can indeed be verified to actually exist, not only for the logistic family, but also for all the other parameterized unimodal functions we have studied. Through direct numerical experimentation, the sizes of the intervals of stability, which sit between consecutive bifurcation values of r, can be estimated. As $m \to \infty$, these bifurcation points and intervals have been found to *pile up* at a certain limiting value of r, beyond which no 2^m-cycle is stable. For the logistic family that r-value has been estimated to be approximately 3.56995.

The infinite sequence of period-doubling bifurcations we have observed here can be conveniently represented in a graph of the stable 2^m-cycle for all $m \geq 0$, as a function of the parameter value. Similar to Figure 3.33, we use the horizontal axis for r and the vertical axis for the locations of the stable fixed point and stable 2^m-cycles of $f_r(x) = rx(1 - x)$. The result is the branching process shown in Figure 3.34.

Each section of this graph branches into two others, since for all m each point of the stable 2^m-cycle bifurcates into two points of the stable 2^{m+1}-cycle. Some of the branches have known equations that we have seen earlier, such as $p(r) = (r - 1)/r$ for $1 \leq r \leq 3$, and the two pieces $p_1(r)$ and $p_2(r)$ for $3 < r < 1 + \sqrt{6}$, whose equations are given in (1). The other sections of the graph, however, can only be approximated.

This period-doubling cascade was discovered by R. M. May as recently as the 1970's, and since then has been observed for most parameterized unimodal families that arise in modeling situations. The specific parameter values where the bifurcations occur are, of course, different for each family. But the overall period-doubling cascade is always the same, there is always a finite limit for the parameter values at which the intervals of stability and bifurcations occur and all have graphs similar to that seen in Figure 3.34.

EXAMPLE 6

For the logistic equation $x_{n+1} = rx_n(1 - x_n)$, find the approximate intervals of stability of the 8-cycles, 16-cycles and 32-cycles.

Solution: Through numerical experimentation one can find

$$\text{stable 8-cycles for} \quad 3.54409 < r < 3.56441$$
$$\text{stable 16-cycles for} \quad 3.56441 < r < 3.56876 \quad \text{and}$$
$$\text{stable 32-cycles for} \quad 3.56876 < r < 3.56969$$

Stable 2^m-cycles for $m \geq 6$ can also be found for slightly larger values of r.   ∎

EXAMPLE 7

Determine whether the threshold model $x_{n+1} = rx_n^2(1 - x_n)$ has stable cycles of periods 1, 2, 4 and 8.

Solution: Again through numerical experimentation, a stable cycle of period 1, 2, 4 or 8 can be found at $r = 5$, 5.5, 5.8 or 5.87 respectively. This model too undergoes an entire period-doubling cascade.   ∎

One may ask: *What happens for parameter values beyond the limit of the period-doubling bifurcations, since none of the 2^m-cycles are stable? Are there stable cycles of other periods, or some other dynamics?* Answering these questions requires the investigation of the set of related complex phenomena known collectively as **chaos,** and an introduction to one of the newest fields of contemporary mathematics, sometimes called **chaos theory.** This is the subject of the next and final section of this chapter.

Exercises 3.7

In Exercises 1–6: (a) Find the interval of stability of the fixed point 0; (b) Find the positive fixed point p(r) *and its intervals of existence and stability; (c) Show that* p(r) *bifurcates from* 0 *at some parameter value* r = r_0.

1. $f_r(x) = rx(5 - 2x)$

2. $f_r(x) = \dfrac{rx}{x + 5}$

3. $f_r(x) = rxe^{-x}$

4. $f_r(x) = rx\sqrt{4 - x}$

5. $f_r(x) = \dfrac{rx}{x^2 + 1}$

6. $f_r(x) = rx(3 - x^2)$

7. Sketch the graph of $p(r)$ from Exercise 3.

8. Sketch the graph of $p(r)$ from Exercise 4.

9. Sketch the graph of $p(r)$ from Exercise 5.

10. Sketch the graph of $p(r)$ from Exercise 6.

In Exercises 11–14: (a) Find the non-negative fixed point p(a) *and its intervals of existence and stability; (b) Find the 2-cycle* $p_1(a)$, $p_2(a)$ *and its intervals of existence and stability; (c) Show that the 2-cycle bifurcates from* p(a) *at some parameter value* a = a_0.

11. $f_a(x) = a - x^2$ (see Exercises 7 and 21 of Section 3.6)

12. $f_a(x) = 0.25 - ax^2$

13. $f_a(x) = \dfrac{-ax}{x^2 + 1}$ (see Exercise 23 of Section 3.6)

14. $f_a(x) = \dfrac{2}{x} + \dfrac{x}{2} - a$ (see Exercise 24 of Section 3.6)

15. Sketch the graphs of $p(a)$, $p_1(a)$, $p_2(a)$ from Exercise 11 together.

16. Sketch the graphs of $p(a)$, $p_1(a)$, $p_2(a)$ from Exercise 12 together.

In another kind of bifurcation two fixed points, one stable and the other unstable, come into existence simultaneously at the same location and for the same parameter value. Show that this occurs for each of the functions in Exercises 17–20.

17. $f_a(x) = a - x^2$

18. $f_r(x) = rx^2(1 - x)$

19. $f_c(x) = \dfrac{1}{x} + 2x - c$

20. $f_c(x) = -c(x^2 - 3x + 2)$

21. (a) Suppose $f_r(x)$ undergoes a period-doubling cascade for $a < r < b$. Does this imply that $f_r^2(x)$ also undergoes a period-doubling cascade for $a < r < b$?

   (b) Suppose $f_r^2(x)$ undergoes a period-doubling cascade for $a < r < b$. Does this imply that $f_r(x)$ also undergoes a period-doubling cascade for $a < r < b$?

Show that the equations in Exercises 22 and 23 undergo period-doubling cascades by using the given substitution to convert each into $x_{n+1} = rx_n(1 - x_n)$ *(which we have already seen has such a cascade).*

22. $y_{n+1} = y_n(a - y_n)$; let $y_n = rx_n$ and $r = a$.

23. $y_{n+1} = a - y_n^2$; let $y_n = r(x_n - 0.5)$ and $r = 1 + \sqrt{4a + 1}$ or $a = r(r - 2)/4$.

Show that the equations in Exercises 24 and 25 undergo period-doubling cascades by using the given substitution to convert each into $x_{n+1} = rx_n e^{-x_n}$ *(which we will see in Computer Project 6 has such a cascade).*

24. $u_{n+1} = u_n e^{a - u_n}$; let $u_n = x_n$ and $r = e^a$.

25. $u_{n+1} = au_n e^{-\sqrt{u_n}}$; let $u_n = 4x_n^2$ and $r = \sqrt{a}$.

Show that the equations in Exercises 26 and 27 undergo period-doubling cascades by using the given substitution to convert each into $x_{n+1} = \dfrac{2}{x_n} + \dfrac{3x_n}{4} - c$ *(which we will see in Computer Project 7 has such a cascade).*

26. $v_{n+1} = \dfrac{1}{v_n} + \dfrac{3v_n}{4} - d$; let $v_n = x_n/\sqrt{2}$ and $c = d\sqrt{2}$.

27. $v_{n+1} = \dfrac{2}{v_n - d} + \dfrac{3v_n}{4}$; let $v_n = x_n + d$ and $c = d/4$.

Show that the equations in Exercises 28 and 29 undergo period-doubling cascades by using the given substitution to convert each into $x_{n+1} = x_n(a - x_n^2)$ *(which we will see in Computer Project 8 has such a cascade).*

28. $y_{n+1} = r y_n(1 - y_n^2)$; let $y_n = x_n/\sqrt{a}$ and $r = a$.

29. $y_{n+1} = r y_n(1 - y_n)^2$; let $y_n = x_n^2/a$ and $r = a^2$.

In Exercises 30–33 show that the given functions DO NOT undergo period-doubling cascades for a > 0.

30. $f_a(x) = ax^2$

31. $f_a(x) = -ax^3$

32. $f_a(x) = a/x^2$

33. $f_a(x) = \dfrac{ax}{x^2 + 1}$

34. A period-doubling bifurcation can be visualized using the example $f_a(x) = a - x^2$, which has one at $a = 3/4$.

   (a) Sketch a careful graph of $y = f_a^2(x) = -x^4 + 2ax^2 + a - a^2$ along with $y = x$ prior to the bifurcation for $a = 1/2$.

   (b) Sketch a careful graph of $y = f_a^2(x)$ along with $y = x$ precisely at the bifurcation for $a = 3/4$.

   (c) Sketch a careful graph of $y = f_a^2(x)$ along with $y = x$ after the bifurcation occurs for $a = 1$.

   (d) Can you explain the mechanism that produced the bifurcation?

Computer Projects 3.7

In Projects 1–4: (a) Use algebra and calculus to find the positive fixed point and its interval of stability. (b) For a sampling of parameter values throughout the interval of stability of the 2-cycle, use direct numerical iteration to find the points of the 2-cycle. (c) Graph the fixed point and points of the 2-cycle together to see if a bifurcation occurs. Note: *The intervals of stability of the 2-cycles were previously found in Computer Projects 2–5 of Section 3.6.*

1. $f_r(x) = rxe^{-x}$

2. $f_c(x) = \dfrac{2}{x} + 0.75x - c$

3. $f_r(x) = rx^2(1 - x)$

4. $f_a(x) = x(a - x^2)$

5. Show that the type of bifurcation described in Exercises 17–20 occurs for $f_c(x) = e^x - c$. The stable fixed point can be found through direct numerical iteration of $f_c(x)$. The unstable fixed point can be found by either using the Newton Root-Finding Method to solve $f_r(x) - x = 0$, or by iterating the inverse of $f_c(x)$, i.e., by iterating $g_c(x) = \ln(x + c)$.

Through numerical experimentation, show that each of the functions in Projects 6–11 undergoes a period-doubling cascade.

6. $f_r(x) = rxe^{-x}$

9. $f_r(x) = rx^2(1 - x)$

7. $f_c(x) = \dfrac{2}{x} + 0.75x - c$

10. $f_r(x) = \dfrac{rx}{(x^2 + 1)^2}$

8. $f_a(x) = x(a - x^2)$

11. $f_r(x) = rxe^{-x^2}$

SECTION

3.8   Chaos

Along with the **period-doubling cascade,** the concept of **chaos** has only recently been recognized and named. Since the 1970's these two related ideas, and their associated geometric counterpart called **fractal geometry,** have been among the most exciting and popular areas of mathematical investigation. Literally scores of books and thousands of research articles have been written on these topics during the past several decades.

In addition to being of interest to mathematicians, chaos has attracted the attention of scientists from many disciplines. Since its inception, chaotic behavior has been experimentally observed in a variety areas, such as meteorology, astrophysics, electrical circuitry, fluid dynamics, neural networks in the brain, heart arrhythmia and pacemaker interaction. The related field of fractal geometry has been used for data compression of computer-generated images, and to model certain biological branching processes, such as plants, trees, the lungs and the cardiovascular system.

Unusual for mathematical topics, these ideas have come to fascinate many in the general public too. There have been a number of popular books written that try to explain the significance of chaos and the beauty of fractal images to readers with minimal mathematical background. At least one of these has achieved best-seller status. Several documentaries exploring these topics have appeared on public TV, and a few theatrical productions inspired by chaos and its implications have been performed on stage. Even a recent blockbuster movie based upon a popular science fiction novel took the inevitability of chaos as its premise (although with a somewhat inaccurate interpretation of the ideas). And computer-generated fractal images have appeared in films, video games, posters, computer screen-savers, calendar art, tee-shirts, coffee mugs etc. Presently, there are copious amounts of software and numerous websites devoted to the study of chaos and fractals.

Although these ideas have certainly gained widespread notoriety among the general public, not many really understand their significance and implications. While fractal geometry will have to wait until after systems of nonlinear equations are treated in Chapter 5, an introduction to chaotic dynamics is appropriate at this point in our investigation.

What Is Chaos?

Chaos may exist for any nonlinear equation, whether it is viewed as being part of a parameterized family $x_{n+1} = f_r(x_n)$ or not $x_{n+1} = f(x_n)$. So in this section, for the sake of generality as well as simplicity, we usually omit the subscript r.

As its name implies, chaos generally involves a complete breakdown of stability, at least for an uncountably infinite set of initial points, sometimes called the **scrambled set.** For example, all stability in the logistic equation is lost for $r > 3.56995$. If the graph

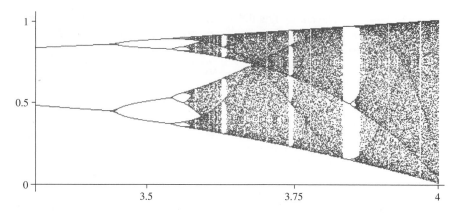

FIGURE 3.35

of Figure 3.34 in Section 3.7 is extended to the right of the period-doubling cascade, the dynamics appear as in Figure 3.35. Rather than a series of 2^m branches dominating the picture, for any $r > 3.56995$ an entire vertical interval or a set of vertical intervals is filled in. These intervals comprise the scrambled set. For some intermittent r values in that chaotic regime, there exist locally stable cycles having periods other than a power of 2, and these cycles undergo their own period-doubling cascades. But the basins of attraction of these cycles are small, and chaos resides in the fragmented scrambled set lying between the points of the cycles.

On the scrambled set, which frequently turns out to include an entire bounded interval or a set of such intervals, the dynamics resemble more a random or stochastic process. This is not to say that given x_0 we cannot compute $x_1, x_2, \ldots$ exactly. Truly stochastic dynamics would imply that iterates could not be computed with 100% certainty, but instead only probabilities could be assigned for their possible values. But here we are still working with a deterministic equation, and consequently given any x_0 we can indeed determine the exact value of each x_n. So in what sense does chaos resemble randomness? To answer this question, we discuss some common characterizations of chaotic dynamics and their implications for the modeling process.

Characterizations of Chaos

The first and simplest property common to all chaotic dynamical systems is the existence of an **infinite number of periodic cycles of different periods.** Based on our findings with regard to the logistic family, we can easily see how an infinite collection of 2^m-cycles would arise. When each 2^m-cycle bifurcates into the next, its stability may be lost, but not its existence. For $r > 3.56995$, since all have bifurcated, the result is an infinite collection of periodic cycles each having a period equal to a power of 2.

But that is only part of the story. When chaos exists, there are generally many more periodic cycles than these, none of which have periods equal to a power of 2 and none of which are stable. In fact, for sufficiently large r the logistic equation, as well as most other parameterized unimodal families, have periodic m-cycles for *every* positive integer period: $m = 1, 2, 3, 4, 5, \ldots.$

In general, finding this infinite collection of periodic cycles can only be done through numerical experimentation. But in some special cases it is possible to determine the locations of periodic points with a few simple calculations.

EXAMPLE 1

Show that $x_{n+1} = 1 - 2|x_n|$ has an m-cycle for any integer $m \geq 1$.

Solution: In this case there are formulas that give the locations of points of period m for any $m \geq 1$. One of these is

$$P(m) = \frac{3 - 2^m}{2^m + 1}$$

For example, a fixed point p corresponds to letting $m = 1$ in that formula, which gives $p = P(1) = 1/3$. For $m = 2$ we get $P(2) = -1/5$, and since

$$1 - 2|P(2)| = 1 - 2|-1/5| = 3/5$$

then letting $p_1 = -1/5$ and $p_2 = 3/5$ yields a 2-cycle. We can similarly find m-cycles for

$m = 3:$ $p_1 = -5/9,$ $p_2 = -1/9,$ $p_3 = 7/9$

$m = 4:$ $p_1 = -13/17,$ $p_2 = -9/17,$ $p_3 = -1/17,$ $p_4 = 15/17$

$m = 5:$ $p_1 = -29/33,$ $p_2 = -25/33,$ $p_3 = -17/33,$ $p_4 = -1/33,$ $p_5 = 31/33$

and all other values of m. The fact that the function being iterated is *piecewise-linear* is the key to making this type of computation possible. In general, no such simple formula exists for finding periodic cycles.   ∎

When chaos exists, not only are there an infinite number of periodic points, but these points are **dense** within the scrambled set. This means that every open interval within the scrambled set, no matter how small, contains a periodic point. An example of a dense collection of points in the interval $0 \leq x \leq 1$ is the collection of all fractional or rational numbers in that interval. They are spread out over the entire interval, and every irrational number has rational numbers so close that they cannot be separated by any minimum distance. Something like this occurs with the countably infinite collection of periodic points within the scrambled set.

Another characteristic of chaotic dynamical systems is the existence of a special type of solution.

DEFINITION

A solution of $x_{n+1} = f(x_n)$ is called a **dense orbit** on some interval I, if the set of iterates x_n are dense within I. In this case $f(x)$ is said to be **transitive.**

As $n \to \infty$ the points x_n of a dense orbit distribute themselves in a random-like pattern over an entire interval I within the scrambled set, getting closer and closer to *all* points of I. In a chaotic process not only does there exist a dense orbit, but in addition the situation is even more complex: there are an uncountably infinite number of different dense orbits. Often, nearly every initial point leads to one. This makes them relatively easy to find.

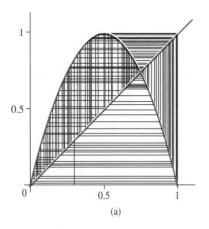

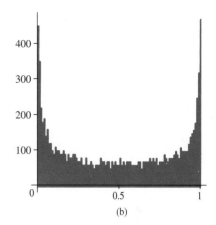

(a)
(b)

FIGURE 3.36

EXAMPLE 2

Show that $x_{n+1} = 4x_n(1 - x_n)$ has a dense orbit.

Solution: The existence of a dense orbit can be demonstrated numerically in several ways. One such way appears in Figure 3.36(a), which shows the result of several hundred steps of cobwebbing for a random choice of x_0. If continued, the web would eventually fill out the entire region between the graphs of $y = 4x(1 - x)$, $y = x$ and the x-axis. This corresponds to the points x_n of the solution being spread out over the interval $0 \le x \le 1$.

Another way to detect a dense orbit is to construct a frequency distribution for the set of points x_n. Figure 3.36(b) shows just such a distribution using the first 10,000 iterates and 100 sub-intervals of $0 \le x \le 1$, each of size 0.01. The nonzero frequency of each sub-interval is another indication of the density of the orbit.   ∎

The feature of chaotic behavior that has perhaps the most important implications for the modeling process is known as **sensitive dependence on initial conditions.** This means that solutions that begin arbitrarily close to one another may lead to very different dynamics. This is what occurs for dense orbits and other points in the scrambled set. Any two such solutions diverge from each other no matter how close together they may begin. They wander seemingly at random and independently of each other, sometimes very close to one another, making it appear as though the two might converge. But when they get too close, they quickly diverge from each other again, only to approach one another again later on.

To see the consequences of this, first recall how much is possible when stability is present. Knowing that a fixed point or periodic cycle is stable means that further investigation is unnecessary. We can predict in advance the long-term or asymptotic behavior of the entire set of solutions x_n for all x_0 in a basin of attraction. But what are the implications if there exists a scrambled set for which solutions remain bounded but never enter any basin of attraction? In this case an uncountably infinite number of *different* asymptotic behaviors must exist. It is as if each solution behaves in its own unique way, independently of all others. No common asymptotic behavior is shared by any significant number of solutions,

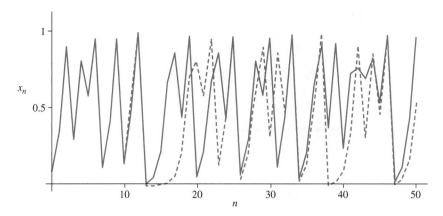

FIGURE 3.37

and so to determine the dynamics for any particular x_0, direct numerical iteration of a large number of x_n would have to be undertaken.

That makes long-term prediction in a modeling situation essentially impossible, since the true initial point x_0 can never really be known. This is, of course, not to mention the fact that the equation itself, as well as any parameters in it, are only approximations to the real-world situation being modeled. All of this means that any specific predictions that are made will become increasingly inaccurate with each step. The model is in effect *too sensitive* to make useful long-term predictions.

EXAMPLE 3

Find several solutions of $x_{n+1} = 4x_n(1 - x_n)$ that exhibit sensitive dependence on initial conditions.

Solution: Figure 3.37 shows time-series graphs of two solutions for slightly different initial values. The *solid* graph corresponds to $x_0 = 0.10000$, and the *dashed* graph to $x_0 = 0.10001$. As would be expected since these initial points are quite close, the respective solutions are at first almost indistinguishable. But they gradually diverge from each other, and after approximately the 15th iteration they are noticeably different. In later iterations they may synchronize once again for a step or two, but they always eventually diverge from each other. The behavior of one solution must be described as quite unlike that of the other, except that both are apparently becoming dense orbits.   ∎

Sensitive dependence is often characterized as the **butterfly effect.** Since our weather is generally believed to be a chaotic system, then in theory the flapping of a butterfly's wings could over time have a significant effect upon weather conditions. Perhaps the butterfly might set in motion a chain reaction that results in a major storm days, weeks or months later. Therefore, to make accurate long-term weather predictions, the behavior of every butterfly (as well as most other living creatures) would have to be known in advance — clearly an impossible task. This is a theoretical consequence of sensitive dependence.

It is possible to quantify somewhat the notion of sensitivity to initial conditions by computing the **Lyapunov exponent** of a solution, especially of a dense orbit. This is a generalization of the idea of determining stability or instability by computing the derivative $f'(x)$ at a fixed point, or the product of the derivatives at the points of an m-cycle. In this case we compute the product of the derivatives at the first N points of a solution:

$$D(N) = f'(x_1) \cdot f'(x_2) \cdot f'(x_3) \cdot \ \cdots \ \cdot f'(x_N) \tag{1}$$

and then ask what happens to the limit of

$$|D(N)|^{1/N} \quad \text{as} \quad N \to \infty \tag{2}$$

To see what this quantity tells us, suppose we compute it for a solution of the linear equation $x_{n+1} = a x_n$ for some a. In this case $f(x) = ax$, and so $f'(x_n) = a$ for all n. This means that

$$D(N) = a \cdot a \cdot a \cdot \ \cdots \ \cdot a = a^N \quad \text{and so} \quad |D(N)|^{1/N} = |a^N|^{1/N} = |a|$$

In other words, for linear equations we get $|a|$, which is a measure of the rate of exponential decay or convergence of solutions if $|a| < 1$, or exponential growth or divergence of solutions if $|a| > 1$.

For nonlinear equations, the limit in (2) may be thought of as a measure of the *average* rate for all n at which a solution x_n asymptotically converges to or diverges from other nearby solutions. Similar to the situation for linear equations, a value between 0 and 1 means x_n may tend to converge to nearby solutions. A value exceeding 1 indicates an average asymptotic divergence from them, which is the meaning of sensitivity to initial conditions.

Since taking the *Nth root* of $D(N)$ in equation (2) is a little too complex, especially since we let $N \to \infty$, believe it or not, the expression can be simplified by taking its natural logarithm. This means we instead compute

$$L = \lim_{N \to \infty} \ln\left(|D(N)|^{1/N}\right) = \lim_{N \to \infty} \frac{\ln|D(N)|}{N}$$

(There are other reasons for doing this that we won't go into.) Replacing $D(N)$ in this equation with its defined value from equation (1) as the product of derivatives and simplifying gives

$$L = \lim_{N \to \infty} \frac{\ln|f'(x_1) \cdot f'(x_2) \cdot \ \cdots \ \cdot f'(x_N)|}{N} = \lim_{N \to \infty} \frac{1}{N} \sum_{n=1}^{N} \ln|f'(x_n)|$$

There is name for that last expression.

DEFINITION

For any solution of $x_{n+1} = f(x_n)$ that satisfies $f'(x_n) \neq 0$ for all $n \geq 0$, the **Lyapunov exponent** is the value of

$$L = \lim_{N \to \infty} \frac{1}{N} \sum_{n=1}^{N} \ln|f'(x_n)|$$

Using this L, the condition for stability now becomes

$$\ln 0 < L < \ln 1 \quad \text{or} \quad -\infty < L < 0$$

and the condition for sensitivity to initial conditions becomes

$$\ln 1 < L < \ln \infty \quad \text{or} \quad 0 < L < +\infty$$

In general, a negative value of L corresponds to some type of asymptotic stability. But for bounded solutions a positive L corresponds to sensitive dependence on initial conditions, as well as other forms of chaotic behavior, and the greater the value of L, the more sensitive and chaotic the dynamics are.

EXAMPLE 4

For $x_{n+1} = 1 - r|x_n|$ with $0 \le r \le 2$, compute the Lyapunov exponent for any solution that does not hit 0.

Solution: Since $f_r(x) = 1 - r|x|$, then although $f_r'(0)$ is undefined,

$$f_r'(x) = +r \quad \text{for} \quad x < 0 \quad \text{and} \quad f_r'(x) = -r \quad \text{for} \quad x > 0$$

This means that as long as $x_n \neq 0$ then

$$|f_r'(x_n)| = |r| = r$$

Similar to the computation above to find the Lyapunov exponent L for a solution of a linear equation, we again have $L = \ln r$. So, stability exists for $0 \le r < 1$, since $L = \ln r < 0$ for these r values. On the other hand, for $1 < r \le 2$ we have $L = \ln r > 0$, and since it can be checked that all solutions remain bounded for these r values, then chaos must exist. Also, since $L = \ln r$, the degree of sensitivity to initial conditions grows as r increases from 1 to 2.   ∎

In this example, it was possible to obtain L exactly as a function of r, since $f_r(x)$ had such a simple piecewise-linear form. Rarely is this the case, however. In general a computer must be used to find a sequence of numerical estimates of L using an increasing number N of iterates. When computing L this way, in general it is best to use a dense orbit. Since such a solution is distributed over an entire interval, these points will provide a *good sample* when computing the average amount of sensitivity that L is intended to measure.

Indicators of Chaos

Now that we have seen the importance of knowing whether a mathematical model is chaotic, it would be beneficial to have some quick indicators that chaos is indeed present, without having to perform extensive numerical iteration before discovering that fact. Using these indicators one would be able to determine in advance that a model is incapable of making precise predictions. We will state without proof some rules for determining the existence of chaos for any continuous function $f(x)$, since the proofs go well beyond the scope of this book.

One rule that predicts that chaos will occur involves finding two closed intervals I and J on the x-axis that satisfy a few simple properties. First, assume I and J don't intersect each other, except perhaps at one end-point. Next, suppose we use the symbol $f(I)$ to represent the *range* of the function $f(x)$ for all x in I, and similar for $f(J)$, i.e.,

$$f(I) = \{f(x) : x \in I\} \quad \text{and} \quad f(J) = \{f(x) : x \in J\}.$$

Since we also assume $f(x)$ is continuous, then both $f(I)$ and $f(J)$ must themselves be closed intervals. Although these two new intervals may at first be viewed as lying on the y-axis, they can be *reflected* back onto the x-axis, as in cobwebbing. If this is done, the following geometric property, which is sometimes given the rather interesting name **horseshoe,** provides a sufficient condition for chaos to occur.

THEOREM *Chaos: Horseshoe*

If $f(I)$ and $f(J)$ each *cover* both I and J, i.e., if

$$I \cup J \subset f(I) \cap f(J)$$

then $x_{n+1} = f(x_n)$ is chaotic. ∎

Although the geometry involved here bears little resemblance to a horseshoe, the generalization of this theorem in higher dimensions, first described by S. Smale in the 1960's, does indeed involve such a shape.

EXAMPLE 5

Show that $x_{n+1} = 4x_n(1 - x_n)$ is chaotic by finding a horseshoe.

Solution: Let I be the interval $0 \le x \le 1/2$ and J be the interval $1/2 \le x \le 1$. Figure 3.38(a) shows that the range of I under $f(x) = 4x(1 - x)$ covers the interval $0 \le x \le 1$, which is $I \cup J$. In fact, $f(I) = I \cup J$. Figure 3.38(b) shows the same for J. The previous theorem therefore implies chaos must exist for this problem. ∎

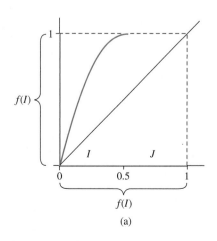
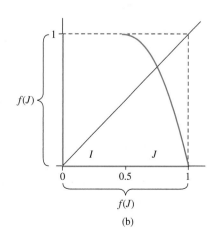

(a)   (b)

FIGURE 3.38

This example is rather untypical. Trying to find a horseshoe is often a tedious task. Instead of searching for two special intervals, the next rule involves finding only a single periodic cycle. The most famous rule in the field of chaos theory is also the simplest to state.

THEOREM   *Chaos: Period 3*

If $f(x)$ has a point of period 3 then $x_{n+1} = f(x_n)$ is chaotic.   ■

This remarkably simple rule, whose discovery by T.-Y. Li and J. A. Yorke in the 1970's almost single-handedly gave birth to the field of chaos theory, indicates that it is enough to demonstrate that a 3-cycle exists, in order to conclude that all the complex dynamics associated with chaos are also present. With some algebra it can be shown that a 3-cycle exists for the logistic equation, as well as most of the other unimodal families we have studied, for all sufficiently large parameter values.

EXAMPLE 6

Prove that $x_{n+1} = 1 - 2|x_n|$ exhibits chaotic behavior.

Solution: According to the previous theorem, it is enough to demonstrate the existence of a 3-cycle. This was previously done in Example 1, where the 3-cycle $-5/9$, $-1/9$ and $7/9$ was found. This equation must therefore possess all the forms of chaotic behavior described above.   ■

Unlike a horseshoe, the existence of a 3-cycle in a higher dimensional setting *does not* imply the existence of chaos, and so the previous theorem does not generalize to functions of two or more variables. In fact, there often exist globally stable 3-cycles when iterating nonlinear functions in two or more dimensions, as we will be doing in Chapter 5.

One last indicator of chaos that we will discuss here is the existence of an unusual kind of solution: one that begins and ends at an unstable fixed (or periodic) point p. If we iterate $x_{n+1} = f(x_n)$, starting with any x_0 close to but distinct from p, then x_n will of course diverge from the repeller p, at least for a while. But because of the nonlinearity of $f(x)$, it is possible for a solution to *snap back* with one such iterate x_N actually hitting p. Since p is a fixed point, then after that we must have $x_n = p$ for all $n \geq N$.

Next, suppose we solve the equation $x_{n+1} = f(x_n)$ for x_n, and iterate the problem *backwards*. The result would look something like

$$x_n = f^{-1}(x_{n+1})$$

although several values may be possible for each x_n since the function $f^{-1}(x)$ is multivalued. Nevertheless, starting with the same x_0 as above, and letting $n = -1, -2, -3, \ldots$, at least one such infinite set of *inverse iterates* $x_{-1}, x_{-2}, x_{-3}, \ldots$ must exist. Further, since p is a repeller for the original problem, it must therefore be an attractor for this inverse problem. (For example, 0 is a repeller for $x_{n+1} = 2x_n$ but an attractor for its inverse $x_n = x_{n+1}/2$.) This means that if x_0 is close enough to p, then a sequence $x_{-1}, x_{-2}, x_{-3}, \ldots$ can be generated that converges to p.

What we have created here is a *doubly-infinite* collection of iterates

$$\dots, \; x_{-3}, \; x_{-2}, \; x_{-1}, \; x_0, \; x_1, \; x_2, \; x_3, \; \dots$$

that satisfies

$$\lim_{n \to -\infty} x_n = p \quad \text{and} \quad \lim_{n \to +\infty} x_n = p$$

In general such a solution is called a **homoclinic orbit,** and p is sometimes called a **snap-back repeller.** Although it might be difficult to see, their presence always implies the existence of a horseshoe, which therefore gives us another rule.

THEOREM   *Chaos: Homoclinic Orbit*

If $f(x)$ has a homoclinic orbit then $x_{n+1} = f(x_n)$ is chaotic.   ∎

For many functions $f(x)$ of the type we have been studying, finding a homoclinic orbit is not as difficult as it might seem. It is often easier than finding horseshoes or 3-cycles. In most cases, the task can be accomplished with a few steps of cobwebbing.

EXAMPLE 7

Show that $x_{n+1} = 4x_n(1 - x_n)$ is chaotic by finding a homoclinic orbit.

Solution: The fixed point 0 can easily be verified to be a repeller. With a little guesswork, as well as some trial and error, a point x_0 can be found that generates a homoclinic orbit. The cobweb graph in Figure 3.39(a) shows a point x_0 close to 0, for which the cobweb at first diverges from 0, but eventually *snaps back* and hits 0 exactly. Figure 3.39(b) shows the cobweb for the entire homoclinic orbit, i.e., as $n \to \pm\infty$. Note how it begins and ends at 0.   ∎

Like horeshoes, the concept of homoclinic orbit can also be extended to higher-dimensional settings, and in those cases too the presence of one is nearly always associated with chaotic behavior.

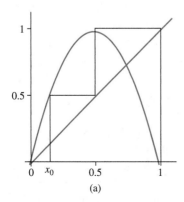
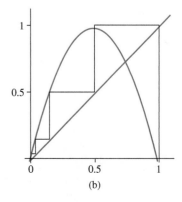

FIGURE 3.39

Summary

In this section we have only scratched the surface of chaos theory. Although we leave the topic for now, many of the same ideas will arise again in Chapter 5 when we investigate systems of nonlinear equations. There, in addition to the chaotic dynamics we have already encountered, we will also see two-dimensional versions of the scrambled set, called **strange attractors,** as well as graphical representations of some of the most bizarre mathematical objects known, called **fractals.**

Exercises 3.8

1. (a) Is the set of irrational numbers in $[0, 1]$ dense in that interval?
   (b) Is the set of all fractional numbers in $[0, 1]$ with denominator equal to a power of 2 dense in that interval?

2. For the function $f(x) = 1 - 2|x|$ that was being iterated in Examples 1 and 6, verify that the formula $P(m) = (3 - 2^m)/(2^m + 1)$ gives a point of period m if (a) $m = 6$; (b) $m = 7$.

3. The function $f(x) = 1 - 2|x|$ that was being iterated in Examples 1 and 6 has another 3-cycle besides the one previously found: $-5/9, -1/9, 7/9$. Show that $1/7$ is part of another 3-cycle.

4. The function $f(x) = \begin{cases} 3x & \text{if } x \leq 1 \\ 6 - 3x & \text{if } x > 1 \end{cases}$ has an infinite number of periodic points. A point $P(m)$ of period m can be found using $P(m) = 6/(3^m + 1)$. Verify this formula for (a) $m = 2$; (b) $m = 3$; (c) $m = 4$; (d) $m = 5$.

5. Suppose that the following points lie on the graph of some continuous function $y = f(x)$: $(3, 8), (7, 1), (4, 4), (5, 1), (1, 9), (8, 3), (11, 11), (9, 5)$.

   (a) Find all fixed points and periodic cycles of $f(x)$ among these points.
   (b) Based on this, is $x_{n+1} = f(x_n)$ a chaotic process? Justify your answer or say that not enough information is given to make that conclusion.

6. Repeat the previous exercise for the set of points $(0, 2), (1, 3), (2, 6), (3, 7), (4, 8),$ $(5, 5), (6, 1), (7, 0)$.

7. Show that $f(x) = 3x$ mod 1 exhibits chaotic behavior by showing:

   (a) All solutions x_n for $0 < x_0 < 1$ remain bounded between 0 and 1.
   (b) The Lyapunov exponent is positive for any solution that does not hit $0, 1/3, 2/3, 1$.

8. Show that $f_r(x) = rx$ mod 1 for any $r > 1$ exhibits chaotic behavior by showing:

   (a) All solutions x_n for $0 < x_0 < 1$ remain bounded between 0 and 1.
   (b) The Lyapunov exponent is positive for any solution x_n for which $f_r'(x_n)$ exists for all $n \geq 0$.

9. (a) Given that $p_1 = 1$ and $p_2 = -2$ constitute a 2-cycle of $f(x) = x^2 - 3$, compute the Lyapunov exponent of the solution x_n that begins at either one of these points.
   (b) Suppose that p_1, p_2 and p_3 constitute a 3-cycle of some differentiable function $f(x)$, and that $f'(p_1) = 1.5$, $f'(p_2) = 0.4$ and $f'(p_3) = -2$. Compute the Lyapunov exponent of the solution x_n that begins at any point of this 3-cycle.

10. (a) Use the Lyapunov exponent to show that $f(x) = \begin{cases} 3x/2 & \text{if } x \le 2/3 \\ 3 - 3x & \text{if } x > 2/3 \end{cases}$ exhibits chaos.

    (b) What bounds must the Lyapunov exponent have?

11. (a) Show graphically that the function $f(x)$ from Exercise 7 has a horseshoe.

    (b) Use cobwebbing to show that this function has a homoclinic orbit.

12. (a) Show graphically that $f_r(x) = rx(1 - x)$ has a horseshoe for all $r \ge 4$.

    (b) Use cobwebbing to show that $f_r(x)$ has a homoclinic orbit for all $r \ge 4$.

13. Prove that the function $f(x) = 1 - 2|x|$, previously investigated in Examples 1 and 6 as well as in Exercises 2 and 3, has a homoclinic orbit by showing:

    (a) $p = -1$ is a repelling fixed point.

    (b) If we define $x_n = -1 + 2^n$ for all $n \le 1$ and $x_n = -1$ for all $n \ge 2$, then this sequence satisfies $x_{n+1} = 1 - 2|x_n|$.

    (c) $\lim\limits_{n \to \pm\infty} x_n = p = -1$.

14. Prove that the function $f(x)$ of Exercise 4 has a homoclinic orbit. *Hint:* Let $x_n = 2 \cdot 3^{n-1}$ for $n \le 1$ and $x_n = 0$ for $n \ge 2$.

In Exercises 15–20: (a) Show graphically that f(x) *has a horseshoe. (b) Use cobwebbing to show that* f(x) *has a homoclinic orbit.*

15. $f(x) = 2 - x^2$

16. $f_r(x) = 2x \bmod 1$

17. $f(x) = 6.75x^2(1 - x)$

18. $f(x) = \pi \sin x$

19. $f(x) = \dfrac{3\sqrt{3}}{2} x(1 - x^2)$

20. $f(x) = 16x^2(1 - x)^2$

▬ Computer Projects 3.8

For each of the functions in Projects 1–8:

(a) *Pick two initial points close together, i.e., that perhaps differ by 0.001 or 0.00001, and perform at least 100 iterations of* $x_{n+1} = f(x_n)$. *Do solutions exhibit sensitive dependence on initial conditions?*

(b) *For several random choices of* x_0 *compute at least 1000 iterates* x_n *and draw a frequency distribution using at least 50 sub-intervals. Do dense orbits appear to exist?*

(c) *Estimate the Lyapunov exponent* L *by picking several random choices of* x_0 *and computing* $\frac{1}{N} \sum_{n=1}^{N} \ln |f'(x_n)|$ *for* N = 1000, 2500, 5000, *etc. Does* L *appear to be positive?*

Note: *When choosing* x_0, *in each case it's best to use a value close but not equal to the local max or min.*

1. $f(x) = 2 - x^2$

2. $f_r(x) = 15xe^{-x}$

3. $f_c(x) = \dfrac{2}{x} + \dfrac{3x}{4} - 2$

4. $f(x) = 6x^2(1 - x)$

5. $f(x) = \pi \sin x$

6. $f(x) = \dfrac{3\sqrt{3}}{2} x(1 - x^2)$

7. $f_r(x) = 7xe^{-x^2}$

8. $f_c(x) = \dfrac{1}{x} + x - 1.8$

It was shown in Exercises 12 and 17 that the functions in Projects 9 and 10 have horseshoes and homoclinic orbits, and hence chaos. Iterate each for random choices of x_0. What happened to the scrambled set?

9. $f(x) = rx(1 - x)$ for $r > 4$

10. $f(x) = 6.75x^2(1 - x)$

4

Modeling with Linear Systems

Thus far, whenever constructing a discrete model we have assumed that the subject of our investigation can be removed from all outside influences and described independently. This approach has led in Chapters 2 and 3 to a consideration of models that are based upon iteration of equations and functions of just *one* variable.

Rarely, however, can a natural process really be isolated this way. For example, with regard to population growth, it is much more common that a variety of different species inhabit the same ecosystem, and that each has some type of effect on the others. Their populations are interwoven, and none can be

predicted accurately without taking all into consideration.

Similar observations can be made with regard to many such processes that arise in the natural and social sciences. Frequently, the dynamics of two or more of them that we wish to investigate interact with one another, and each influences the evolution over time of all the others. As a result, the discrete models we develop to study such processes must inevitably involve iteration of *systems* of equations, or equivalently, iteration of functions of *several* variables. In this chapter we commence an investigation of these types of models by considering the dynamics associated with iteration of systems of linear equations.

4.1   Some Linear Systems Models

We first provide some motivation for the investigations we will be undertaking in this chapter by constructing several models involving iteration of systems of two or more linear equations. As in the previous chapter, we consider only those that are *autonomous*. That is, the variable n does not explicitly appear in the system except as the subscript of the primary variables involved. Once again, these are the only types of problems for which a comprehensive theory can be established.

Prey–Predator Systems

Iterated systems are frequently used to model the population dynamics of several interacting species inhabiting the same ecosystem. One such model describes the population growth of two species in which one, the **prey,** provides sustenance for the other, the **predator.** If we disregard density dependence and assume that independent of the other each would either grow or decay exponentially, then the prey population P_n and predator population Q_n might in that case be modeled separately by

$$P_{n+1} = r_1 P_n \quad \text{and} \quad Q_{n+1} = r_2 Q_n$$

respectively, where r_1 and r_2 are their intrinsic growth rates.

However, since we assume the predator species consumes the prey, then the next prey population P_{n+1} will be diminished to a degree which we might assume is directly proportional to the size of the present predator population Q_n. Conversely, the next predator population Q_{n+1} will be enhanced by an amount directly proportional to the present prey population P_n. If these two ideas are combined we arrive at a model.

LINEAR PREY-PREDATOR MODEL I

The linear system

$$\begin{aligned}
P_{n+1} &= r_1 P_n - s_1 Q_n \\
Q_{n+1} &= s_2 P_n + r_2 Q_n
\end{aligned} \tag{1}$$

where all parameters are non-negative, models a prey–predator relationship between two interacting species. P_n represents the prey and Q_n the predator.

The ecological scheme we have constructed here, which is also sometimes called a **host–parasite** system, involves the iteration of two equations for two interacting quantities or variables that evolve over time, in this case P_n and Q_n. In such models it is nearly always true that the number of equations matches the number of variables, with each equation giving the next value of one of them. To iterate this system, the pair of initial populations P_0 and Q_0 must be known. Both of these are needed to compute P_1 and Q_1, and this is accomplished by letting $n = 0$ in (1). Successively letting $n = 1, 2, 3, \ldots$ there allows computation of both P_n and Q_n, one-by-one for all $n > 0$. For example,

$$P_1 = r_1 P_0 - s_1 Q_0 \qquad P_2 = r_1 P_1 - s_1 Q_1$$
$$\text{and}$$
$$Q_1 = s_2 P_0 + r_2 Q_0 \qquad Q_2 = s_2 P_1 + r_2 Q_1$$

This prey–predator system is classified as *linear* since each of its two equations is a linear function of its two variables. In other words, $z = r_1 x - s_1 y$ and $z = s_2 x + r_2 y$ both describe planes in three-dimensional (x, y, z) space.

But this is not the only type of linear system that might be used to model the populations of such species. Suppose, for example, that at least one of the populations undergoes immigration, migration or harvesting at a constant level. We previously saw examples of these processes in Chapter 2. To incorporate immigration, a constant was added to the equation describing the next population, and for migration or harvesting a constant was subtracted. To include these ideas in our present ecological model, a constant could be added to or subtracted from either equation or both. So the model might become:

LINEAR PREY-PREDATOR MODEL II

The linear system

$$P_{n+1} = r_1 P_n - s_1 Q_n + k_1$$
$$Q_{n+1} = s_2 P_n + r_2 Q_n + k_2$$

where r_1, r_2, s_1, s_1 are non-negative, models a prey–predator relationship with constant immigration for positive k_1 or k_2, or constant migration or harvesting for negative k_1 or k_2. Again, P_n represents prey and the Q_n the predator.

Competing Species Systems

Rather than having a prey–predator relationship, two species may instead be competitors for the same life-giving resources: food, water, habitat etc. In this case, each may grow or decay exponentially on its own, but be diminished by an amount directly proportional to the size of the other. If we call the sizes of the two species P_n and Q_n, this leads to the following model.

LINEAR COMPETITION MODEL I

The linear system

$$P_{n+1} = r_1 P_n - s_1 Q_n$$
$$Q_{n+1} = -s_2 P_n + r_2 Q_n$$

where all parameters are non-negative, models the populations of two competing species.

Of course, we can introduce the following modification, as we did for the prey–predator model:

LINEAR COMPETITION MODEL II

The linear system

$$P_{n+1} = r_1 P_n - s_1 Q_n + k_1$$
$$Q_{n+1} = -s_2 P_n + r_2 Q_n + k_2$$

where r_1, r_2, s_1, s_1 are non-negative, models the populations of two competing species, with constant immigration for positive k_1 or k_2, or constant migration or harvesting for negative k_1 or k_2.

Overlapping-Generations Systems

Models involving the interaction of different species are not the only types of population schemes that lead to systems of equations. A system can also result from modeling the population of just a single species — a species in which several generations overlap. Suppose we call P_n the population of the nth generation of a species for which two consecutive generations simultaneously contribute to the population of the generation that follows them. For example, both P_0 and P_1 may contribute to P_2, although at perhaps different rates owing to different maturity levels. In general, P_{n+1} may depend on both P_n and P_{n-1} according to a linear equation such as

$$P_{n+1} = r P_n + s P_{n-1} \qquad (2)$$

where $r, s \geq 0$. To iterate this equation, we must assume that the populations of the present and previous generations are known: P_0 and P_{-1}, respectively.

An equation such as (2) is usually classified as **second-order** since the difference between the highest and lowest subscripts in the equation is $(n + 1) - (n - 1) = 2$. Although this does not appear to be a system, it can easily be converted into one if desired. Suppose we call $Q_n = P_{n-1}$, which implies $Q_{n+1} = P_n$ for all $n \geq 0$. Combining these with (2) gives the following:

LINEAR OVERLAPPING-GENERATIONS MODEL I

The linear system

$$P_{n+1} = r P_n + s Q_n$$
$$Q_{n+1} = P_n$$

for $r, s \geq 0$ models the population growth of a species with overlapping generations. The system is equivalent to the second-order equation (2) with $Q_n = P_{n-1}$.

Since $Q_n = P_{n-1}$ and we have assumed that P_{-1} and P_0 are given, then letting $n = 0$ gives $Q_0 = P_{-1}$. This means that both P_0 and Q_0 are known, and so iteration of the system can begin by letting $n = 0$.

As with the prey–predator and competition models above, immigration, migration or harvesting effects could be introduced into the system. However, this time only a single constant is needed, since we are really modeling a single species. For convenience, we choose to place that constant in the first equation.

LINEAR OVERLAPPING-GENERATIONS MODEL II

The linear system

$$P_{n+1} = r\,P_n + s\,Q_n + k$$
$$Q_{n+1} = P_n$$

for $r, s \geq 0$ models the population growth of a species with overlapping generations and constant immigration for $k > 0$, or constant migration or harvesting for $k < 0$.

Economic Systems

Iterated systems also arise in economic models describing the interplay between the supply, demand and price of a product. Recall that equation (3) of Section 3.1 says that the next price P_{n+1} depends on the present price P_n, as well as the present supply S_n and demand D_n according to

$$P_{n+1} = P_n + (a_1\,D_n - b_1) - (a_2\,S_n - b_2) \tag{3}$$

where all parameters are non-negative. Another economic principle states that a high price generally causes the demand to decrease and the supply to increase from their present levels. On the other hand, the demand may increase and the supply decrease when the price is low. If we use linear functions to represent these notions, then we could say

$$D_{n+1} = D_n - c_1\,P_n + k_1 \quad \text{and} \quad S_{n+1} = S_n + c_2\,P_n - k_2$$

where parameters are once again non-negative. Combining these with the equation for P_{n+1}, and letting $a_3 = b_2 - b_1$, yields the following model.

LINEAR PRICE-DEMAND-SUPPLY MODEL

The linear system

$$P_{n+1} = P_n + a_1\,D_n - a_2\,S_n + a_3$$
$$D_{n+1} = -c_1\,P_n + D_n + k_1$$
$$S_{n+1} = c_2\,P_n + S_n - k_2$$

where all parameters (except perhaps a_3) are non-negative, describes the interaction between the price P_n, the demand D_n and the supply S_n.

Unlike the population models above, this economic model involves iterating three rather than two linear equations. As such, it would prove to be too unwieldy to consider here. It would therefore be wise to somehow reduce the size of the problem, if that is possible. One way would be to assume instead that supply either remains constant or reacts immediately to price, so that

$$S_n = c_2 P_n + k_2, \quad \text{which means (3) may be written} \quad P_{n+1} = d_1 P_n + a_1 D_n + h_1,$$

where $d_1 = 1 - a_2 c_2$ and $h_1 = b_2 - b_1 - a_2 k_2$. Combining this new equation for P_{n+1} with the previous one for D_{n+1} yields the simplified model:

LINEAR PRICE-DEMAND MODEL

The linear system

$$P_{n+1} = d_1 \, P_n + a_1 \, D_n + h_1$$
$$D_{n+1} = -c_1 \, P_n + D_n + k_1$$

where a_1, c_1, $k_1 \geq 0$, $d_1 \leq 1$ and any value h_1, describes the interaction between the price P_n and the demand D_n.

Other Iterated Systems Models

The above models provide just a small sample of the areas in which iterated systems have application. Other disciplines in which such systems arise include computer graphics and animation; Markov processes; not to mention some purely mathematical uses, such as finding approximate solutions to systems of differential equations and sets of simultaneous equations. Some of these applications will be discussed in the exercises. Others will be introduced in later sections of this chapter.

Exercises 4.1 ──────────────────────────

1. For the prey–predator model $\begin{cases} P_{n+1} = 1.5P_n - Q_n \\ Q_{n+1} = 0.4P_n + 1.2Q_n \end{cases}$ with $P_0 = 400$ and $Q_0 = 150$:

    (a) Compute P_1 and Q_1; (b) Compute P_2 and Q_2.

2. Identify r_1, r_2, s_1, s_2 in each of the following prey–predator models:

    (a) $P_{n+1} = P_n - 1.5Q_n$ (b) $P_{n+1} = 0.8P_n - 0.2Q_n$

            $Q_{n+1} = 2P_n + 0.9Q_n$ $Q_{n+1} = 1.1Q_n$

3. Construct the linear prey–predator model with the following parameters:

    (a) $r_1 = 1.3, r_2 = 0.9, s_1 = 0.2, s_2 = 0.5, k_1 = k_2 = 0$.

    (b) $r_1 = 2, r_2 = 1, s_1 = 0, s_2 = 1.2, k_1 = 3000, k_2 = -2000$.

4. Construct the linear prey–predator model with $k_1 = k_2 = 0$, satisfying:

    (a) $r_1 = 1, r_2 = 1, P_0 = 2000, Q_0 = 1000, P_1 = 1800, Q_1 = 2400$.

    (b) $s_1 = 1, s_2 = 1, P_0 = 400, Q_0 = 200, P_1 = 300, Q_1 = 420$.

5. Construct the linear prey–predator model with no immigration, migration or harvesting, and:

   (a) The prey's growth rate is 1.2 and the predator's is 1.3; the prey population is diminished by 0.2 times that of the predator; the predator population is increased by 0.3 times that of the prey.

   (b) On their own each population would remain fixed, but the prey population is diminished by twice the predator population, and predator population increased by twice the prey population.

6. Identify r_1, r_2, s_1, s_2 in each of the following competition models:

   (a) $P_{n+1} = 1.2P_n - 0.7Q_n$

   $Q_{n+1} = -P_n + Q_n$

   (b) $P_{n+1} = 3(1.1P_n - 0.5Q_n)$

   $Q_{n+1} = (Q_n - P_n)/2$

7. Construct the linear competition model that satisfies:

   (a) $r_1 = 1.35, r_2 = 0.75, s_1 = 0.1, s_2 = 0.05, k_1 = k_2 = 0,$
   (b) $r_1 = 1, r_2 = 2, s_1 = 1, s_2 = 0, k_1 = -100, k_2 = 150.$

8. Construct the linear competition model that satisfies:

   (a) The growth rate of the first species is 1.5, and of the second is 1.25; each is diminished by 0.4 times the population of the other.

   (b) In addition to the characteristics of (a), the first species undergoes immigration at 2500 per step, and the second migrates at 1200 per step.

9. Construct the linear competition model with no immigration, migration or harvesting that satisfies:

   (a) $r_1 = 2, r_2 = 3/2, P_0 = 1000, Q_0 = 1600, P_1 = 1200, Q_1 = 800.$
   (b) $P_0 = 100, Q_0 = 200, P_1 = 150, Q_1 = 100, P_2 = 250, Q_2 = 25.$

10. For the overlapping-generations model $P_{n+1} = 1.2P_n + 0.5P_{n-1}$ with $P_{-1} = 1200$ and $P_0 = 1500$, compute $P_1, P_2,$ and P_3.

11. Convert the second-order equation of the previous exercise into a system and identify the initial conditions.

12. Show that $\begin{cases} P_{n+1} = rP_n + sQ_n \\ Q_{n+1} = P_n \end{cases}$ and $\begin{cases} P_{n+1} = rP_n + Q_n \\ Q_{n+1} = sP_n \end{cases}$ describe the same model for P_n.

13. The following are overlapping-generations models that have been converted into systems. Find the original equation for P_n:

    (a) $P_{n+1} = 1.8P_n + 0.3Q_n$

    $Q_{n+1} = P_n$

    (b) $P_{n+1} = 0.7P_n + Q_n$

    $Q_{n+1} = P_n + 1000$

14. Construct the linear overlapping-generations model that satisfies:

    (a) The population of the next generation equals 3/4 of the present population, plus 1/2 of the previous population.

    (b) In addition to that described in (a), there is migration at 2500 per step.

15. Construct the linear overlapping-generations model with no immigration, migration or harvesting, that satisfies:

    (a) $r = 1/2, P_{-1} = 5000, P_0 = 6000, P_1 = 6500.$
    (b) $P_{-1} = 100, P_0 = 400, P_1 = 600, P_2 = 2000.$

16. The equation $F_{n+1} = F_n + F_{n-1}$ is called the *Fibonacci Equation*.

    (a) If $F_{-1} = 0$ and $F_0 = 1$, compute $F_1, \ldots, F_{10}$.

    (b) Convert this equation into a system.

17. For the price-supply-demand system
$$\begin{cases} P_{n+1} = P_n + 2D_n - S_n + 3 \\ D_{n+1} = -P_n + D_n + 1 \\ S_{n+1} = 2P_n + S_n - 2 \end{cases} \quad \text{with } P_0 = 1,$$

    $D_0 = 10$, $S_0 = 15$:

    (a) Compute P_1, D_1, S_1;   (b) Compute P_2, D_2, S_2.

18. For the price–demand model $\begin{cases} P_{n+1} = P_n + D_n + h_1 \\ D_{n+1} = -2P_n + D_n + k_1 \end{cases}$:

    (a) when $D_0 = 100$ what is the minimum value of h_1 that will make $P_1 \geq P_0$?

    (b) when $P_0 = 20$ what is the maximum value of k_1 that will make $D_1 \leq D_0$?

19. Construct a linear price–supply model by assuming instead that demand is either constant or reacts immediately to price: $D_n = k_1 - c_1 P_n$.

20. The system $\begin{cases} x_{n+1} = ax_n - by_n \\ y_{n+1} = bx_n + ay_n \end{cases}$ where $a = \cos\theta$ and $b = \sin\theta$ for some angle θ, has application in computer animation.

    (a) Construct the model for $\theta = \pi/6$.

    (b) What θ corresponds to the system $\begin{cases} x_{n+1} = (x_n - y_n)/\sqrt{2} \\ y_{n+1} = (x_n + y_n)/\sqrt{2} \end{cases}$?

21. Many of the systems we have seen involve linear equations of the form $z = ax + by$. To obtain such a model from a collection of data points $(x_1, y_1, z_1), \ldots, (x_N, y_N, z_N)$, a form of regression can be used to find the values of a and b that minimize $\sum_{i=1}^{N}(ax_i + by_i - z_i)^2$.

    (a) Derive a set of *normal equations* that a and b must satisfy in this case.

    (b) Use your result from (a) and the data points $(1, 1, 2)$, $(1, 2, 3)$, $(2, 1, 1)$, $(3, 2, 1)$, to find a least-squares model of the form $z = ax + by$.

22. Suppose the populations (P_n, Q_n) (in thousands) of two interacting species over a 4 year period are: $(10, 10)$, $(10, 15)$, $(15, 15)$, $(15, 20)$, $(20, 20)$. Use Exercise 21 to find the least-squares model of the form $\begin{cases} P_{n+1} = aP_n + bQ_n \\ Q_{n+1} = cP_n + dQ_n \end{cases}$.

Computer Projects 4.1

1. Use Exercise 21 and a computer to find the least-squares equation of the form $z = ax + by$ for the data points $(100, 200, 200)$, $(150, 150, 200)$, $(200, 100, 150)$, $(200, 250, 300)$, $(250, 250, 350)$, $(250, 150, 200)$, $(300, 350, 400)$, $(350, 350, 300)$, $(350, 225, 150)$, $(400, 400, 400)$.

2. Suppose the populations (P_n, Q_n) of two interacting species over a 10-year period are $(1000, 1000)$, $(900, 1500)$, $(850, 1600)$, $(800, 1650)$, $(750, 1800)$, $(700, 1850)$,

(600, 1900), (650, 2000), (600, 2200), (500, 2250), (500, 2250). Use Exercise 21 and a computer to find the least-squares model of the form $\begin{cases} P_{n+1} = a P_n + b Q_n \\ Q_{n+1} = c P_n + d Q_n \end{cases}$.

4.2 Linear Systems and Their Dynamics

Having seen in the previous section some important areas of application of iterated linear systems, we now begin an investigation of their dynamics. As we shall see in Chapter 5, the tools developed here will serve also as a starting place in the analysis of nonlinear systems as well.

Homogeneous and Non-Homogeneous Systems

We first introduce some general terminology often used to classify linear systems.

D E F I N I T I O N

The general form of an **autonomous linear system** for x_n and y_n is

$$x_{n+1} = a\,x_n + b\,y_n + h$$
$$y_{n+1} = c\,x_n + d\,y_n + k \tag{1}$$

where the parameters a, b, c, d, h and k are any real constants. When initial values are provided for x_0 and y_0, a unique solution exists for x_n and y_n for all $n \geq 0$. If $h = 0$ and $k = 0$, the system is called **homogeneous.** Otherwise, it is **non-homogeneous.**

As before, we generally omit the word *autonomous* since this is the only type of problem we consider. It is easy to see that all of the two-variable models introduced in the previous section are of the form (1). Additionally, the population models for which no immigration, migration or harvesting occurs are also homogeneous.

Although in theory a formula for the exact solution of a linear system can always be found, the process can at times be rather tedious. So instead, we will adopt the precedent of the previous chapter and concentrate primarily on techniques for determining the qualitative behavior of solutions. This, after all, is the more realistic approach. Since most models, as we have previously pointed out, are only approximations to reality, what purpose would it serve to find their *exact* solutions? In the end, the qualitative dynamics we uncover for these systems will yield about the same amount of useful information concerning the true nature of the processes being modeled as would the exact solutions.

Second-Order Equations

To more easily investigate solutions of (1) it is sometimes wise to **uncouple** the system, i.e., to find an independent equation for each of its two variables. Recall from the last section that Linear Overlapping-Generations Model I began as a single second-order equation describing the population growth of just one species. By introducing another variable to represent the previous generation, this was converted into a linear system of equations. We would now like to do the reverse — to begin with a general linear system, and convert it into a single second-order equation for (either) one of its two variables. Here, we demonstrate the technique for homogeneous systems. The steps for the non-homogeneous case are similar.

To uncouple the general homogeneous linear system

$$x_{n+1} = a\,x_n + b\,y_n \tag{2}$$

$$y_{n+1} = c\,x_n + d\,y_n \tag{3}$$

first note that (2) implies

$$x_{n+2} = a\,x_{n+1} + b\,y_{n+1} \tag{4}$$

Also, multiplying (3) by b gives

$$b\,y_{n+1} = b\,c\,x_n + b\,d\,y_n \tag{5}$$

and multiplying (2) by $-d$ gives

$$-d\,x_{n+1} = -a\,d\,x_n - b\,d\,y_n \tag{6}$$

Finally, suppose we add together the three equations (4), (5) and (6). This yields

$$x_{n+2} + b\,y_{n+1} - d\,x_{n+1} = a\,x_{n+1} + b\,y_{n+1} + b\,c\,x_n + b\,d\,y_n - a\,d\,x_n - b\,d\,y_n$$

which after some cancellation reduces to

$$x_{n+2} - d\,x_{n+1} = a\,x_{n+1} + b\,c\,x_n - a\,d\,x_n$$

Combining terms, this can be written more simply as

$$x_{n+2} = (a + d)\,x_{n+1} + (b\,c - a\,d)\,x_n \tag{7}$$

which is an uncoupled equation involving the variable x_n alone.

This equation is second order since once again the difference between its highest and lowest subscripts is $(n + 2) - n = 2$. Also, since we presume the initial values x_0 and y_0 are given for the system, we can let $n = 0$ in (2) to obtain

$$x_1 = a\,x_0 + b\,y_0$$

This means that both x_0 and x_1 are available, so that (7) can be iterated to find x_2, x_3, x_4, etc.

It is interesting to note that a similar set of calculations would result in the exact same second-order equation for y_n, i.e.,

$$y_{n+2} = (a + d)\,y_{n+1} + (b\,c - a\,d)\,y_n$$

for which both y_0 and y_1 could be known. This allows us to determine the nature of solutions for y_n as well. Alternatively, once the x_n have been evaluated using (7), the solution for y_n could instead be computed by writing (2) as $y_n = (x_{n+1} - a x_n)/b$, assuming $b \neq 0$.

EXAMPLE 1

Uncouple the linear predator–prey model

$$P_{n+1} = 1.5P_n - 0.5Q_n$$
$$Q_{n+1} = 0.5P_n + 2.5Q_n$$

Solution: The first equation of the system can be written

$$P_{n+2} = 1.5P_{n+1} - 0.5Q_{n+1}$$

Multiplying the second equation of the system by -0.5 and the first by -2.5 gives

$$-0.5Q_{n+1} = -0.25P_n - 1.25Q_n \quad \text{and} \quad -2.5P_{n+1} = -3.75P_n + 1.25Q_n$$

respectively. Adding these last three equations yields

$$P_{n+2} = 4P_{n+1} - 4P_n \quad \text{or} \quad P_{n+2} - 4P_{n+1} + 4P_n = 0$$

which is uncoupled. The same equation can be derived for Q_n.   ∎

Some Numerical Solutions

Before developing any additional tools for the analysis of linear systems, it may be worth first seeing some examples of the kinds of dynamics we will have to deal with. Through direct numerical iteration, a solution can easily be obtained for any system, and it can then be displayed in the same ways we're already accustomed to. Computer techniques for generating such solutions, and for displaying them numerically and graphically, are discussed in the Appendix.

EXAMPLE 2

Use a computer to generate a table and time-series graphs for the first 20 iterates of the prey–predator model

$$P_{n+1} = 1.2P_n - 0.1Q_n$$
$$Q_{n+1} = 0.1P_n + Q_n$$

beginning with $P_0 = 1200$ and $Q_0 = 1000$.

Solution: The data is shown in Table 4.1, the graph of P_n in Figure 4.1(a) and the graph of Q_n in Figure 4.1(b). From these it is clear that both prey and predator populations grow without bound. Further analysis shows that these growth patterns are asymptotically exponential with the same base. This can be seen by computing the ratios P_n/P_{n-1} and Q_n/Q_{n-1} as n gets large. Each of these ratios approaches the same constant $1.100\ldots$.   ∎

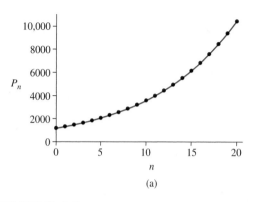

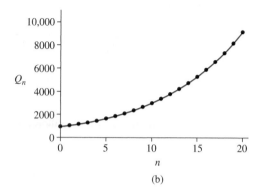

FIGURE 4.1

TABLE 4.1

n	P_n	n	P_n
0	1200		
1	1340	11	3994
2	1496	12	4451
3	1670	13	4959
4	1863	14	5524
5	2079	15	6152
6	2319	16	6851
7	2586	17	7628
8	2884	18	8492
9	3215	19	9452
10	3584	20	10519

(a)

n	Q_n	n	Q_n
0	1000		
1	1120	11	3424
2	1254	12	3823
3	1404	13	4268
4	1571	14	4764
5	1757	15	5316
6	1965	16	5932
7	2197	17	6617
8	2455	18	7380
9	2744	19	8229
10	3065	20	9174

(b)

EXAMPLE 3

Create a table and time-series graphs for the first 20 iterates of the competition model

$$P_{n+1} = 0.7P_n - 0.05Q_n$$
$$Q_{n+1} = -0.05P_n + 0.8Q_n$$

beginning with $P_0 = 5000$ and $Q_0 = 2000$.

Solution: The data is shown in Table 4.2 and the graphs in Figures 4.2(a) and 4.2(b). This time the populations appear to decay to 0 exponentially, with the ratios P_n/P_{n-1} and Q_n/Q_{n-1} both approaching $0.8207\ldots$ as $n \to \infty$.   ■

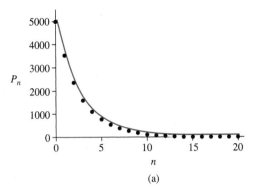

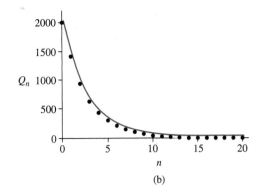

(a)   (b)

FIGURE 4.2

TABLE 4.2

n	P_n	n	P_n		n	Q_n	n	Q_n
0	5000				0	2000		
1	3400	11	74		1	1350	11	22
2	2313	12	50		2	910	12	14
3	1573	13	35		3	612	13	9
4	1071	14	24		4	411	14	5
5	729	15	16		5	275	15	3
6	496	16	11		6	184	16	2
7	338	17	8		7	122	17	1
8	231	18	5		8	81	18	0
9	157	19	4		9	53	19	0
10	108	20	3		10	35	20	0

       (a)   (b)

For the simple behavior exhibited by solutions in these two examples, tables and time-series graphs may be sufficient. But to truly comprehend the dynamics of more general linear systems, the following additional tool is needed.

Solution Space Graphs

A simple graphical technique that in some sense replaces the now obsolete cobweb graph is called a **solution space graph.** This is a two-dimensional plot in the (x, y)-plane of the set of solution points (x_n, y_n) for an iterated system. Consecutive points of the graph are usually connected by lines or curves in order to more easily follow the path or **trajectory** of the solution and thereby determine its asymptotic behavior, which is the ultimate goal.

Solution space graphs are also sometimes called **phase plane graphs,** although the latter is reserved more for continuous solutions of differential equations.

A solution space graph eliminates the need for two separate time-series graphs, one for each of x_n and y_n. It also allows the *geometry* of a trajectory to be easily identified and classified. As we shall see, only a small handful of such geometries are possible for the trajectory of a linear system, and each corresponds to a particular qualitative feature of the system.

The following are some examples of solution space graphs. Computer techniques for creating such graphs appear in the Appendix.

EXAMPLE 4

For the prey–predator model with constant immigration

$$P_{n+1} = 0.8P_n - 0.3Q_n + 8000$$
$$Q_{n+1} = 0.2P_n + 0.9Q_n + 2000$$

create a solution space graph for the first 50 iterates with $P_0 = 1000$ and $Q_0 = 15{,}000$.

Solution: The graph is shown in Figure 4.3. The set of points (P_n, Q_n) of the solution assume the shape of an inward-directed spiral. It appears that this trajectory spirals inward to a particular point (2500, 25,000). Further analysis would show a similar behavior for solutions starting at all other initial points. ∎

EXAMPLE 5

Create a solution space graph for the first 50 iterates of the price–demand model

$$P_{n+1} = P_n + 0.4D_n - 20$$
$$D_{n+1} = -0.3P_n + D_n + 5$$

for $P_0 = 15$ and $D_0 = 50$.

Solution: The graph is shown in Figure 4.4. This time the set of points (P_n, D_n) of the solution assume the shape of an outward-directed spiral. If continued, it appears that

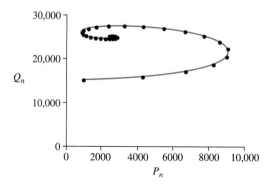

FIGURE 4.3

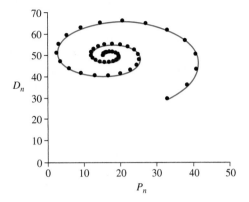

FIGURE 4.4

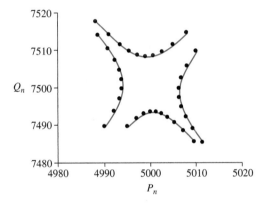

FIGURE 4.5

this trajectory would spiral out to ∞, although the model breaks down as soon as either coordinate becomes negative. It can be checked that (almost) all other solutions of the system have this same spiral structure. ∎

EXAMPLE 6

For the model of two competing species with constant immigration

$$P_{n+1} = P_n - 0.2Q_n + 1500$$
$$Q_{n+1} = -0.4P_n + Q_n + 2000$$

create solution space graphs for this system for several different pairs of initial values.

Solution: Figure 4.5 shows the result of plotting together in the same coordinate system the solution space graphs for four different initial points. Note the hyperbolic nature of the trajectories. This geometry, as well as the spiral structures of the previous examples, would not be apparent from corresponding time-series graphs. This is one of the advantages of solution space graphing. ∎

Fixed Points

It may not yet be apparent, but fixed points played a key role in the dynamics we just observed in the previous numerical examples. In the present setting, a fixed point should be viewed as a point (p, q) in the plane representing the solution space that remains unchanged under iteration of the system (1). The following is a more formal characterization.

DEFINITION

A point (p, q) is a **fixed point** of the linear system (1) if it satisfies

$$p = a\,p + b\,q + h$$
$$q = c\,p + d\,q + k \qquad (8)$$

Letting $(x_0, y_0) = (p, q)$ will make $(x_n, y_n) = (p, q)$ for all $n \geq 0$, and so (p, q) is an **equilibrium point** of the model.

Any linear system that corresponds to a useful mathematical model very likely has exactly one fixed point (p, q), and if such a system is homogeneous, that point must be $(p, q) = (0, 0)$. If the system is non-homogeneous, then solving the set of two simultaneous linear equations

$$x = a\,x + b\,y + h \qquad \text{or equivalently} \qquad (1 - a)\,x - b\,y = h$$
$$y = c\,x + d\,y + k \qquad\qquad\qquad -c\,x + (1 - d)\,y = k$$

will generally yield a unique fixed point (p, q) that is not equal to $(0, 0)$.

EXAMPLE 7

Find the equilibrium point of the price–demand model from Example 5.

Solution: Substituting x for P_n and y for D_n, that system becomes

$$x = x + 0.4y - 20$$
$$y = -0.3x + y + 5$$

It is immediately evident that the unique solution is $x = 50/3$, $y = 50$, which means the model is in equilibrium when the price is approximately $p = \$16.67$ and the demand is $q = 50$.   ∎

EXAMPLE 8

Find the equilibrium population levels for the model of Example 4.

Solution: Here we must solve the system of simultaneous equations

$$x = 0.8x - 0.3y + 8000 \qquad\qquad 0.2x + 0.3y = 8000$$
$$y = 0.2x + 0.9y + 2000 \qquad \text{which is equivalent to} \qquad -0.2x + 0.1y = 2000$$

The unique solution is quickly found to be $x = 2500$, $y = 25{,}000$. The fixed point is $(p, q) = (2500, 25{,}000)$, which means the model is in equilibrium when the prey population is $p = 2500$ and the predator population is $q = 25{,}000$.   ∎

Sinks, Sources and Saddles

To see the role of fixed points in the dynamics of linear systems, we need look no further than the numerical examples above. The solution obtained in Example 3 converges to the fixed point $(0, 0)$ as $n \to \infty$. All other solutions of that system behave similarly. The fixed point of Example 4, which was found to be $(2500, 25{,}000)$ in Example 8, again appears to attract all solutions. As in the previous chapters, such behavior indicates that these fixed points are stable attractors. For linear systems a stable equilibrium is sometimes called a **sink,** since the trajectories of solutions resemble water flowing toward the drain of a sink (see Fig. 4.6). In the case of Example 4 there is a spiral flow into the drain, further suggesting the sink image.

For the linear systems considered in Examples 2 and 5 above, the solutions obtained there increase with n and diverge to ∞. Further numerical experimentation would show that this occurs for all solutions of those systems, unless of course the fixed point is used as the initial point. In both cases the fixed point repels all solutions, and so appears to be an unstable equilibrium. This type of unstable fixed point is called a **source,** since solutions flow away from it like a faucet. There are no other fixed (or periodic) points of these systems, and so these flows must continue outward to ∞ (see Fig. 4.7).

There is one additional type of geometry that we encountered above. Example 6 shows a system in which trajectories appear hyperbolic. There exist two straight lines through the same point $(5000, 7500)$, which can be verified to be a fixed point, such that solutions diverge from one of the lines and approach the other. This makes trajectories resemble hyperbolas,

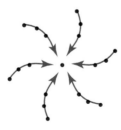

FIGURE 4.6

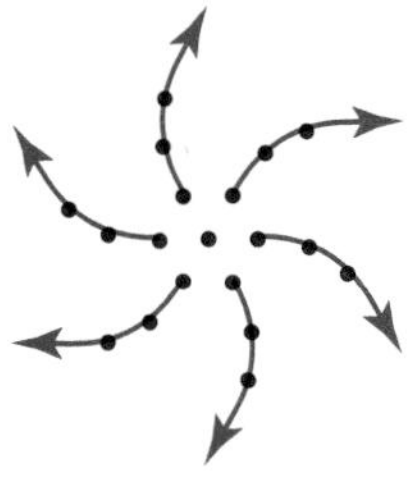

FIGURE 4.7

FIGURE 4.8

and the two straight lines resemble asymptotes (see Fig. 4.8). When this happens, such a fixed point is unstable, but it is not a source since most solutions do not flow directly away from it. Rather, it is called a **saddle,** due to the three-dimensional saddle-surface interpretation that it is sometimes given.

Fixed points of linear systems almost always fall into one of these three categories: sinks, sources and saddles. And the collective behavior of all solutions of such a linear system depends upon what type of fixed point it has. Therefore, to determine the overall dynamics of virtually any linear model, one needs only to find its unique fixed point (if it has one), and then decide which type it is.

While the technique associated with the first step of that process, finding the fixed point, has already been demonstrated, we have not yet seen how to determine whether that fixed point is a sink, source or saddle — unless of course we are willing to perform extensive numerical iteration. A relatively simple method of classifying fixed points does indeed exist, and will be developed later in this chapter.

Exercises 4.2

1. Identify the parameters a, b, c, d in each of the following homogeneous systems:

    (a) $x_{n+1} = (2.5x_n - 1.8y_n)/2$

    $y_{n+1} = 1.5(x_n + 2y_n)$

    (b) $x_{n+1} = 2.8x_n - 0.5(x_n + y_n)$

    $y_{n+1} = 0.3(y_n - x_n)$

2. Identify the parameters a, b, c, d, h, k in each of the following non-homogeneous systems:

    (a) $x_{n+1} = 5x_n - (y_n + 1000)$

    $y_{n+1} = 2y_n - 3(x_n + 500)$

    (b) $x_{n+1} = 2y_n - 600$

    $y_{n+1} = 0.5(x_n + y_n) + 0.75(x_n - y_n)$

In Exercises 3–6 convert each of the given systems into a second-order equation for each of its variables.

3. $x_{n+1} = x_n + y_n$

    $y_{n+1} = x_n - y_n$

4. $x_{n+1} = y_n - 10$

    $y_{n+1} = x_n + 8$

5. $P_{n+1} = 2P_n - Q_n$

    $Q_{n+1} = 3P_n + 4Q_n$

6. $P_{n+1} = 2P_n - 3Q_n + 1000$

    $Q_{n+1} = 1.5P_n + 2Q_n + 500$

In Exercises 7–10 find the unique fixed point of the given system.

7. $P_{n+1} = P_n - 2Q_n + 100$

    $Q_{n+1} = -5P_n + Q_n + 200$

9. $P_{n+1} = 5P_n - 4Q_n$

    $Q_{n+1} = P_n + 2Q_n - 1000$

8. $x_{n+1} = 0.75x_n + 0.5y_n + 20$

    $y_{n+1} = 4x_n - y_n + 40$

10. $x_{n+1} = 2x_n + y_n - 10$

    $y_{n+1} = -x_n + 2y_n - 5$

For each of the systems given in Exercises 11 and 12 explain why a unique fixed point does not exist.

11. $x_{n+1} = 2x_n - y_n$

    $y_{n+1} = x_n$

12. $x_{n+1} = 2x_n + y_n + 2$

    $y_{n+1} = 3x_n + 4y_n + 9$

In Exercises 13–16 find a formula for the exact solution of the given system.

13. $x_{n+1} = a x_n$

    $y_{n+1} = d y_n$

15. $x_{n+1} = a x_n + h$

    $y_{n+1} = d y_n + k$

14. $x_{n+1} = ax_n$

    $y_{n+1} = cx_n$

16. $x_{n+1} = a x_n + h$

    $y_{n+1} = cx_n + k$

In Exercises 17–20 find the fixed point of the given system and determine whether it is a sink, source or saddle using the results of Exercises 13–16.

17. $x_{n+1} = \frac{2}{3} x_n + 40$

    $y_{n+1} = 2x_n - 80$

19. $x_{n+1} = \frac{3}{2}x_n - 2500$

    $y_{n+1} = \frac{3}{4}y_n + 700$

18. $x_{n+1} = 2.5x_n$

    $y_{n+1} = 0.8y_n + 2000$

20. $x_{n+1} = \frac{1}{3}y_n - 10$

    $y_{n+1} = 2y_n + 5$

In Exercises 21 and 22 determine whether the origin is a sink, source or saddle of the given system by iterating and graphing $(x_0, y_0), \ldots, (x_4, y_4)$ in solution space for several choices of initial conditions.

21. $x_{n+1} = \frac{1}{4} x_n + \frac{1}{8} y_n$

    $y_{n+1} = \frac{1}{2} x_n + \frac{1}{4} y_n$

22. $x_{n+1} = 3x_n + y_n$

    $y_{n+1} = 2x_n + 3y_n$

23. In the computer animation model described in Exercise 20 of Section 4.1, compute $(x_1, y_1), \ldots, (x_5, y_5)$ for $\theta = \pi/2$ and $(x_0, y_0) = (1, 0)$, and graph these points in solution space. Describe what happens to (x_n, y_n) as $n \to \infty$. Explain the relevance to computer animation.

24. Repeat the previous exercise for (a) $\theta = \pi/3$; (b) $\theta = \pi/4$.

■ Computer Projects 4.2

In Projects 1–4 find the fixed point and determine whether it is a sink, source or saddle by iterating and graphing in solution space the first few iterates for several choices of initial conditions.

1. $x_{n+1} = x_n + y_n$

    $y_{n+1} = x_n - y_n$

2. $x_{n+1} = x_n - y_n + 30$

    $y_{n+1} = x_n + y_n - 20$

3. $P_{n+1} = 0.9P_n - 0.2Q_n + 40$
   $Q_{n+1} = 0.5P_n + 0.8Q_n - 5$

4. $P_{n+1} = 0.2P_n - 0.5Q_n + 5$
   $Q_{n+1} = -0.8x_n + 0.4y_n + 6$

5. The second-order equation $x_{n+1} = 0.36x_n - 0.63x_{n-1}$ can be converted into the two different systems below. For each system compute and graph in solution space the first few iterates for several choices of initial conditions. Compare the graphs and determine whether the fixed point of each is a sink, source or saddle.

$$x_{n+1} = 0.36x_n - 0.63y_n$$
$$y_{n+1} = x_n$$

and

$$x_{n+1} = 0.36x_n - y_n$$
$$y_{n+1} = 0.63x_n$$

6. Convert the second-order equation $P_{n+1} = 0.55P_n - 0.44P_{n-1} + 20$ into a system in two different ways and repeat the previous project for these systems.

7. Through numerical experimentation find all values of a and b for which the origin is a sink for the system $\begin{cases} x_{n+1} = ax_n + by_n \\ y_{n+1} = bx_n - ay_n \end{cases}$.

4.3 Some Vector and Matrix Arithmetic

As one may have noticed, the linear systems we have been investigating have a rather cumbersome notation. It would therefore be advantageous if we could abbreviate the statement of the problem by adopting vector and matrix notation from linear algebra. In this section we introduce some of those ideas.

Recognizing the relationship between the iteration of linear systems and the algebra of vectors and matrices is more than just a notational convenience, however. Its use will ultimately provide a means of determining the dynamics of a linear system without the need for direct iteration.

Vectors in the Plane

A **vector** in the plane is an ordered pair of numbers x and y, usually written vertically and surrounded by square brackets: $\begin{bmatrix} x \\ y \end{bmatrix}$. This type of vector may be visualized as an *arrow* in the plane with its tail at the origin and its tip at the point (x, y) (see Fig. 4.9).

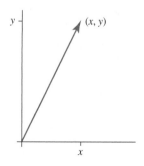

FIGURE 4.9

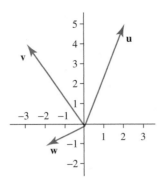

FIGURE 4.10

For convenience, vectors are often given names, such as

$$\mathbf{u} = \begin{bmatrix} 2 \\ 5 \end{bmatrix}, \quad \mathbf{v} = \begin{bmatrix} -3 \\ 4 \end{bmatrix} \quad \text{and} \quad \mathbf{w} = \begin{bmatrix} -2 \\ -1 \end{bmatrix}$$

Boldface is commonly used to denote a vector name. The vectors $\mathbf{u}$, $\mathbf{v}$ and $\mathbf{w}$ given above are shown in Figure 4.10. The zero vector is usually written as $\mathbf{0} = \begin{bmatrix} 0 \\ 0 \end{bmatrix}$. Such names make it easier to work symbolically with vectors, i.e., to manipulate them in equations as we will be doing below.

Often, the point (x, y) and the vector $\begin{bmatrix} x \\ y \end{bmatrix}$ are thought of as being interchangeable, since there is a one-to-one correspondence between the two. Since it takes two coordinates to denote any particular vector in the plane, they are *two-dimensional*. Vectors may be of any positive dimension, however. For example,

$$\begin{bmatrix} 3 \\ 5 \\ -2 \end{bmatrix} \quad \text{is three-dimensional, and} \quad \begin{bmatrix} 1 \\ 0 \\ 7 \\ -6 \end{bmatrix} \quad \text{is four-dimensional}$$

Although such vectors are frequently needed in mathematics, we will restrict the discussion here to just the two-dimensional case.

There is entire arithmetic associated with vectors. For example, any two vectors $\mathbf{u}$ and $\mathbf{v}$ may be added to get a new vector $\mathbf{u} + \mathbf{v}$. When adding vectors, the computations are done *component-wise*. This means that the first coordinate of $\mathbf{u}$ is added to the first coordinate of $\mathbf{v}$ to get the first coordinate of $\mathbf{u} + \mathbf{v}$. A similar computation gives the second coordinate. For example,

$$\begin{bmatrix} 4 \\ -3 \end{bmatrix} + \begin{bmatrix} 2 \\ 7 \end{bmatrix} = \begin{bmatrix} 4+2 \\ -3+7 \end{bmatrix} = \begin{bmatrix} 6 \\ 4 \end{bmatrix} \quad \text{and} \quad \begin{bmatrix} 1 \\ 8 \end{bmatrix} + \begin{bmatrix} 3 \\ 0 \end{bmatrix} = \begin{bmatrix} 1+3 \\ 8+0 \end{bmatrix} = \begin{bmatrix} 4 \\ 8 \end{bmatrix}$$

Any two vectors $\mathbf{u}$ and $\mathbf{v}$ may also be subtracted to get $\mathbf{u} - \mathbf{v}$, and again the computations are done component-wise. For example,

$$\begin{bmatrix} 1 \\ 0 \end{bmatrix} - \begin{bmatrix} 0 \\ 5 \end{bmatrix} = \begin{bmatrix} 1-0 \\ 0-5 \end{bmatrix} = \begin{bmatrix} 1 \\ -5 \end{bmatrix} \quad \text{and} \quad \begin{bmatrix} 5 \\ 7 \end{bmatrix} - \begin{bmatrix} 2 \\ 9 \end{bmatrix} = \begin{bmatrix} 5-2 \\ 7-9 \end{bmatrix} = \begin{bmatrix} 3 \\ -2 \end{bmatrix}$$

Although two vectors cannot be multiplied or divided by one another, each can be multiplied by a constant, which is often called a **scalar.** If the vector **v** is multiplied by the scalar c the result is the vector $c\mathbf{v}$, which is obtained by multiplying both components of **v** by c. For example,

$$3\begin{bmatrix} 2 \\ 5 \end{bmatrix} = \begin{bmatrix} 3 \cdot 2 \\ 3 \cdot 5 \end{bmatrix} = \begin{bmatrix} 6 \\ 15 \end{bmatrix} \quad \text{and} \quad -1/2\begin{bmatrix} 4 \\ -9 \end{bmatrix} = \begin{bmatrix} -\frac{1}{2} \cdot 4 \\ -\frac{1}{2} \cdot (-9) \end{bmatrix} = \begin{bmatrix} -2 \\ 4.5 \end{bmatrix}$$

It is worth noting the effect that scalar multiplication has on a vector. Suppose, for example, we let $\mathbf{v} = \begin{bmatrix} 3 \\ 1 \end{bmatrix}$, which means $2\mathbf{v} = \begin{bmatrix} 6 \\ 2 \end{bmatrix}$. If these are plotted in the plane, it is apparent that $2\mathbf{v}$ lies in the same direction as **v**, but has twice the length. This can be seen by using the distance formula for each:

$$\sqrt{6^2 + 2^2} = \sqrt{40} = 2\sqrt{10} = 2\sqrt{3^2 + 1^2}$$

Similarly, $\frac{1}{2}\mathbf{v} = \begin{bmatrix} 3/2 \\ 1/2 \end{bmatrix}$ lies in the same direction but has one-half the length (see Fig. 4.11(a)). On the other hand, $-2\mathbf{v} = \begin{bmatrix} -6 \\ -2 \end{bmatrix}$ also has twice the length of **v**, but this time it points in the exact opposite direction as **v**. Similarly, $-0.8\mathbf{v} = \begin{bmatrix} -0.24 \\ -0.8 \end{bmatrix}$ points in the opposite direction and has 4/5 the length of **v** (see Fig. 4.11(b)).

In general, multiplying a vector **v** by a scalar c always causes a stretching or shrinking of a vector and/or a reversing of its direction (unless of course $c = 1$). If $c > 1$ then $c\mathbf{v}$ is a vector pointing in the same direction as **v** but elongated by a factor of c. If $0 < c < 1$, then $c\mathbf{v}$ again has the same direction as **v** but is now shortened by the factor c. If $c < 0$, this stretching or shrinking once again occurs, but additionally the direction is reversed.

The special case $-1 \cdot \mathbf{v}$, written more simply as $-\mathbf{v}$, is called the **negative** of **v**. It has the same length as **v** but the opposite direction. Subtracting may actually be thought of as adding the negative, i.e., $\mathbf{u} - \mathbf{v} = \mathbf{u} + (-\mathbf{v})$.

It can easily be checked that many of the commonly known arithmetic rules for numbers, such as the commutative, associative and distributive laws, also work for vectors.

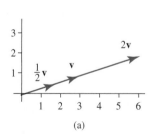

(a)

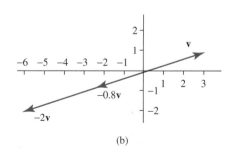

(b)

FIGURE 4.11

EXAMPLE 1

If $\mathbf{u} = \begin{bmatrix} 10 \\ -8 \end{bmatrix}$, $\mathbf{v} = \begin{bmatrix} 4 \\ 5 \end{bmatrix}$ and $\mathbf{w} = \begin{bmatrix} -6 \\ 0 \end{bmatrix}$, show that: (a) $\mathbf{u} + \mathbf{v} = \mathbf{v} + \mathbf{u}$; (b) $(\mathbf{u} + \mathbf{v}) + \mathbf{w} = \mathbf{u} + (\mathbf{v} + \mathbf{w})$.

Solution: (a) $\mathbf{u} + \mathbf{v} = \begin{bmatrix} 10 \\ -8 \end{bmatrix} + \begin{bmatrix} 4 \\ 5 \end{bmatrix} = \begin{bmatrix} 14 \\ -3 \end{bmatrix}$, and $\mathbf{v} + \mathbf{u} = \begin{bmatrix} 4 \\ 5 \end{bmatrix} + \begin{bmatrix} 10 \\ -8 \end{bmatrix} = \begin{bmatrix} 14 \\ -3 \end{bmatrix}$. So the two are equal.

(b) Since $\mathbf{u} + \mathbf{v} = \begin{bmatrix} 14 \\ -3 \end{bmatrix}$, then adding this with $\mathbf{w} = \begin{bmatrix} -6 \\ 0 \end{bmatrix}$ gives $\begin{bmatrix} 8 \\ -3 \end{bmatrix}$. On the other hand, $\mathbf{v} + \mathbf{w} = \begin{bmatrix} 4 \\ 5 \end{bmatrix} + \begin{bmatrix} -6 \\ 0 \end{bmatrix} = \begin{bmatrix} -2 \\ 5 \end{bmatrix}$. Adding this with $\mathbf{u} = \begin{bmatrix} 10 \\ -8 \end{bmatrix}$ gives $\begin{bmatrix} 8 \\ -3 \end{bmatrix}$, and so (b) is shown. ■

EXAMPLE 2

Using $\mathbf{u}$ and $\mathbf{v}$ from the previous example, show that: (a) $5(\mathbf{u} + \mathbf{v}) = 5\mathbf{u} + 5\mathbf{v}$; (b) $2\mathbf{v} - 6\mathbf{v} = -4\mathbf{v}$.

Solution: (a) Since $\mathbf{u} + \mathbf{v} = \begin{bmatrix} 14 \\ -3 \end{bmatrix}$ then $5(\mathbf{u} + \mathbf{v}) = \begin{bmatrix} 70 \\ -15 \end{bmatrix}$. Also, $5\mathbf{u} = \begin{bmatrix} 50 \\ -40 \end{bmatrix}$ and $5\mathbf{v} = \begin{bmatrix} 20 \\ 25 \end{bmatrix}$, which means $5\mathbf{u} + 5\mathbf{v} = \begin{bmatrix} 70 \\ -15 \end{bmatrix}$. This shows (a).

(b) $2\mathbf{v} = \begin{bmatrix} 8 \\ 10 \end{bmatrix}$ and $6\mathbf{v} = \begin{bmatrix} 24 \\ 30 \end{bmatrix}$. Subtracting these gives $2\mathbf{v} - 6\mathbf{v} = \begin{bmatrix} -16 \\ -20 \end{bmatrix}$. Also, $-4\mathbf{v} = \begin{bmatrix} -16 \\ -20 \end{bmatrix}$. Since the two are equal, (b) is shown. ■

2 by 2 Matrices

A **matrix** is a rectangular array of numbers. These numbers are laid out in horizontal rows and vertical columns, and usually surrounded by square brackets. In general, a matrix may have any number of rows and columns, such as

$$\begin{bmatrix} 6 & 0 & -4 \\ 1 & 2 & -1 \\ 0 & 5 & 5 \end{bmatrix}, \quad \begin{bmatrix} -8 & 15 & 1/2 \\ 3 & -6 & 0 \end{bmatrix} \quad \text{and} \quad \begin{bmatrix} 5 & 12 \\ -4 & 2 \end{bmatrix}$$

A matrix with m rows and n columns is called an **m by n matrix,** and if $m = n$ the matrix is called **square.** The first example above is a square 3 by 3 matrix, and the second is 2 by 3 but obviously not square. We will be concerned here only with those square matrices having exactly two rows and two columns, such as the third example above. The general form of a 2 by 2 matrix is $\begin{bmatrix} a & b \\ c & d \end{bmatrix}$ where a, b, c and d are any real numbers.

Like vectors, matrices are often given names to make them easier to work with. However, capital letters instead of boldface is the conventional notation for them. For example, we could say

$$A = \begin{bmatrix} 4 & 10 \\ 8 & -9 \end{bmatrix} \quad \text{or} \quad B = \begin{bmatrix} 5 & 0 \\ 0 & 6 \end{bmatrix}$$

Also like vectors, there is an arithmetic associated with matrices that allows them to be added or subtracted (if they have the same size), or multiplied by a scalar. As before, when adding or subtracting, the computations are done component-wise. For example,

$$\begin{bmatrix} 7 & 0 \\ -2 & 4 \end{bmatrix} - \begin{bmatrix} 5 & 8 \\ -3 & 10 \end{bmatrix} = \begin{bmatrix} 7-5 & 0-8 \\ -2-(-3) & 4-10 \end{bmatrix} = \begin{bmatrix} 2 & -8 \\ 1 & -6 \end{bmatrix}$$

And when multiplying by a scalar, each entry of the matrix is multiplied by that scalar. For example,

$$3 \begin{bmatrix} 4 & 2/3 \\ -6 & 0 \end{bmatrix} = \begin{bmatrix} 3 \cdot 4 & 3 \cdot 2/3 \\ 3 \cdot (-6) & 3 \cdot 0 \end{bmatrix} = \begin{bmatrix} 12 & 2 \\ -18 & 0 \end{bmatrix}$$

As with vectors, the usual arithmetic rules involving the commutative, associative and distributive laws work when adding or subtracting matrices, or when multiplying by a scalar.

Matrix Multiplication

Besides multiplication by a scalar, there is another useful form of multiplication involving 2 by 2 matrices. A vector in the plane can be multiplied by a 2 by 2 matrix, and the product will be another vector in the plane. Also, any two such matrices can be multiplied together, and the product this time will be another 2 by 2 matrix.

To multiply the vector $\mathbf{x} = \begin{bmatrix} x \\ y \end{bmatrix}$ by the matrix $A = \begin{bmatrix} a & b \\ c & d \end{bmatrix}$ we follow the rule

$$A\mathbf{x} = A \cdot \mathbf{x} = \begin{bmatrix} a & b \\ c & d \end{bmatrix} \cdot \begin{bmatrix} x \\ y \end{bmatrix} = \begin{bmatrix} ax + by \\ cx + dy \end{bmatrix}$$

Note above that $ax + by$ and $cx + dy$ are each single quantities, which means $A\mathbf{x}$ is a two-dimensional vector.

EXAMPLE 3

If $A = \begin{bmatrix} 2 & -3 \\ 5 & 1 \end{bmatrix}$, compute $A\mathbf{x}$ if: (a) $\mathbf{x} = \begin{bmatrix} 4 \\ 2 \end{bmatrix}$; (b) $\mathbf{x} = \begin{bmatrix} -1 \\ 3 \end{bmatrix}$.

Solution: (a) $A\mathbf{x} = \begin{bmatrix} 2 & -3 \\ 5 & 1 \end{bmatrix} \cdot \begin{bmatrix} 4 \\ 2 \end{bmatrix} = \begin{bmatrix} 2 \cdot 4 + (-3) \cdot 2 \\ 5 \cdot 4 + 1 \cdot 2 \end{bmatrix} = \begin{bmatrix} 2 \\ 22 \end{bmatrix}$.

(b) $A\mathbf{x} = \begin{bmatrix} 2 & -3 \\ 5 & 1 \end{bmatrix} \cdot \begin{bmatrix} -1 \\ 3 \end{bmatrix} = \begin{bmatrix} 2 \cdot (-1) + (-3) \cdot 3 \\ 5 \cdot (-1) + 1 \cdot 3 \end{bmatrix} = \begin{bmatrix} -11 \\ -2 \end{bmatrix}$.   ∎

It is important to note that while $A \cdot \mathbf{x}$ is defined, $\mathbf{x} \cdot A$ is not. In other words, multiplying a vector by a matrix (written to the left of it) is defined, but not the other way around.

There is a special 2 by 2 matrix $I = \begin{bmatrix} 1 & 0 \\ 0 & 1 \end{bmatrix}$ that has a unique property. If we let

$\mathbf{x} = \begin{bmatrix} x \\ y \end{bmatrix}$ then

$$I\mathbf{x} = \begin{bmatrix} 1 \cdot x + 0 \cdot y \\ 0 \cdot x + 1 \cdot y \end{bmatrix} = \begin{bmatrix} x \\ y \end{bmatrix} = \mathbf{x}$$

Since $I\mathbf{x} = \mathbf{x}$, this means that 2 by 2 vectors are always left unchanged when multiplied by I. The matrix I acts like the number 1 in regular arithmetic, and for this reason I is called the **identity matrix**.

Multiplying one matrix by another is similar to multiplying a vector by a matrix. In essence, one just treats each column of the second matrix as if it is a vector. If $A = \begin{bmatrix} a_1 & b_1 \\ c_1 & d_1 \end{bmatrix}$

and $B = \begin{bmatrix} a_2 & b_2 \\ c_2 & d_2 \end{bmatrix}$, then

$$AB = A \cdot B = \begin{bmatrix} a_1 & b_1 \\ c_1 & d_1 \end{bmatrix} \cdot \begin{bmatrix} a_2 & b_2 \\ c_2 & d_2 \end{bmatrix} = \begin{bmatrix} a_1 a_2 + b_1 c_2 & a_1 b_2 + b_1 d_2 \\ c_1 a_2 + d_1 c_2 & c_1 b_2 + d_1 d_2 \end{bmatrix}$$

EXAMPLE 4

If $A = \begin{bmatrix} 3 & 2 \\ 0 & 4 \end{bmatrix}$, and $B = \begin{bmatrix} 1 & 6 \\ -3 & 5 \end{bmatrix}$ compute: (a) AB; (b) BA.

Solution: (a) $AB = \begin{bmatrix} 3 & 2 \\ 0 & 4 \end{bmatrix} \cdot \begin{bmatrix} 1 & 6 \\ -3 & 5 \end{bmatrix} = \begin{bmatrix} 3 \cdot 1 + 2 \cdot (-3) & 3 \cdot 6 + 2 \cdot 5 \\ 0 \cdot 1 + 4 \cdot (-3) & 0 \cdot 6 + 4 \cdot 5 \end{bmatrix}$

$= \begin{bmatrix} -3 & 28 \\ -12 & 20 \end{bmatrix}$

(b) $BA = \begin{bmatrix} 1 & 6 \\ -3 & 5 \end{bmatrix} \cdot \begin{bmatrix} 3 & 2 \\ 0 & 4 \end{bmatrix} = \begin{bmatrix} 1 \cdot 3 + 6 \cdot 0 & 1 \cdot 2 + 6 \cdot 4 \\ -3 \cdot 3 + 5 \cdot 0 & -3 \cdot 2 + 5 \cdot 4 \end{bmatrix} = \begin{bmatrix} 3 & 26 \\ -9 & 14 \end{bmatrix}$

In the previous example since $AB \neq BA$, this shows that in general matrix multiplication is **not** commutative. However, multiplication is commutative if the identity matrix $I = \begin{bmatrix} 1 & 0 \\ 0 & 1 \end{bmatrix}$ is involved. Not only are AI and IA equal, but both equal A itself, and so again I acts like the number 1.

Matrix multiplication is always associative, however. That is, $A(BC) = (AB)C$ for any 2 by 2 matrices A, B and C. One implication of this is that $A^2 = A \cdot A$, $A^3 = A \cdot A \cdot A$, etc., can be uniquely defined. We will need that fact in the next section.

Simultaneous Equations

One of the uses of matrix multiplication is to abbreviate the notation when solving a pair of simultaneous linear equations. Suppose, for example, we wish to solve

$$2x - 3y = 2$$
$$x + y = 6$$

(1)

for the unknowns x and y. Using vectors and matrices, this can be written

$$\begin{bmatrix} 2 & -3 \\ 1 & 1 \end{bmatrix} \begin{bmatrix} x \\ y \end{bmatrix} = \begin{bmatrix} 2 \\ 6 \end{bmatrix}$$

If we call $A = \begin{bmatrix} 2 & -3 \\ 1 & 1 \end{bmatrix}$, $\mathbf{x} = \begin{bmatrix} x \\ y \end{bmatrix}$ and $\mathbf{b} = \begin{bmatrix} 2 \\ 6 \end{bmatrix}$, then the problem could be written more concisely as $A\mathbf{x} = \mathbf{b}$.

Any pair of two linear equations involving two unknowns can be abbreviated this way. The general system

$$\begin{matrix} ax + by = h \\ cx + dy = k \end{matrix} \quad \text{becomes} \quad \begin{bmatrix} a & b \\ c & d \end{bmatrix} \begin{bmatrix} x \\ y \end{bmatrix} = \begin{bmatrix} h \\ k \end{bmatrix}$$

and so can be written in the form $A\mathbf{x} = \mathbf{b}$ with $A = \begin{bmatrix} a & b \\ c & d \end{bmatrix}$, $\mathbf{x} = \begin{bmatrix} x \\ y \end{bmatrix}$ and $\mathbf{b} = \begin{bmatrix} h \\ k \end{bmatrix}$.

Notice the similarity of the vector equation $A\mathbf{x} = \mathbf{b}$ to the single linear equation $ax = b$. Just as in elementary algebra, where one is interested in solving $ax = b$ for the unknown number x when the constants a and b are given, in linear algebra one is interested in solving $A\mathbf{x} = \mathbf{b}$ for the unknown vector $\mathbf{x}$, when the **coefficient matrix** A and vector $\mathbf{b}$ are given.

Although the two equations have obvious similarities, there is one significant difference, however. The equation $ax = b$ can be solved by dividing both sides by a to get $x = b/a$. But, the equation $A\mathbf{x} = \mathbf{b}$ cannot be solved by dividing, since matrix division does not really exist.

There is a substitute for division however. Suppose we could find another 2 by 2 matrix A^{-1} that satisfies $A^{-1} \cdot A = I$, where I is the identity matrix. If so, then we could multiply this A^{-1} on both sides of the equation $A\mathbf{x} = \mathbf{b}$ to obtain

$$A^{-1} \cdot A\mathbf{x} = A^{-1} \cdot \mathbf{b}$$

But since $A^{-1}A = I$ then

$$I\mathbf{x} = A^{-1}\mathbf{b}$$

and since $I\mathbf{x} = \mathbf{x}$ then $\mathbf{x} = A^{-1}\mathbf{b}$. In other words,

$$\text{the solution of} \quad A\mathbf{x} = \mathbf{b} \quad \text{is} \quad \mathbf{x} = A^{-1}\mathbf{b}$$

The matrix A^{-1}, if it exists, is called the **inverse** of A. Although inverses do not always exist, they usually do. For example,

$$\text{if} \quad A = \begin{bmatrix} 2 & -3 \\ 1 & 1 \end{bmatrix} \quad \text{then} \quad A^{-1} = \begin{bmatrix} 1/5 & 3/5 \\ -1/5 & 2/5 \end{bmatrix}$$

since $A^{-1}A = I$. This means the solution of (1) above can be solved using $\mathbf{x} = A^{-1}\mathbf{b}$ to obtain

$$\begin{bmatrix} x \\ y \end{bmatrix} = \begin{bmatrix} 1/5 & 3/5 \\ -1/5 & 2/5 \end{bmatrix} \begin{bmatrix} 2 \\ 6 \end{bmatrix} = \begin{bmatrix} 4 \\ 2 \end{bmatrix}$$

So, the solution of (1) is $x = 4$, $y = 2$.

There is a formula for the inverse of any 2 by 2 matrix that has one:

INVERSE FORMULA

The inverse of the matrix $A = \begin{bmatrix} a & b \\ c & d \end{bmatrix}$, if it exists, is given by $A^{-1} = \dfrac{1}{ad - bc} \begin{bmatrix} d & -b \\ -c & a \end{bmatrix}$.

It is easily verified that, with A^{-1} defined this way, $A^{-1}A = I$ as well as $AA^{-1} = I$.

EXAMPLE 5

Use an inverse matrix to solve the linear system

$$3x + 4y = 3$$
$$x - 2y = 11$$

Solution: Here, the coefficient matrix is $A = \begin{bmatrix} 3 & 4 \\ 1 & -2 \end{bmatrix}$, which makes $ad - bc = 3 \cdot (-2) - 4 \cdot 1 = -10$, and so

$$A^{-1} = \frac{1}{-10} \begin{bmatrix} -2 & -4 \\ -1 & 3 \end{bmatrix} = \begin{bmatrix} -2/(-10) & -4/(-10) \\ -1/(-10) & 3/(-10) \end{bmatrix} = \begin{bmatrix} 1/5 & 2/5 \\ 1/10 & -3/10 \end{bmatrix}$$

The solution can then be found by computing

$$\begin{bmatrix} x \\ y \end{bmatrix} = \begin{bmatrix} 1/5 & 2/5 \\ 1/10 & -3/10 \end{bmatrix} \begin{bmatrix} 3 \\ 11 \end{bmatrix} = \begin{bmatrix} 5 \\ -3 \end{bmatrix}$$

The solution of the linear system is therefore $x = 5$, $y = -3$.   ■

The formula for A^{-1} has several implications. First, it indicates which 2 by 2 matrices A have inverses and which don't. Since division by 0 is impossible, then A^{-1} exists if and only if $ad - bc \neq 0$. Because the existence of A^{-1} is determined by this value, the number $ad - bc$ is called the **determinant** of A.

A less obvious implication of the formula for A^{-1} is that it indicates how many solutions the **homogeneous** system $Ax = 0$ has. Obviously, $\mathbf{x} = \mathbf{0}$ is always one solution. This is called the **trivial solution.** The following rule, which will be useful in the next section, describes how the determinant can be used to quickly determine if there are any others. Its proof is left as an exercise.

THEOREM   *Determinant and Nontrivial Solutions*

A homogeneous system of simultaneous linear equations has nontrivial solutions if and only if the determinant of its corresponding coefficient matrix is 0.   ■

■ **EXAMPLE 6**

Determine whether the linear system

$$6x - 4.5y = 0$$
$$4x - 3y = 0$$

has only the solution $x = 0$, $y = 0$.

Solution: If we call $A = \begin{bmatrix} 6 & -4.5 \\ 4 & -3 \end{bmatrix}$ then the determinant is $6(-3) - 4(-4.5) = 0$. So there are nontrivial solutions, such as $x = 3$, $y = 4$ and $x = -6$, $y = -8$. In fact, there are an infinite number of solutions since the equations of the system are multiples of one another. For 2 by 2 matrices this is always the case when the determinant is 0. ■

Exercises 4.3 ─────────────────────────────────

In Exercises 1 and 2 compute: (a) $3\mathbf{u}$; (b) $\mathbf{u} + \mathbf{v}$; (c) $-\frac{1}{2}\mathbf{v}$; (d) $2\mathbf{u} - \mathbf{v}$; (e) $\frac{1}{3}\mathbf{u} + 3\mathbf{v}$.

1. $\mathbf{u} = \begin{bmatrix} 3 \\ -2 \end{bmatrix}$, $\mathbf{v} = \begin{bmatrix} -1 \\ 2 \end{bmatrix}$

2. $\mathbf{u} = \begin{bmatrix} -6 \\ 0 \end{bmatrix}$, $\mathbf{v} = \begin{bmatrix} 5 \\ 3 \end{bmatrix}$

In Exercises 3 and 4 verify that: (a) $\mathbf{u} + \mathbf{v} = \mathbf{v} + \mathbf{u}$ and (b) $(\mathbf{u} + \mathbf{v}) - \mathbf{w} = \mathbf{u} + (\mathbf{v} - \mathbf{w})$.

3. $\mathbf{u} = \begin{bmatrix} 2.5 \\ -4 \end{bmatrix}$, $\mathbf{v} = \begin{bmatrix} 4 \\ 6 \end{bmatrix}$, $\mathbf{w} = \begin{bmatrix} -1.5 \\ 1.75 \end{bmatrix}$

4. $\mathbf{u} = \begin{bmatrix} -3 \\ 4.5 \end{bmatrix}$, $\mathbf{v} = \begin{bmatrix} 2 \\ 0 \end{bmatrix}$, $\mathbf{w} = \begin{bmatrix} -2 \\ -5 \end{bmatrix}$

In Exercises 5 and 6 verify that: (a) $-3(\mathbf{u} + \mathbf{v}) = -3\mathbf{u} - 3\mathbf{v}$ and (b) $\frac{1}{2}\mathbf{u} + \frac{5}{2}\mathbf{u} = 3\mathbf{u}$.

5. $\mathbf{u} = \begin{bmatrix} -2/3 \\ 7 \end{bmatrix}$, $\mathbf{v} = \begin{bmatrix} 5 \\ -4 \end{bmatrix}$

6. $\mathbf{u} = \begin{bmatrix} 8.5 \\ 2.25 \end{bmatrix}$, $\mathbf{v} = \begin{bmatrix} -5.5 \\ 0 \end{bmatrix}$

7. For each of the following, graph $\mathbf{v}$, $2\mathbf{v}$, $-3\mathbf{v}$, $\frac{1}{2}\mathbf{v}$ in the same coordinate system and compute the lengths of each:

(a) $\mathbf{v} = \begin{bmatrix} 7 \\ 4 \end{bmatrix}$;

(b) $\mathbf{v} = \begin{bmatrix} -2 \\ 0 \end{bmatrix}$

8. If $\mathbf{u} = \begin{bmatrix} 1 \\ 2 \end{bmatrix}$ and $\mathbf{v} = \begin{bmatrix} 5 \\ 1 \end{bmatrix}$:

(a) Compute $\mathbf{u} + \mathbf{v}$.

(b) Graph $\mathbf{u}$, $\mathbf{v}$, $\mathbf{u} + \mathbf{v}$ in the same coordinate system.

(c) Can you explain how $\mathbf{u} + \mathbf{v}$ could have been obtained graphically from $\mathbf{u}$ and $\mathbf{v}$?

(d) Try this again for two other vectors of your choosing.

In Exercises 9 and 10 compute (a) $5A$; (b) $A + B$; (c) $4A - 2B$.

9. $A = \begin{bmatrix} 2 & 0 \\ 1 & -3 \end{bmatrix}$, $B = \begin{bmatrix} 4 & 1 \\ -5 & 0 \end{bmatrix}$

10. $A = \begin{bmatrix} 2.75 & 1.5 \\ 2.5 & -1 \end{bmatrix}$, $B = \begin{bmatrix} -4 & -3 \\ 1.5 & 2 \end{bmatrix}$

In Exercises 11 and 12 compute (a) A**u** *and (b)* A**v**.

11. $A = \begin{bmatrix} 3 & 4 \\ 2 & 6 \end{bmatrix}$, $\mathbf{u} = \begin{bmatrix} 5 \\ 1 \end{bmatrix}$, $\mathbf{v} = \begin{bmatrix} 0 \\ 2 \end{bmatrix}$

12. $A = \begin{bmatrix} 2 & 0 \\ 5 & 1 \end{bmatrix}$, $\mathbf{u} = \begin{bmatrix} 3 \\ 7 \end{bmatrix}$, $\mathbf{v} = \begin{bmatrix} 0 \\ -2 \end{bmatrix}$

In Exercises 13 and 14 verify that (a) $A(\mathbf{u} - \mathbf{v}) = A\mathbf{u} - A\mathbf{v}$; *(b)* $(A + B)\mathbf{u} = A\mathbf{u} + B\mathbf{u}$.

13. $A = \begin{bmatrix} 2 & 0 \\ -2 & 1 \end{bmatrix}$, $B = \begin{bmatrix} -3 & 7 \\ 9 & 2 \end{bmatrix}$ $\mathbf{u} = \begin{bmatrix} 4 \\ 5 \end{bmatrix}$, $\mathbf{v} = \begin{bmatrix} 8 \\ -1 \end{bmatrix}$.

14. $A = \begin{bmatrix} 5 & -1 \\ -3 & 4 \end{bmatrix}$, $B = \begin{bmatrix} 0 & -2 \\ 3 & 0 \end{bmatrix}$ $\mathbf{u} = \begin{bmatrix} 3/5 \\ -6 \end{bmatrix}$, $\mathbf{v} = \begin{bmatrix} 0 \\ 3/2 \end{bmatrix}$.

In Exercises 15 and 16 compute (a) AB; *(b)* BA; *(c)* A^2; *(d)* A^3.

15. *A and B from Exercise 9.*   16. *A and B from Exercise 10.*

In Exercises 17 and 18 verify that: (a) $A(B + C) = AB + AC$; *(b)* $(A + B)C = AC + BC$; *(c)* $AB \neq BA$.

17. $A = \begin{bmatrix} 1 & 3 \\ -1 & 2 \end{bmatrix}$, $B = \begin{bmatrix} 4 & -1 \\ 0 & 1 \end{bmatrix}$, $C = \begin{bmatrix} 2 & 0 \\ -2 & 3 \end{bmatrix}$

18. $A = \begin{bmatrix} 0 & -4 \\ 2 & 1 \end{bmatrix}$, $B = \begin{bmatrix} 8 & -1 \\ 1 & 0 \end{bmatrix}$, $C = \begin{bmatrix} 3 & -2 \\ -2 & 0 \end{bmatrix}$

19. Verify that the matrices from Exercise 17 satisfy $A(BC) = (AB)C$.

20. Verify that the matrices from Exercise 18 satisfy $A(BC) = (AB)C$.

In Exercises 21–24 compute the determinants.

21. $\begin{bmatrix} 2 & -3 \\ 1 & 6 \end{bmatrix}$ 22. $\begin{bmatrix} 5 & 2 \\ -8 & -4 \end{bmatrix}$ 23. $\begin{bmatrix} 2/3 & -5/6 \\ -6 & 9 \end{bmatrix}$ 24. $\begin{bmatrix} 7.25 & 0.5 \\ -1.8 & 0 \end{bmatrix}$

In Exercises 25–28 find the inverse A^{-1} *of the given matrix* A. *Verify that* $A^{-1}A = I$ *and* $AA^{-1} = I$.

25. $\begin{bmatrix} 8 & 2 \\ 7 & 2 \end{bmatrix}$ 26. $\begin{bmatrix} 0 & -1 \\ 2 & -5 \end{bmatrix}$ 27. $\begin{bmatrix} 0 & -1/4 \\ 2 & -5 \end{bmatrix}$ 28. $\begin{bmatrix} 6 & 1/2 \\ -3 & 1/3 \end{bmatrix}$

In Exercises 29–32 write the given system in vector/matrix form and then use the inverse to solve.

29. $6x - 3y = 3$
    $2x + y = 9$

30. $3x - 6y = 0$
    $5x - 9y = 1$

31. $4x - 7y = 1$
    $x - 2y = 2$

32. $\frac{1}{2}x + 3y = 1$
    $3x + 9y = 9$

33. (a) If $A = \begin{bmatrix} -2 & 5 \\ 2 & 1 \end{bmatrix}$, compute the determinant of $A - 3I$ and of $A + 4I$.

(b) If $B = \begin{bmatrix} 6 & 8 \\ 2 & 0 \end{bmatrix}$, find the two values of r that make the determinant of $B - rI$ equal to 0.

34. (a) Prove that the determinant of a matrix is 0 if and only if one row of the matrix is a multiple of the other.

(b) Prove that the determinant of a matrix is 0 if and only if one column of the matrix is a multiple of the other.

35. For any matrix $A = \begin{bmatrix} a & b \\ c & d \end{bmatrix}$, prove that its powers $A^n = \begin{bmatrix} a_n & b_n \\ c_n & d_n \end{bmatrix}$ may be computed by iterating the system

$$a_{n+1} = a\,a_n + b\,c_n$$
$$b_{n+1} = a\,b_n + b\,d_n$$
$$c_{n+1} = c\,a_n + d\,c_n$$
$$d_{n+1} = c\,b_n + d\,d_n$$

with $a_0 = d_0 = 1$ and $b_0 = c_0 = 0$.

▬▬▬▬ Computer Projects 4.3

In Projects 1–4 compute A^n for $n = 1, \ldots, 20$ (see Exercise 35). Can you predict what will happen as $n \to \infty$?

1. $A = \begin{bmatrix} 1/2 & 1/3 \\ 1/2 & 2/3 \end{bmatrix}$

3. $A = \begin{bmatrix} 1/2 & 1/4 \\ 1/4 & 1/2 \end{bmatrix}$

2. $A = \begin{bmatrix} 0.9 & 0.3 \\ 0.1 & 0.7 \end{bmatrix}$

4. $A = \begin{bmatrix} 1 & 1/2 \\ 1/2 & 1 \end{bmatrix}$

*For a matrix A, the maximum length of $A\mathbf{v}$ for all vectors $\mathbf{v}$ of length 1 is called the **norm** of A. By computing $A\mathbf{v}$ for a large sample of vectors $\mathbf{v} = \begin{bmatrix} x \\ y \end{bmatrix}$ satisfying $\sqrt{x^2 + y^2} = 1$, estimate the norms of the matrices given in Projects 5 and 6.*

5. $A = \begin{bmatrix} 1/3 & 2/3 \\ 2/3 & 1/3 \end{bmatrix}$

6. $A = \begin{bmatrix} 1 & -1 \\ 1 & 1 \end{bmatrix}$

SECTION

4.4   Stability and Eigenvalues

Having familiarized ourselves in the previous section with the algebra of vectors and matrices, we are now in a position to make use of that algebra to help establish stability criteria for iterated linear systems. In this section we take the first steps of a search, which will occupy much of the remainder of this chapter, for a means of quickly identifying whether the fixed point of any such system is a sink, a source or a saddle. Since the general behavior of all solutions depends upon this, the overall dynamics of the system will then be known.

Homogeneous Systems

To simplify the discussion, for now we confine our attention to homogeneous systems only. We shall see later that the stability criteria we derive for this case applies to non-homogeneous systems as well.

Using vector and matrix notation introduced in the previous section, we now see that any homogeneous system

$$x_{n+1} = a\,x_n + b\,y_n$$
$$y_{n+1} = c\,x_n + d\,y_n$$

can be written more concisely as

$$\begin{bmatrix} x_{n+1} \\ y_{n+1} \end{bmatrix} = \begin{bmatrix} a & b \\ c & d \end{bmatrix} \begin{bmatrix} x_n \\ y_n \end{bmatrix}$$

Suppose we let $\mathbf{x}_n = \begin{bmatrix} x_n \\ y_n \end{bmatrix}$ be the **solution vector** of that system. If we also let the **coefficient matrix** be $A = \begin{bmatrix} a & b \\ c & d \end{bmatrix}$, then the system can be abbreviated as

$$\mathbf{x}_{n+1} = A\mathbf{x}_n$$

Note how similar this is to the homogeneous linear equation $x_{n+1} = a x_n$ treated in Chapter 2.

EXAMPLE 1

Write the general Linear Prey–Predator Model I from Section 4.1

$$P_{n+1} = r_1\,P_n - s_1\,Q_n$$
$$Q_{n+1} = s_2\,P_n + r_2\,Q_n$$

in vector/matrix form, and identify the coefficient matrix.

Solution: The system may be written $\begin{bmatrix} P_{n+1} \\ Q_{n+1} \end{bmatrix} = \begin{bmatrix} r_1 & -s_1 \\ s_2 & r_2 \end{bmatrix} \begin{bmatrix} P_n \\ Q_n \end{bmatrix}$, and also as $\mathbf{x}_{n+1} = A\mathbf{x}_n$, where $\mathbf{x}_n = \begin{bmatrix} P_n \\ Q_n \end{bmatrix}$ and the coefficient matrix is $A = \begin{bmatrix} r_1 & -s_1 \\ s_2 & r_2 \end{bmatrix}$. ∎

This notation allows several important observations to be made. First, iteration of homogeneous linear systems is equivalent to starting with a vector in the plane $\mathbf{x}_0 = \begin{bmatrix} x_0 \\ y_0 \end{bmatrix}$, multiplying it by the matrix A to obtain another vector $\mathbf{x}_1 = \begin{bmatrix} x_1 \\ y_1 \end{bmatrix}$, and then multiplying this in turn by A to obtain the next vector $\mathbf{x}_2 = \begin{bmatrix} x_2 \\ y_2 \end{bmatrix}$, etc. Each step of the iteration process is equivalent to multiplying the present vector $\mathbf{x}_n = \begin{bmatrix} x_n \\ y_n \end{bmatrix}$ by A to obtain the next one $\mathbf{x}_{n+1} = \begin{bmatrix} x_{n+1} \\ y_{n+1} \end{bmatrix}$.

Consequently, since $\mathbf{x}_1 = A\mathbf{x}_0$ and $\mathbf{x}_2 = A\mathbf{x}_1$, then

$$\mathbf{x}_2 = A\mathbf{x}_1 = A \cdot A\mathbf{x}_0 = A^2\mathbf{x}_0$$

Similarly, since $\mathbf{x}_3 = A\mathbf{x}_2$, then

$$\mathbf{x}_3 = A\mathbf{x}_2 = A \cdot A^2\mathbf{x}_0 = A^3\mathbf{x}_0$$

If this process is continued, we see that

$$\mathbf{x}_n = A^n\mathbf{x}_0 \quad \text{for all } n > 0$$

Just as $\mathbf{x}_{n+1} = A\mathbf{x}_n$ resembles $x_{n+1} = ax_n$ from Chapter 2, its solution $\mathbf{x}_n = A^n\mathbf{x}_0$ resembles the solution $x_n = a^nx_0$ of that previous equation. Seeing connections like this between linear equations of one variable and linear systems of several variables is one nice advantage of using vectors and matrices.

Matrix notation allows another observation to easily be made with regard to solutions. Suppose $\mathbf{y}_n$ and $\mathbf{z}_n$ are solutions of $\mathbf{x}_{n+1} = A\mathbf{x}_n$ for different initial points $\mathbf{y}_0$ and $\mathbf{z}_0$. This means that $\mathbf{y}_{n+1} = A\mathbf{y}_n$ and $\mathbf{z}_{n+1} = A\mathbf{z}_n$. For any constants h and k we must have

$$h\,\mathbf{y}_{n+1} = h\,A\mathbf{y}_n = A(h\,\mathbf{y}_n) \quad \text{and} \quad k\,\mathbf{z}_{n+1} = k\,A\mathbf{y}_n = A(k\,\mathbf{y}_n)$$

Adding these equations gives

$$(h\,\mathbf{y}_{n+1} + k\,\mathbf{z}_{n+1}) = A(h\,\mathbf{y}_n) + A(k\,\mathbf{z}_n) = A(h\,\mathbf{y}_n + k\,\mathbf{z}_n)$$

This implies that $h\,\mathbf{y}_n + k\,\mathbf{z}_n$ is another solution of $\mathbf{x}_{n+1} = A\mathbf{x}_n$, the solution whose initial point is $h\,\mathbf{y}_0 + k\,\mathbf{z}_0$. Since h and k are constants, the sum $h\,\mathbf{y}_n + k\,\mathbf{z}_n$ is called a **linear combination** of $\mathbf{y}_n$ and $\mathbf{z}_n$, which are functions of n. What we have shown here is that any linear combination of solutions of a homogeneous linear system $\mathbf{x}_{n+1} = A\mathbf{x}_n$ must also be a solution. This will be important in later sections.

Although we saw earlier that the solution of $\mathbf{x}_{n+1} = A\mathbf{x}_n$ may be written in the form $\mathbf{x}_n = A^n\mathbf{x}_0$, since this does not really provide values for the individual components x_n and y_n of the solution vector $\mathbf{x}_n = \begin{bmatrix} x_n \\ y_n \end{bmatrix}$, that alone will not allow us to determine the behavior of solutions. To uncover this behavior, one needs to look further.

Eigenvectors and Eigenvalues

To determine the dynamics of solutions of $\mathbf{x}_{n+1} = A\mathbf{x}_n$, we begin with what may appear to be just a hypothetical situation. Suppose there exists a non-zero vector $\mathbf{v}$ in the plane such that

$$A\mathbf{v} = r\mathbf{v}$$

for some real number r. That is, multiplying $\mathbf{v}$ by A would have the same effect as if multiplying $\mathbf{v}$ by the scalar r. Graphically, this would mean a stretching or shrinking of that vector $\mathbf{v}$ (if $|r| \neq 1$), and/or a reversing of its direction (if $r < 0$). If such a vector exists, it is called an **eigenvector**, and its corresponding real scalar r is an **eigenvalue** (see Figs. 4.12(a) and 4.12(b)).

Eigenvectors and eigenvalues do indeed exist for many matrices, as the following example shows.

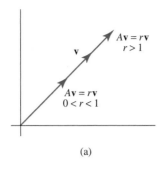

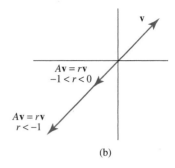

(a)

(b)

FIGURE 4.12

EXAMPLE 2

Show that $\mathbf{v} = \begin{bmatrix} 2 \\ 1 \end{bmatrix}$ and $\mathbf{u} = \begin{bmatrix} 16 \\ 4 \end{bmatrix}$ are eigenvectors of the coefficient matrix of the competition model

$$P_{n+1} = P_n - Q_n$$
$$Q_{n+1} = \tfrac{1}{8} P_n + \tfrac{1}{4} Q_n,$$

and determine each one's corresponding eigenvalue.

Solution: Since $A = \begin{bmatrix} 1 & -1 \\ 1/8 & 1/4 \end{bmatrix}$,   then

$$A\mathbf{v} = \begin{bmatrix} 1 & -1 \\ 1/8 & 1/4 \end{bmatrix} \begin{bmatrix} 2 \\ 1 \end{bmatrix} = \begin{bmatrix} 1 \\ 1/2 \end{bmatrix} = 1/2 \begin{bmatrix} 2 \\ 1 \end{bmatrix} = \tfrac{1}{2}\mathbf{v}$$

and so $\mathbf{v}$ is an eigenvector with eigenvalue $r = 1/2 = 0.5$. Similarly,

$$A\mathbf{u} = \begin{bmatrix} 1 & -1 \\ 1/8 & 1/4 \end{bmatrix} \begin{bmatrix} 16 \\ 4 \end{bmatrix} = \begin{bmatrix} 12 \\ 3 \end{bmatrix} = 3/4 \begin{bmatrix} 16 \\ 4 \end{bmatrix} = \tfrac{3}{4}\mathbf{u}$$

which means $\mathbf{u}$ is an eigenvector with eigenvalue $s = 3/4 = 0.75$.   ∎

A simple computation shows that if $\mathbf{v}$ is an eigenvector of A with eigenvalue r, then for any constant c

$$A \cdot (c\mathbf{v}) = c \cdot A\mathbf{v} = c \cdot r\mathbf{v} = r \cdot (c\mathbf{v})$$

which means that any multiple $c\mathbf{v}$ of an eigenvector $\mathbf{v}$ is also an eigenvector corresponding to the same eigenvalue r. Since, as we saw in the last section, $c\mathbf{v}$ always amounts to either a stretching or shrinking of $\mathbf{v}$, then the set of all eigenvectors associated with any particular real eigenvalue of a matrix generally constitutes a line through the origin.

To see what all this has to do with stability, suppose we first let $\mathbf{x}_0 = \mathbf{v}$, where $\mathbf{v}$ is an eigenvector with eigenvalue r, and then begin iterating the system $\mathbf{x}_{n+1} = A\mathbf{x}_n$. For $n = 0$ this gives

$$\mathbf{x}_1 = A\mathbf{x}_0 = A\mathbf{v} = r\mathbf{v}$$

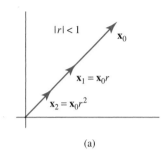
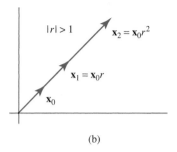

(a)   (b)

FIGURE 4.13

and so $x_1 = rv$. For $n = 1$ this in turn implies

$$x_2 = Ax_1 = A \cdot rv = r \cdot Av = r \cdot rv$$

which means $x_2 = r^2 v$. A pattern may be seen emerging that can be verified for all subsequent iterates, namely $x_n = r^n v$, which we prefer to write as $x_n = vr^n$.

With this it is possible to determine the asymptotic behavior of x_n. Since all $x_n = vr^n = x_0 r^n$, then the length of the solution vector x_n either goes to 0 if $|r| < 1$, or goes to ∞ if $|r| > 1$ (see Figs. 4.13(a) and 4.13(b)). Note the similarity with the behavior of solutions $x_n = a^n x_0$ of the single variable equation $x_{n+1} = ax_n$ from Chapter 2. (As before, the rare borderline case $r = \pm 1$ will not be considered here.) Thus it is really the magnitude of the eigenvalue, and not the eigenvector, that is important when determining the asymptotic behavior.

EXAMPLE 3

For the competition model of Example 2 determine the asymptotic behavior of the solution with $P_0 = 4000$ and $Q_0 = 2000$.

Solution: Here we will call $x_n = \begin{bmatrix} P_n \\ Q_n \end{bmatrix}$ for all $n \geq 0$. In Example 2 it was found that $v = \begin{bmatrix} 2 \\ 1 \end{bmatrix}$ is an eigenvector of the coefficient matrix A for this system with corresponding eigenvalue $r = 0.5$. Since any multiple of an eigenvector is again an eigenvector, this means that

$$x_0 = \begin{bmatrix} P_0 \\ Q_0 \end{bmatrix} = \begin{bmatrix} 4000 \\ 2000 \end{bmatrix} = 2000 \begin{bmatrix} 2 \\ 1 \end{bmatrix} = 2000v$$

must also be an eigenvector associated with $r = 0.5$. So the solution can be written $x_n = x_0(0.5)^n$. Since $|0.5| < 1$, this implies x_n converges to 0 as $n \to \infty$, which means that both populations P_n and Q_n will decay to 0. We remark that, although a certain long-term asymptotic behavior is predicted in this example, this may not really reflect the true dynamics if either population eventually becomes negative.   ∎

Two Real Eigenvalues

The technique we have described so far for determining the asymptotic behavior of solutions of $\mathbf{x}_{n+1} = A\mathbf{x}_n$ applies only to the special case in which $\mathbf{x}_0$ is an eigenvector. But most initial points do not correspond to eigenvectors. So one may ask: *How do we determine the dynamics of most solutions?*

This may be partially answered by first observing that the coefficient matrix from Example 2 has two different real eigenvalues. This is common for 2 by 2 matrices, although not always the case. For now, let us assume that the matrix A does have two distinct real eigenvalues r and s. Although we shall not prove so here, it will always be true that any vector $\mathbf{x}_0$ in the plane can consequently be written as the sum of at most two corresponding eigenvectors. That is, $\mathbf{x}_0 = \mathbf{v} + \mathbf{u}$, where either $\mathbf{v}$ and $\mathbf{u}$ are eigenvectors corresponding to r and s respectively, or at least one of them equals $\mathbf{0}$. Further, if they are both eigenvectors, they must lie along different lines. That is, neither is a multiple of the other.

Since $\mathbf{x}_0 = \mathbf{v} + \mathbf{u}$, then we must have

$$\mathbf{x}_1 = A\mathbf{x}_0 = A(\mathbf{v} + \mathbf{u}) = A\mathbf{v} + A\mathbf{u} = r\,\mathbf{v} + s\,\mathbf{u}$$

and so we see that $\mathbf{x}_1 = r\mathbf{v} + s\mathbf{u}$. Similarly,

$$\mathbf{x}_2 = A\mathbf{x}_1 = A(r\mathbf{v} + s\mathbf{u}) = A(r\mathbf{v}) + A(s\mathbf{u})$$

$$= r(A\mathbf{v}) + s(A\mathbf{u}) = r \cdot r\,\mathbf{v} + s \cdot s\,\mathbf{u}$$

So, $\mathbf{x}_2 = r^2\mathbf{v} + s^2\mathbf{u}$. From this a general formula for the exact solution may be seen: $\mathbf{x}_n = r^n\mathbf{v} + s^n\mathbf{u}$ for all $n \geq 0$, which we prefer to write as

$$\mathbf{x}_n = \mathbf{v}\,r^n + \mathbf{u}\,s^n \tag{1}$$

We now have a way of determining the stability of the fixed point $\mathbf{0}$ of any homogeneous linear system $\mathbf{x}_{n+1} = A\mathbf{x}_n$ when A has two real and distinct eigenvalues.

THEOREM *Stability: Two Real Eigenvalues*

Suppose the coefficient matrix of a homogeneous linear system has two real distinct eigenvalues r and s. Then the fixed point $(0, 0)$ of the system is

(i) a sink if both $|r| < 1$ and $|s| < 1$, or
(ii) a source if both $|r| > 1$ and $|s| > 1$, or
(iii) a saddle if $|r| < 1$ and $|s| > 1$, or if $|r| > 1$ and $|s| < 1$.

Proof: The conclusion of part *(i)* is fairly evident. If $|r| < 1$ and $|s| < 1$, then both $r^n, s^n \to 0$ as $n \to \infty$. Consequently, since $\mathbf{v}$ and $\mathbf{u}$ are constant vectors, all solutions $\mathbf{x}_n = \mathbf{v}\,r^n + \mathbf{u}\,s^n$ must approach $\mathbf{0}$ as $n \to \infty$, indicating that $\mathbf{0}$ is a sink.

For part *(ii)*, since $|r| > 1$ and $|s| > 1$, then both $|r|^n, |s|^n \to \infty$ as $n \to \infty$. If either $\mathbf{v}$ or $\mathbf{u}$ equals $\mathbf{0}$ then $\mathbf{x}_0$ is itself an eigenvector and, as seen earlier, the length of the solution vector $\mathbf{x}_n$ approaches ∞ (except for the fixed-point solution $\mathbf{x}_n = \mathbf{0}$). On the other hand, if both $\mathbf{v} \neq \mathbf{0}$ and $\mathbf{u} \neq \mathbf{0}$, assume $|s| \leq |r|$. (The reverse is argued similarly.) In this case,

$$\mathbf{x}_n = \mathbf{v}\,r^n + \mathbf{u}\,s^n = r^n(\mathbf{v} + \mathbf{u}(s/r)^n)$$

Since $(s/r)^n$ approaches either 0 or ± 1 as $n \to \infty$, then $\mathbf{v} + \mathbf{u}(s/r)^n$ approaches either $\mathbf{v}$ or $\mathbf{v} \pm \mathbf{u}$. None of these three vectors can equal $\mathbf{0}$ since we assumed $\mathbf{v} \neq \mathbf{0}$, and $\mathbf{v} \pm \mathbf{u} = \mathbf{0}$ would imply that $\mathbf{v} = \mp\mathbf{u}$ or that $\mathbf{v}$ is a multiple of $\mathbf{u}$, which cannot happen if r and s are distinct. And so as $n \to \infty$, the length of $\mathbf{x}_n$ approaches ∞. Since the lengths of all nonzero solutions approach ∞, $\mathbf{0}$ must therefore be a source.

Part *(iii)* is a bit trickier to see. Suppose we assume $|r| < 1$ and $|s| > 1$. (The reverse is argued similarly.) In this case $\mathbf{x}_n = \mathbf{v}\,r^n + \mathbf{u}\,s^n$ has one part $\mathbf{v}\,r^n$ converging to $\mathbf{0}$ while the length of the other $\mathbf{u}\,s^n$ diverges to ∞. In such a situation, solutions diverge from the line through $\mathbf{0}$ along which $\mathbf{v}$ lies, and approach the line through $\mathbf{0}$ along which $\mathbf{u}$ lies. All solutions have hyperbolic trajectories except for those exactly on those lines, which are the asymptotes. This makes $\mathbf{0}$ a saddle. ∎

The following example demonstrates the existence of these hyperbolic trajectories.

EXAMPLE 4

Graph the trajectory of $\mathbf{x}_n = \mathbf{v}\,r^n + \mathbf{u}\,s^n$ if (a) $r = 1.5$, $s = 0.9$, $\mathbf{v} = \begin{bmatrix} 1 \\ 1 \end{bmatrix}$ and $\mathbf{u} = \begin{bmatrix} -100 \\ 100 \end{bmatrix}$;

(b) $r = -1.25$, $s = 0.75$, $\mathbf{v} = \begin{bmatrix} 1 \\ 0 \end{bmatrix}$ and $\mathbf{u} = \begin{bmatrix} 0 \\ 8 \end{bmatrix}$

Solution: The trajectory for (a), shown in Figure 4.14(a), has the shape of a hyperbola with center at the origin and asymptotes $y = \pm x$. This resembles what we saw earlier in Example 6 of Section 4.2, where the origin was a saddle. The trajectory for (b), shown in Figure 4.14(b), is slightly different. Because of the negative eigenvalue $r = -1.25$, the solution jumps back and forth between two adjacent hyperbolas as it diverges from one asymptote, the y axis, and converges to the other, the x axis. ∎

EXAMPLE 5

For the system from Example 2: (a) Find the unique solution that satisfies $P_0 = 1000$ and $Q_0 = 300$. (b) Find the asymptotic behavior of all solutions.

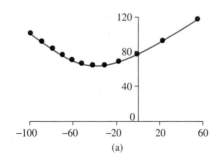

(a)

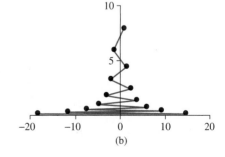

(b)

FIGURE 4.14

Solution: (a) Although we are given $P_0 = 1000$ and $Q_0 = 300$, to find a formula for the unique solution we will also need P_1 and Q_1, which can be determined by letting $n = 0$ in the original system:

$$P_1 = P_0 - Q_0 = 700 \quad \text{and} \quad Q_1 = \frac{1}{8}P_0 + \frac{1}{4}Q_0 = 200$$

Next, since we previously found that $r = 0.5$ and $s = 0.75$ are the eigenvalues of this system, then from (1) all solutions $\mathbf{x}_n = \begin{bmatrix} P_n \\ Q_n \end{bmatrix}$ have the form

$$\mathbf{x}_n = \mathbf{v}\,(0.5)^n + \mathbf{u}\,(0.75)^n \quad \text{or equivalently} \qquad \begin{aligned} P_n &= v_1(0.5)^n + u_1(0.75)^n \\ Q_n &= v_2(0.5)^n + u_2(0.75)^n \end{aligned}$$

where $\mathbf{v} = \begin{bmatrix} v_1 \\ v_2 \end{bmatrix}$ and $\mathbf{u} = \begin{bmatrix} u_1 \\ u_2 \end{bmatrix}$ are eigenvectors. To find the constants v_1 and u_1, note that letting $n = 0$ and $n = 1$ in $P_n = v_1(0.5)^n + u_1(0.75)^n$ respectively gives

$$P_0 = v_1 + u_1 \quad \text{and} \quad P_1 = 0.5v_1 + 0.75u_1$$

But since we were given that $P_0 = 1000$ and we found above that $P_1 = 700$, then

$$v_1 + u_1 = 1000 \quad \text{and} \quad 0.5v_1 + 0.75u_1 = 700$$

This is a set of two simultaneous equations for the unknowns v_1 and u_1, which can easily be solved giving $v_1 = 200$ and $u_1 = 800$. The solution for P_n is therefore

$$P_n = v_1(0.5)^n + u_1(0.75)^n = 200(0.5)^n + 800(0.75)^n$$

A similar set of calculations would show that v_2 and u_2 satisfy

$$v_2 + u_2 = 300 \quad \text{and} \quad 0.5v_2 + 0.75u_2 = 200$$

whose solution is $v_2 = 100$ and $u_2 = 200$, making

$$Q_n = v_2(0.5)^n + u_2(0.75)^n = 100(0.5)^n + 200(0.75)^n$$

The unique solution of the system that satisfies $P_0 = 1000$ and $Q_0 = 300$ is therefore

$$P_n = 200(0.5)^n + 800(0.75)^n \quad \text{and} \quad Q_n = 100(0.5)^n + 200(0.75)^n$$

(b) From (1) the general form of all solutions is $\mathbf{x}_n = \mathbf{v}\,(1/2)^n + \mathbf{u}\,(3/4)^n$, where $\mathbf{v}$ and $\mathbf{u}$ are eigenvectors. Since $|1/2| < 1$ and $|3/4| < 1$ then all solutions $\mathbf{x}_n \to \mathbf{0}$ as $n \to \infty$. The origin is a sink. ∎

One may notice from part (b) of this example how much quicker it is to determine the dynamics of *all* solutions of a linear system, when compared to finding in part (a) just *one* exact solution. Because so much more is gained with much less effort, the dynamical-systems approach described here is often the preferred method of analysis of such systems.

Computing Eigenvalues

Although we have been discussing eigenvalues and their importance in determining stability, we have not yet demonstrated a method for computing them. Using a theorem from the previous section, however, this turns out to be relatively easy.

Recall from the definition of an eigenvalue r of the matrix A, we assume $A\mathbf{v} = r\mathbf{v}$ for some *non-zero* eigenvector $\mathbf{v}$. Using some matrix algebra and properties of the identity matrix I, this equation can be written

$$A\mathbf{v} - r\mathbf{v} = \mathbf{0} \quad \text{or} \quad A\mathbf{v} - rI\mathbf{v} = \mathbf{0}$$

If we factor $\mathbf{v}$ from that last equation, then we have

$$(A - rI)\mathbf{v} = \mathbf{0}$$

This equation now represents a homogeneous system of simultaneous equations. According to the theorem from Section 4.3, a non-zero vector solution (in this case an eigenvector $\mathbf{v}$) exists if and only if the determinant of the corresponding coefficient matrix $A - rI$ is equal to 0.

To see what this implies, first note that if $A = \begin{bmatrix} a & b \\ c & d \end{bmatrix}$ then

$$A - rI = \begin{bmatrix} a & b \\ c & d \end{bmatrix} - r\begin{bmatrix} 1 & 0 \\ 0 & 1 \end{bmatrix} = \begin{bmatrix} a & b \\ c & d \end{bmatrix} - \begin{bmatrix} r & 0 \\ 0 & r \end{bmatrix} = \begin{bmatrix} a-r & b \\ c & d-r \end{bmatrix}$$

Since the determinant of $A - rI$ is $(a - r)(d - r) - bc$, the eigenvalues of A are the values of r that satisfy the quadratic equation $(a - r)(d - r) - bc = 0$.

We have derived the following.

EIGENVALUE FORMULA

The eigenvalues of a matrix $A = \begin{bmatrix} a & b \\ c & d \end{bmatrix}$ are the values of r that make the determinant of $A - rI = \begin{bmatrix} a-r & b \\ c & d-r \end{bmatrix}$ equal to 0. They are found by solving $(a - r)(d - r) - bc = 0$.

EXAMPLE 6

Find the eigenvalues of $A = \begin{bmatrix} 1 & -1 \\ -1 & 2 \end{bmatrix}$ and use them to determine the stability of $\mathbf{x}_{n+1} = A\mathbf{x}_n$.

Solution: Letting the determinant of $A - rI = \begin{bmatrix} 1-r & -1 \\ -1 & 2-r \end{bmatrix}$ equal 0 gives

$$(1 - r)(2 - r) - 1 = 0 \quad \text{or} \quad r^2 - 3r + 1 = 0$$

The solutions are $r = (3 \pm \sqrt{5})/2$, or approximately 0.382 and 2.618. Since these eigenvalues satisfy $|0.382| < 1$ and $|2.618| > 1$, then according to the above theorem the origin must be a saddle. ■

■■■   **EXAMPLE 7**

Show that $A = \begin{bmatrix} 1 & -1 \\ 2 & 1 \end{bmatrix}$ has no real eigenvalues.

Solution: Since $A - rI = \begin{bmatrix} 1-r & -1 \\ 2 & 1-r \end{bmatrix}$, then letting its determinant equal 0 yields $(1-r)^2 + 2 = 0$ or $r^2 - 2r + 3 = 0$. Using the quadratic formula this time gives $(2 \pm \sqrt{-8})/2$, which indicates there are no real solutions and therefore no real eigenvalues.   ■

■■■   **EXAMPLE 8**

Determine the stability of the fixed point $(0, 0)$ for the competition model

$$P_{n+1} = 0.7P_n - 0.05Q_n$$
$$Q_{n+1} = -0.05P_n + 0.8Q_n$$

which was investigated in Example 3 of Section 4.2.

Solution: Here the coefficient matrix is $A = \begin{bmatrix} 0.7 & -0.05 \\ -0.05 & 0.8 \end{bmatrix}$, which means that $A - rI = \begin{bmatrix} 0.7-r & -0.05 \\ -0.05 & 0.8-r \end{bmatrix}$. Letting the determinant equal 0 gives

$$(0.7-r)(0.8-r) - 0.0025 = 0 \quad \text{or} \quad r^2 - 1.5r + 0.5575 = 0$$

Since both solutions $(1.5 \pm \sqrt{0.02})/2 \approx 0.68, \ 0.82$ of this quadratic equation are between -1 and $+1$, these eigenvalues indicate that the origin is a sink. Both populations P_n and Q_n must therefore always decay to 0, confirming the dynamics indicated in Table 4.2 and Figure 4.2 (in Section 4.2). We remark once again, however, that in general the model breaks down if either population becomes negative.   ■

Exercises 4.4 _____

*In Exercises 1–4 verify that **u** and **v** are eigenvectors of the given matrix A and find their corresponding eigenvalues.*

1. $A = \begin{bmatrix} -1 & 5 \\ 0.5 & 0.5 \end{bmatrix}$, $\mathbf{u} = \begin{bmatrix} 5 \\ -1 \end{bmatrix}$, $\mathbf{v} = \begin{bmatrix} 2 \\ 1 \end{bmatrix}$

2. $A = \begin{bmatrix} 0.8 & 1.2 \\ 1 & 1 \end{bmatrix}$, $\mathbf{u} = \begin{bmatrix} 6 \\ -5 \end{bmatrix}$, $\mathbf{v} = \begin{bmatrix} 3 \\ 3 \end{bmatrix}$

3. $A = \begin{bmatrix} 4 & 3 \\ 1 & 2 \end{bmatrix}$, $\mathbf{u} = \begin{bmatrix} 2 \\ -2 \end{bmatrix}$, $\mathbf{v} = \begin{bmatrix} 1 \\ 1/3 \end{bmatrix}$

4. $A = \begin{bmatrix} 5 & 3 \\ -2 & -0.5 \end{bmatrix}$, $\mathbf{u} = \begin{bmatrix} 3 \\ -4 \end{bmatrix}$, $\mathbf{v} = \begin{bmatrix} -1 \\ 0.5 \end{bmatrix}$

In Exercises 5 and 6, given the matrix A and initial vector x_0: (a) Use $x_{n+1} = Ax_n$ to compute x_1 and x_2; (b) Find a formula for x_n.

5. $A = \begin{bmatrix} 4.5 & -1 \\ -1.5 & 2 \end{bmatrix}$, $x_0 = \begin{bmatrix} 1000 \\ 3000 \end{bmatrix}$ 　　6. $A = \begin{bmatrix} 1.75 & -1 \\ -0.5 & 1.25 \end{bmatrix}$, $x_0 = \begin{bmatrix} 2500 \\ 2500 \end{bmatrix}$

In Exercises 7–10 compute the eigenvalues of the given matrix.

7. $\begin{bmatrix} 8 & -1 \\ 2 & 5 \end{bmatrix}$ 　　8. $\begin{bmatrix} 2 & 2 \\ 1 & 2 \end{bmatrix}$ 　　9. $\begin{bmatrix} 2 & 3 \\ 1 & 1 \end{bmatrix}$ 　　10. $\begin{bmatrix} 1.5 & 0.5 \\ -0.5 & 0.25 \end{bmatrix}$

In Exercises 11–18, given the matrix A, determine whether the origin is a sink, source or saddle of $x_{n+1} = Ax_n$.

11. The matrix A from Exercise 1. 　　15. $A = \begin{bmatrix} 2.25 & 2.5 \\ 0.75 & 2.0 \end{bmatrix}$

12. The matrix A from Exercise 2. 　　16. $A = \begin{bmatrix} 1/4 & -1/8 \\ -1/2 & 3/4 \end{bmatrix}$

13. The matrix A from Exercise 7. 　　17. $A = \begin{bmatrix} 0.5 & -0.75 \\ -0.25 & 0.5 \end{bmatrix}$

14. The matrix A from Exercise 8. 　　18. $A = \begin{bmatrix} 1 & 1 \\ 3 & 1 \end{bmatrix}$

19. Determine all positive values of the constant t for which the origin is a sink for the system $\begin{cases} x_{n+1} = 0.5x_n + ty_n \\ y_{n+1} = x_n \end{cases}$

20. Repeat the previous exercise for the system $\begin{cases} x_{n+1} = tx_n + 0.18y_n \\ y_{n+1} = 2x_n + ty_n \end{cases}$

The eigenvectors of a matrix A are all solutions $v = \begin{bmatrix} x \\ y \end{bmatrix}$ of $(A - rI)v = 0$, where r is an eigenvalue. For each of the matrices A given in Exercises 21–24 find the eigenvalues and the set of eigenvectors associated with each. Graph each set of eigenvectors.

21. $A = \begin{bmatrix} 2 & 1 \\ 1 & 2 \end{bmatrix}$ 　　22. $A = \begin{bmatrix} 6 & 1 \\ -2 & 3 \end{bmatrix}$ 　　23. $A = \begin{bmatrix} 1 & 2 \\ 1 & 1 \end{bmatrix}$ 　　24. $A = \begin{bmatrix} 0.5 & 5 \\ 1 & 0 \end{bmatrix}$

For each of the matrices A given in Exercises 25–28: (a) Find the general form of all solutions $x_n = \begin{bmatrix} x_n \\ y_n \end{bmatrix}$ of $x_{n+1} = Ax_n$; (b) Find the unique solution x_n that satisfies the given initial values x_0 and y_0; (c) Compute $\lim_{n \to \infty} x_n$ and $\lim_{n \to \infty} y_n$ if they exist.

25. The matrix A from Exercise 1 with $x_0 = 0$, $y_0 = 2$.

26. The matrix A from Exercise 6 with $x_0 = 800$, $y_0 = 400$.

27. The matrix A from Exercise 7 with $x_0 = 250$, $y_0 = 200$.

28. The matrix A from Exercise 24 with $x_0 = 20$, $y_0 = 10$.

29. Suppose that a matrix A has nonzero eigenvalues r and s. Determine the eigenvalues of (a) cA for any constant c; (b) A^{-1}.

30. (a) Show that a matrix has 0 as an eigenvalue if and only if its determinant is 0.

    (b) Show that a matrix has 0 as an eigenvalue if and only if it has no inverse.

Computer Projects 4.4

The form of the solution (1) implies that x_n and y_n grow or decay exponentially as $n \to \infty$ *with base equal to the larger of $|r|$ or $|s|$. Verify this for each of the coefficient matrices A given in Projects 1 and 2, by computing $\left|\frac{x_{n+1}}{x_n}\right|$ and $\left|\frac{y_{n+1}}{y_n}\right|$ for $n = 0, \ldots, 50$ for several choices of (x_0, y_0). Compare these ratios with the eigenvalues r and s.*

1. The matrix A from Exercise 2.

2. $A = \begin{bmatrix} 0.25 & 0.5 \\ 0.5 & -0.25 \end{bmatrix}$

Verify that the origin is a saddle of the system $\boldsymbol{x}_{n+1} = A\boldsymbol{x}_n$ for each of the matrices A given in Projects 3 and 4 by computing the eigenvalues. Then show that hyperbolic trajectories exist by generating and plotting on the same graph (x_n, y_n) for $n = 0, \ldots, 4$ for each of the given initial points.

3. $A = \begin{bmatrix} 1.5 & 0.5 \\ 2 & 1.5 \end{bmatrix}$ for (x_0, y_0) equal to (a) $(-10, 22)$; (b) $(-11, 20)$; (c) $(10, -22)$;
   (d) $(11, -20)$.

4. $A = \begin{bmatrix} 2 & -1 \\ -2 & 1.5 \end{bmatrix}$ for (x_0, y_0) equal to (a) $(11, 21)$; (b) $(14, 21)$; (c) $(-11, -21)$;
   (d) $(-14, -21)$.

The origin is a saddle of the system $\boldsymbol{x}_{n+1} = A\boldsymbol{x}_n$ for each of the matrices A given in Projects 5–7, but now A has at least one negative eigenvalue. Verify this, and then generate and graph (x_n, y_n) for $n = 0, \ldots, 9$ for each of the given initial points. Are trajectories still hyperbolic?

5. $A = \begin{bmatrix} -1 & 2 \\ 1 & -1 \end{bmatrix}$ for (x_0, y_0) equal to (a) $(4.25, 3)$; (b) $(-4.25, -3)$.

6. $A = \begin{bmatrix} 3/8 & 9/8 \\ 9/8 & 3/8 \end{bmatrix}$ for (x_0, y_0) equal to (a) $(-10, 11)$; (b) $(10, -11)$.

7. $A = \begin{bmatrix} -1 & -0.25 \\ -0.25 & -1 \end{bmatrix}$ for (x_0, y_0) equal to (a) $(10, -8)$; (b) $(-8, 10)$.

SECTION
4.5 Repeated Real Eigenvalues

In the previous section we learned the importance of eigenvalues in determining the stability of linear systems. When the coefficient matrix corresponding to a homogeneous linear system has two real and distinct eigenvalues, the nature of its fixed point can be completely determined using the stability theorem derived there. But that theorem does not apply to all homogeneous systems. It is common for a coefficient matrix to have only one real eigenvalue, or perhaps even none at all. So how are the equilibria of such systems to be dealt with?

In this and the next several sections, we attempt to answer that question by developing stability criteria for these more complicated situations. Here, we consider systems in which the coefficient matrix has only one eigenvalue.

Repeated Eigenvalues

Recall that when two real and distinct eigenvalues r and s exist for a matrix A, then any initial vector $\mathbf{x}_0$ in the plane can be written as the sum of two corresponding eigenvectors: $\mathbf{x}_0 = \mathbf{v} + \mathbf{u}$. The formula for the exact solution of the system $\mathbf{x}_{n+1} = A\mathbf{x}_n$ can then be obtained: $\mathbf{x}_n = \mathbf{v}r^n + \mathbf{u}s^n$. In this way, the dynamics of *all* solutions of the system can be determined.

The problem we are confronted with now, however, is that sometimes when we use the Eigenvalue Formula of the previous section, we find that equating the determinant of $A - rI$ to 0 yields only one solution r. This happens when the corresponding quadratic equation we solve has a single *repeated* real root r, representing a unique eigenvalue. The difficulty stems from the fact that now most vectors $\mathbf{x}_0$ in the plane cannot be written as the sum of eigenvectors, and so the dynamics of most solutions cannot be determined as before. To establish these dynamics, a whole different approach is called for.

Before taking on this problem, it is worth first recognizing the special relationships that the components of $A = \begin{bmatrix} a & b \\ c & d \end{bmatrix}$ must satisfy when only one real eigenvalue r exists. Since the determinant of $A - rI$ must be 0 then $(a - r)(d - r) - bc = 0$, or

$$r^2 - (a + d)r + ad - bc = 0 \tag{1}$$

Although the quadratic formula therefore gives us

$$r = \frac{a + d \pm \sqrt{(a + d)^2 - 4(ad - bc)}}{2}$$

for this to have only one solution it must be that $(a + d)^2 - 4(ad - bc) = 0$, which means

$$r = \frac{a + d}{2} \quad \text{or} \quad 2r - (a + d) = 0 \tag{2}$$

EXAMPLE 1

Show that $A = \begin{bmatrix} 4 & -1 \\ 1 & 2 \end{bmatrix}$ has a single repeated eigenvalue, and that (1) and (2) are true.

Solution: Letting the determinant of $A - rI$ equal 0 gives

$$(4 - r)(2 - r) - 1 \cdot (-1) = 0 \quad \text{or} \quad r^2 - 6r + 9 = 0$$

which corresponds to (1). The unique solution and eigenvalue is $r = 3$, which also satisfies (2).   ∎

We remark that in the event that the repeated eigenvalue is $r = 0$, it can be shown that all solutions satisfy $\mathbf{x_n} = \mathbf{0}$ for all $n \geq 2$. This case need not be discussed further since it implies that the origin must always be a stable sink for the system $\mathbf{x}_{n+1} = A\mathbf{x}_n$.

Solving the Uncoupled Equation

The most convenient approach to investigating the behavior of solutions of $\mathbf{x}_{n+1} = A\mathbf{x}_n$ when $A = \begin{bmatrix} a & b \\ c & d \end{bmatrix}$ has a single real eigenvalue is to uncouple the system. Recall from Section 4.2 that the first component x_n of the solution vector $\mathbf{x_n} = \begin{bmatrix} x_n \\ y_n \end{bmatrix}$ must satisfy a second-order equation that may be written

$$x_{n+2} - (a + d) x_{n+1} + (a d - b c) x_n = 0 \tag{3}$$

Before determining the dynamics of x_n we must first find what form solutions of this equation take when A has only one real eigenvalue. Some motivation might be found by observing the form that solutions have when two real eigenvalues r and s exist. We previously found in this case that all solutions of $\mathbf{x}_{n+1} = A\mathbf{x}_n$ have the form $\mathbf{x}_n = \mathbf{v}r^n + \mathbf{u}s^n$, where $\mathbf{v}$ and $\mathbf{u}$ are corresponding eigenvectors, or

$$\mathbf{x}_n = \begin{bmatrix} x_n \\ y_n \end{bmatrix} = \begin{bmatrix} v_1 \\ v_2 \end{bmatrix} r^n + \begin{bmatrix} u_1 \\ u_2 \end{bmatrix} s^n = \begin{bmatrix} v_1 r^n + u_1 s^n \\ v_2 r^n + u_2 s^n \end{bmatrix}$$

where v_1, v_2, u_1, u_2 are constants. Looking at just the first component of this vector equation gives the general form that all solutions of (3) must have in this case:

$$x_n = v_1 r^n + u_1 s^n$$

Because v_1 and u_1 are constants, this expression for x_n is once again called a *linear combination* of r^n and s^n, which are functions of n.

In the present situation, however, only one eigenvalue r exists. One might nevertheless assume that the term $v_1 r^n$ still describes at least some solutions x_n of (3), but as there is no other eigenvalue s this time, what term replaces $u_1 s^n$ in that linear combination? Or is another term even needed?

It is immediately evident that unless $r = 0$ a term other than $v_1 r^n$ is essential to truly describe all solutions of (3). This can be seen by observing that letting $n = 0$ and $n = 1$ in $x_n = v_1 r^n$ would imply respectively that

$$x_0 = v_1 r^0 = v_1 \quad \text{and} \quad x_1 = v_1 r^1 = v_1 r$$

which leads to $x_1 = r x_0$. Consequently, only those rare solutions of (3) whose initial conditions satisfy $x_1 = r x_0$ can be written as $x_n = v_1 r^n$. So this is not the form of most solutions.

To uncover the correct form, we must engage in some creative *guesswork,* and make a tentative assumption: Since some solutions of (3) are found by multiplying r^n by a constant v_1 to obtain $v_1 r^n$, maybe others can be found by multiplying this same r^n by something else — perhaps a simple function of n. The simplest of these is a linear function $u_1 n$ where u_1 is a constant, which would produce new solutions of the form $u_1 n r^n$. This method of generating additional solutions is sometimes called **variation of parameters.**

If we construct linear combinations using our new proposed set of solutions $u_1 n r^n$ along with the previous ones $v_1 r^n$, we arrive at the general form we might expect *all* solutions of (3) to have when only one real non-zero eigenvalue r exists:

$$x_n = v_1 r^n + u_1 n r^n$$

where v_1 and u_1 are constants. However, at present this is just an assumption and must be borne out before proceeding. To prove this assertion we first show that individually both $x_n = r^n$ and $x_n = n r^n$ are solutions of (3). Since we learned in the previous section that all linear combinations of solutions are again solutions, then $x_n = v_1 r^n + u_1 n r^n$ would consequently also be a solution for any constants v_1 and u_1.

To show that $x_n = r^n$ is a solution of (3), we first compute $x_{n+1} = r^{n+1}$ and $x_{n+2} = r^{n+2}$, and then substitute all three expressions into (3). This yields

$$r^{n+2} - (a + d) r^{n+1} + (a d - b c) r^n = 0$$

Dividing by r^n gives

$$r^2 - (a + d) r + (a d - b c) = 0$$

which is obviously true since r is the eigenvalue that satisfies (1).

Next, to see that $x_n = n r^n$ also solves (3), we substitute that expression along with $x_{n+1} = (n + 1) r^{n+1}$ and $x_{n+2} = (n + 2) r^{n+2}$ into (3), which then gives

$$(n + 2) r^{n+2} - (a + d)(n + 1) r^{n+1} + (a d - b c) n r^n = 0$$

Dividing by r^n, this becomes

$$(n + 2) r^2 - (a + d)(n + 1) r + (a d - b c) n = 0$$

Using the distributive law and then rearranging some terms, this can be written

$$n r^2 - (a + d) n r + (ad - bc) n + 2 r^2 - (a + d) r = 0$$

and factoring finally yields

$$n (r^2 - (a + d) r + ad - bc) + r (2r - (a + d)) = 0$$

Again using (1) and this time also (2), which says that $2r - (a + d) = 0$ when the eigenvalue r is repeated, we see that both terms on the left-hand side of this equation equal 0, making the equation true.

This confirms that both $x_n = r^n$ and $x_n = n r^n$ are solutions of (3), and therefore so too are all linear combinations of them $x_n = v_1 r^n + u_1 n r^n$. Although we shall not do so here, it can be proven that *all* solutions of (3) must in fact have this form. So we have successfully *guessed* the correct form of all solutions to that equation.

Since y_n satisfies the same second-order equation as x_n, its solution must also be a linear combination of r^n and $n r^n$, although most likely with different constants. This means that y_n has the form

$$y_n = v_2 r^n + u_2 n r^n$$

where v_2 and u_2 are constants.

EXAMPLE 2

For the prey–predator model from Example 1 of Section 4.2

$$P_{n+1} = 1.5 P_n - 0.5 Q_n$$
$$Q_{n+1} = 0.5 P_n + 2.5 Q_n$$

(a) Find the form of all solutions for P_n and Q_n.
(b) Find the unique solution that satisfies $P_0 = 2000$ and $Q_0 = 1000$.

Solution: (a) It can be checked that the coefficient matrix of the system $A = \begin{bmatrix} 1.5 & -0.5 \\ 0.5 & 2.5 \end{bmatrix}$
has a unique eigenvalue $r = 2$. So the solution must take the form

$$P_n = v_1 2^n + u_1 n\, 2^n \quad \text{and} \quad Q_n = v_2 2^n + u_2 n\, 2^n$$

where v_1, v_2, u_1, u_2 are constants.
   (b) To find the unique solution for $P_0 = 2000$ and $Q_0 = 1000$, we first compute

$$P_1 = 1.5 P_0 - 0.5 Q_0 = 2500 \quad \text{and} \quad Q_1 = 0.5 P_0 + 2.5 Q_0 = 3500$$

using the original system. We then find the constants v_1 and u_1 in the equation $P_n = v_1 2^n + u_1 n\, 2^n$ that satisfy $P_0 = 2000$ and $P_1 = 2500$. Letting $n = 0$ gives $v_1 = 2000$, and letting $n = 1$ gives $P_1 = 2v_1 + 2u_1$. Using $P_1 = 2500$ and $v_1 = 2000$, this equation becomes $2500 = 4000 + 2 u_1$, which means $u_1 = -750$. With $v_1 = 2000$ and $u_1 = -750$ the exact solution for P_n must be

$$P_n = v_1 2^n + u_1 n\, 2^n = 2000 \cdot 2^n - 750 \cdot n\, 2^n$$

Using similar reasoning, we find $v_2 = 1000$ and $u_2 = 750$, and so

$$Q_n = v_2 2^n + u_2 n\, 2^n = 1000 \cdot 2^n + 750 \cdot n\, 2^n$$ ∎

Repeated Eigenvalues and Stability

We can now return to the central question of the section: *How do we determine stability for* $\mathbf{x}_{n+1} = A\mathbf{x}_n$ *when A has only one eigenvalue r?* Since we have seen that the components of $\mathbf{x}_n = \begin{bmatrix} x_n \\ y_n \end{bmatrix}$ have the forms

$$x_n = v_1 r^n + u_1 n\, r^n \quad \text{and} \quad y_n = v_2 r^n + u_2 n\, r^n$$

where v_1, v_2, u_1, u_2 are constants, the solution vector $\mathbf{x}_n$ has the form

$$\mathbf{x}_n = \begin{bmatrix} x_n \\ y_n \end{bmatrix} = \begin{bmatrix} v_1 r^n + u_1 n\, r^n \\ v_2 r^n + u_2 n\, r^n \end{bmatrix} = \begin{bmatrix} v_1 \\ v_2 \end{bmatrix} r^n + \begin{bmatrix} u_1 \\ u_2 \end{bmatrix} n\, r^n$$

Thus the solution of $\mathbf{x}_{n+1} = A\mathbf{x}_n$ when only one non-zero eigenvalue r exists can always be written

$$\mathbf{x}_n = \mathbf{v} r^n + \mathbf{u} n\, r^n$$

where $\mathbf{v} = \begin{bmatrix} v_1 \\ v_2 \end{bmatrix}$ and $\mathbf{u} = \begin{bmatrix} u_1 \\ u_2 \end{bmatrix}$ are vectors, although not necessarily eigenvectors as before.

As a result of all this, stability criteria for the equilibrium of any homogeneous linear system can now easily be established for this case.

THEOREM   *Stability: One Repeated Real Eigenvalue*

Suppose the coefficient matrix of a homogeneous linear system has only one repeated real eigenvalue r. Then the fixed point $(0, 0)$ of the system is:

(i) a sink if $|r| < 1$; or
(ii) a source if $|r| > 1$.

Proof: For part *(i)* since $|r| < 1$ then we know that $r^n \to 0$ as $n \to \infty$. But now we also make use of the fact that $nr^n \to 0$ as well, when $|r| < 1$. (L'Hopital's Rule is helpful in seeing this.) Consequently, for any solution $\mathbf{x}_n = \mathbf{v}\,r^n + \mathbf{u}\,n\,r^n$, each of its terms approaches $\mathbf{0}$ as $n \to \infty$. So $\mathbf{x}_n$ must also decay to $\mathbf{0}$, implying that $\mathbf{0}$ is a sink.

For part *(ii)* suppose we write

$$\mathbf{x}_n = \mathbf{v}\,r^n + \mathbf{u}\,n\,r^n = (\mathbf{v} + \mathbf{u}\,n)\,r^n$$

Since $|r| > 1$ then $r^n \to \pm\infty$ as $n \to \infty$. Also, the length of $\mathbf{v} + \mathbf{u}\,n$ approaches ∞, unless $\mathbf{u} = \mathbf{0}$, in which case its length is constant. So the length of the solution vector $\mathbf{x}_n$, which is the product of these two, must become infinite as $n \to \infty$, unless both $\mathbf{v} = \mathbf{0}$ and $\mathbf{u} = \mathbf{0}$, which makes $\mathbf{x}_n$ the zero solution. This means $\mathbf{0}$ is a source.   ∎

EXAMPLE 3

Determine the dynamics of solutions of the prey–predator model considered in Example 2.

Solution: It was previously found that P_n and Q_n have the forms

$$P_n = v_1\,2^n + u_1\,n\,2^n \quad \text{and} \quad Q_n = v_2\,2^n + u_2\,n\,2^n$$

and so the solution vector $\mathbf{x}_n = \begin{bmatrix} P_n \\ Q_n \end{bmatrix}$ can be written in the form $\mathbf{x}_n = \mathbf{v}\,2^n + \mathbf{u}\,n\,2^n$. Since $|2| > 1$, all non-zero solutions diverge to $\pm\infty$, making $\mathbf{0}$ a source.   ∎

EXAMPLE 4

Determine the dynamics of the prey–predator model

$$P_{n+1} = P_n - 0.25 Q_n$$
$$Q_{n+1} = 0.25 P_n + 0.5 Q_n$$

Solution: The only eigenvalue of the coefficient matrix $A = \begin{bmatrix} 1 & -0.25 \\ 0.25 & 0.5 \end{bmatrix}$ is $r = 3/4$.

According to the stability theorem for this case, both the prey population P_n and the predator population Q_n must decay to 0 as n increases.   ∎

We remark once again that, although a certain long-term asymptotic behavior is predicted in these examples, this may not really reflect the true dynamics after either population becomes negative.

Exercises 4.5

In Exercises 1–4 find the unique eigenvalue of the given matrix A.

1. $A = \begin{bmatrix} 1 & 1 \\ -1 & 3 \end{bmatrix}$

2. $A = \begin{bmatrix} 0 & -0.5 \\ 0.5 & -1 \end{bmatrix}$

3. $A = \begin{bmatrix} 1 & -0.2 \\ 0.2 & 0.6 \end{bmatrix}$

4. $A = \begin{bmatrix} 1 & -2.5 \\ 2.5 & -4 \end{bmatrix}$

For each of the matrices A given in Exercises 5–12 determine whether the origin is a sink or source of $x_{n+1} = Ax_n$.

5. The matrix A from Exercise 1.

6. The matrix A from Exercise 2.

7. The matrix A from Exercise 3.

8. The matrix A from Exercise 4.

9. $A = \begin{bmatrix} 3/2 & -1/2 \\ 1/8 & 1 \end{bmatrix}$

10. $A = \begin{bmatrix} 2/3 & -2 \\ 8/9 & -2 \end{bmatrix}$

11. $A = \begin{bmatrix} -1 & 1/8 \\ -1/2 & -1/2 \end{bmatrix}$

12. $A = \begin{bmatrix} 1 & -2 \\ 8 & -7 \end{bmatrix}$

For each of the second-order equations given in Exercises 13–16: (a) Use direct substitution to find the value of r that makes $x_n = v_1 r^n + u_1 n r^n$ a solution; (b) Find the unique solution that satisfies the given initial values x_0 and x_1.

13. $x_{n+2} - 6x_{n+1} + 9x_n = 0$, $x_0 = 2000$, $x_1 = 3000$.

14. $25x_{n+2} - 40x_{n+1} + 16x_n = 0$, $x_0 = 300$, $x_1 = 400$.

15. $16x_{n+2} - 24x_{n+1} + 9x_n = 0$, $x_0 = 6000$, $x_1 = 0$.

16. $64x_{n+2} - 80x_{n+1} + 25x_n = 0$, $x_0 = 10$, $x_1 = -5$.

For each of the systems given in Exercises 17–20: (a) Find the general form of all solutions; (b) Find the unique solution that satisfies the given initial values; (c) Compute its limit as $n \to \infty$ if it exists.

17. $x_{n+1} = 2x_n - 4y_n$
    $y_{n+1} = 4x_n - 6y_n$ $x_0 = 2$, $y_0 = 1$

18. $P_{n+1} = \sqrt{8}P_n - Q_n$
    $Q_{n+1} = 2P_n$ $P_0 = 5000$, $Q_0 = 10{,}000$

19. $P_{n+1} = \frac{3}{4}P_n - \frac{1}{8}Q_n$
    $Q_{n+1} = \frac{1}{2}P_n + \frac{1}{4}Q_n$ $P_0 = 400$, $Q_0 = 600$

20. $P_{n+1} = P_n - 0.1Q_n$
    $Q_{n+1} = 0.1P_n + 0.8Q_n$ $P_0 = 7000$, $Q_0 = 5000$

21. Since 1 is the only eigenvalue of the coefficient matrix $A = \begin{bmatrix} 2 & -1 \\ 1 & 0 \end{bmatrix}$, the stability theorem does not apply. For each of the following initial conditions (x_0, y_0), find the unique solution for x_n and y_n, and compute $\lim_{n \to \infty} x_n$ and $\lim_{n \to \infty} y_n$ if they exist. (a) (1, 1); (b) (1, 0); (c) (2, 2); (d) (−1, −1); (e) (0, 1).

22. Repeat the previous exercise for $A = \begin{bmatrix} -2 & -1 \\ -1 & 0 \end{bmatrix}$ which has the unique eigenvalue −1.

▰▰▰▰▰ Computer Projects 4.5

For each of the coefficient matrices A *given in Projects 1 and 2 determine the limiting values as* n → ∞ *that the form of the solution implies for* $\left| \frac{x_{n+1}}{x_n} \right|$ *and* $\left| \frac{y_{n+1}}{y_n} \right|$. *Then verify this by computing these ratios for* n = 0, ..., 50 *for several choices of initial condition* (x_0, y_0).

1. The matrix A from Exercise 17.   2. The matrix A from Exercise 18.

For each of the matrices A *given in Projects 3 and 4 find the unique eigenvalue* r *and then solve* $(A - rI)\boldsymbol{v} = \boldsymbol{0}$ *to find the set of eigenvectors* $\boldsymbol{v} = \begin{bmatrix} x \\ y \end{bmatrix}$. *Graph the trajectories for a large sample of initial points* (x_0, y_0) *both on and off the eigenvectors. How would you describe the shapes of the trajectories?*

3. The matrix A from Exercise 3.   4. The matrix A from Exercise 9.

4.6   Complex Numbers and Their Arithmetic

In order to establish a comprehensive stability theory based upon eigenvalues, as we have been attempting to do in the previous sections, one last case remains to be investigated. It is possible for the coefficient matrix of a linear system to have no real eigenvalues, and in fact it is quite common for this to occur for many types of models. Of the three cases we must consider, the others involving either one or two real eigenvalues, this offers the greatest challenge. Since it implies the eigenvalues must be complex numbers, we are inevitably led into the rather strange workings of the Complex Plane.

In this section we introduce some elementary properties of complex numbers and the arithmetic associated with them. With this background, we will then be able to complete our stability analysis of linear systems.

Imaginary Numbers

The set of imaginary numbers is an abstract creation of the human mind, even more so than is the real number system. Unlike real numbers, these don't actually count or measure quantities that exist in the *real world*. But they are often useful nonetheless. In fact they are quite valuable in the analysis of many natural and scientific processes — even indispensable, one might say.

To see how this could be, the place to begin may be to see how they might naturally arise in simple algebra. Suppose, for example, we would like to solve the quadratic equation $x^2 + 1 = 0$. Since this doesn't factor, one may first write $x^2 = -1$ and then take the square root of both sides, which gives us $x = \pm\sqrt{-1}$. The most common, and usually most appropriate, response at this point would be to say that no solution exists, as it is impossible to take the square root of any negative number. In other words, no number exists whose square equals -1.

But suppose that, to overcome this insufficiency of numbers, we create a new one whose square does equal -1. Certainly, we could not think of it as being a number in the same sense that real numbers like 2, -5 and 24.75 are, and it could not be placed on the real number line since it is not one of them. Such a number may not actually exist, but it can be *imagined*, as was probably just done by you as you read this. For this reason, such a number is called **imaginary.** The imaginary number $\sqrt{-1}$ plays a central role in all this, and so is identified by a special name i, which is, of course, short for *imaginary*.

With this i it is now possible to say that the equation $x^2 + 1 = 0$ does have solutions. In fact it has two different solutions: $x = i$ and $x = -i$. This is true since if $x = \pm i$ then

$$x^2 = (\pm i)^2 = (\pm\sqrt{-1})^2 = -1$$

Since both satisfy $x^2 = -1$, or equivalently $x^2 + 1 = 0$, they must both be solutions.

There are many imaginary numbers besides i that are useful in mathematics — in fact, an infinite number of them. But there is no need to make up names for those others. If we adopt the same rules of arithmetic and algebra for imaginary numbers as for real numbers, then additional names are unnecessary, since all other imaginary numbers can be expressed in terms of i. For example,

$$\sqrt{-4} = \sqrt{-1 \cdot 4} = \sqrt{-1} \cdot \sqrt{4} = \sqrt{-1} \cdot 2 = 2\sqrt{-1} = 2i$$

All imaginary numbers can be written in the form βi where β is real. To conform to the same properties as do real numbers, certain arithmetic rules must be true with regard to imaginary numbers. For example, if β_1 and β_2 are any real numbers, then we can add $\beta_1 i$ and $\beta_2 i$ (in any order) or subtract them:

$$\beta_1 i + \beta_2 i = \beta_2 i + \beta_1 i = (\beta_1 + \beta_2) i \quad \text{and} \quad \beta_1 i - \beta_2 i = (\beta_1 - \beta_2) i$$

We can also multiply them (again, in any order) or divide them:

$$\beta_1 i \cdot \beta_2 i = \beta_2 i \cdot \beta_1 i = \beta_1 \beta_2 i^2 = \beta_1 \beta_2 (-1) = -\beta_1 \beta_2 \quad \text{and} \quad \frac{\beta_1 i}{\beta_2 i} = \frac{\beta_1}{\beta_2}$$

which in both cases yields an answer that is real.

EXAMPLE 1

Compute and simplify: (a) $3i - 8i$; (b) $-5i \cdot 6i$; (c) i^3; (d) $\dfrac{7}{2i}$.

Solution: (a) $3i - 8i = (3 - 8)i = -5i$.

(b) $-5i \cdot 6i = (-5 \cdot 6)i^2 = -30 \cdot (-1) = 30$.

(c) $i^3 = i^2 \cdot i = -1 \cdot i = -i$.

(d) $\dfrac{7}{2i} = \dfrac{7}{2i} \cdot \dfrac{i}{i} = \dfrac{7i}{2i^2} = \dfrac{7i}{-2} = -\dfrac{7}{2}i$. ∎

Complex Numbers

Although these computations seem to indicate that combining any two imaginary numbers produces a number that is either real or imaginary, that is not always the case. For example, if we add or subtract a real number and an imaginary one, a new type emerges. A number like $2 + 3i$ does not reduce to either a real or an imaginary number. It is of neither type exactly but has components of both, and so is given a different name: **complex number.** The quantities $6 - 5i$, $2.5 + 7i$ and $\sqrt{2} + i\sqrt{3}$ are all complex numbers.

The general form of any complex number is $\alpha + \beta i$, where both α and β are real numbers. The number α is called the **real part** of the complex number, and β is called the **imaginary part,** even though it is also a real number. It is the fact that β is multiplied by i that makes the term imaginary. For example, the real part of $4 - 6i$ is $\alpha = 4$ and its imaginary part is $\beta = -6$.

The complex number system includes within it all real numbers. Each can be obtained by letting $\beta = 0$ and having α assume a real value. It also includes all the imaginary numbers described earlier, which can be generated by letting $\alpha = 0$ and β be an arbitrary real number. Like real and imaginary ones, complex numbers can also be added, subtracted, multiplied or divided. The result will always be another complex number, although in some cases it takes a bit of algebra to write the answer once again in the form $\alpha + \beta i$.

EXAMPLE 2

Compute and simplify: (a) $(4 - 2i) + (-1 + i)$; (b) $5i \cdot (\sqrt{3} + 4i)$; (c) $(2 - 3i) \cdot (7 + 2i)$;

(d) $\dfrac{4 - i}{2 + 3i}$.

Solution: (a) $(4 - 2i) + (-1 + i) = 4 - 1 - 2i + i = 3 - i$.

(b) $5i \cdot (\sqrt{3} + 4i) = 5i \cdot \sqrt{3} + 5i \cdot 4i = 5i\sqrt{3} + 20i^2 = 5i\sqrt{3} - 20 = -20 + 5\sqrt{3}i$.

(c) $(2 - 3i) \cdot (7 + 2i) = 2 \cdot 7 + 2 \cdot 2i - 3i \cdot 7 - 3i \cdot 2i = 14 + 4i - 21i - 6i^2 = 14 - 17i + 6 = 20 - 17i$.

(d) $\dfrac{4 - i}{2 + 3i} = \dfrac{4 - i}{2 + 3i} \cdot \dfrac{2 - 3i}{2 - 3i} = \dfrac{8 - 12i - 2i + 3i^2}{4 - 6i + 6i - 9i^2} = \dfrac{8 - 14i - 3}{4 + 9} = \dfrac{5 - 14i}{13}$. ∎

Within the complex number system, it is always possible in theory to solve *any* algebraic equation. That is, roots of equations always exist, provided of course we allow them to be complex valued. No additional types of numbers ever need to be introduced in order to express the solution of any algebraic equation. This applies, in particular, to quadratic equations.

EXAMPLE 3

Solve $2x^2 - 2x + 5 = 0$.

Solution: The quadratic formula gives $x = \dfrac{2 \pm \sqrt{-36}}{4} = \dfrac{2 \pm 6i}{4}$. So the solutions are $(1 + 3i)/2$ and $(1 - 3i)/2$.   ∎

In this example the roots, which may also be written as $\frac{1}{2} + \frac{3}{2}i$ and $\frac{1}{2} - \frac{3}{2}i$, have a special relationship. The first has the form $\alpha + \beta i$ and the second $\alpha - \beta i$, where α and β are real numbers. These are called **complex conjugates.** The quadratic formula indicates that complex solutions of any quadratic equation (with real coefficients) always come in complex conjugate pairs. It can readily be seen therefore that every such quadratic equation always has either two real and distinct roots, a single real but repeated root, or two distinct complex conjugate roots. For the eigenvalue problem, this means that we indeed have only the one last case of complex conjugate roots to consider.

It is often necessary to perform algebraic manipulations on variables that represent complex numbers. In this case the variable z is typically used in the same way that x and y usually stand for real numbers. If z designates a particular complex number $\alpha + \beta i$, then $\bar{z}$ often represents its complex conjugate $\bar{z} = \alpha - \beta i$. For example, if $z = 9 - 5i$ then $\bar{z} = 9 + 5i$.

A complex number z may also be used as the *argument* of a trigonometric, exponential or logarithmic function. For example, $\sin z$, $\cos z$, e^z and $\ln z$ all have meaning for complex z. Although their values are complex and can again be written in the form $\alpha + \beta i$, this is often a tedious task, and so we will not go into those methods here. Such techniques are discussed in a standard undergraduate course in Complex Variables. The notions of derivative and integral based upon complex valued variables z and functions $f(z)$ are also typically developed in such a course.

However, one important formula involving complex valued trigonometric and exponential functions is worth mentioning at this point. For any real number θ it can be shown that

$$e^{i\theta} = \cos \theta + i \sin \theta \tag{1}$$

This is called **Euler's Formula.** An outline of its proof involving Taylor series expansions appears in the exercises. Among the important implications of this formula is one of the most interesting and beautiful equations in mathematics. Suppose we let $\theta = \pi$. In this case Euler's Formula becomes

$$e^{i\pi} = \cos \pi + i \sin \pi = -1 + i \, 0 = -1$$

The preferred way of writing this is

$$e^{i\pi} + 1 = 0$$

since it now states that there is a relationship — an amazingly simple one — between what are perhaps the five most fundamental numbers in mathematics: $0, 1, e, \pi, i$. Each of these appears exactly once in that equation, and are joined together using the three most fundamental arithmetic operations: addition, multiplication and exponentiation, used exactly once each. (Recall that subtraction and division can be thought of as adding and multiplying negatives and reciprocals respectively.) Even the equal sign is used exactly once. There is probably no other equation in mathematics that matches $e^{i\pi} + 1 = 0$ in its simple beauty.

As interesting as this may be, the implication of (1) that is most valuable to us right now, however, is that it provides an easy way to compute a certain quantity we will need later on. Because of (1) we must have

$$e^{in\theta} = (e^{i\theta})^n = (\cos\theta + i\,\sin\theta)^n$$

But replacing θ by $n\theta$ in (1) gives another expression for $e^{in\theta}$:

$$e^{in\theta} = \cos n\theta + i\,\sin n\theta$$

Equating the two expressions for $e^{in\theta}$ yields

$$(\cos\theta + i\,\sin\theta)^n = \cos n\theta + i\,\sin n\theta \tag{2}$$

This is called **de Moivre's Formula** and will be of great importance to us below.

EXAMPLE 4

Compute: (a) $(\cos 2\pi/5 + i\sin 2\pi/5)^5$; (b) $(\cos\pi/3 + i\sin\pi/3)^8$.

Solution: (a) $(\cos 2\pi/5 + i\sin 2\pi/5)^5 = \cos(5\cdot 2\pi/5) + i\sin(5\cdot 2\pi/5) = \cos(2\pi) + i\sin(2\pi) = 1$.

(b) $(\cos\pi/3 + i\sin\pi/3)^8 = \cos(8\cdot\pi/3) + i\sin(8\cdot\pi/3) = \cos(2\pi + 2\pi/3) + i\sin(2\pi + 2\pi/3) = \cos(2\pi/3) + i\sin(2\pi/3) = \dfrac{-1 + i\sqrt{3}}{2}$. ∎

Complex Plane

Although complex numbers cannot be placed on the real number line (unless the imaginary part is 0), there is a simple way to graphically represent them. Suppose we create a two-dimensional coordinate system using the horizontal axis to represent the real part α of any complex number $\alpha + \beta i$, and the vertical axis for the imaginary part β. In essence, we are actually just plotting the point (α, β) as we would in the (x, y) plane (see Fig. 4.15(a)). This coordinate system is called the **Complex Plane.** Here, the horizontal axis is called the **Real Axis** and the vertical is the **Imaginary Axis.** Some examples are shown in Figure 4.15(b).

The Complex Plane has many uses, but only one interests us here. Graphing a complex number will allow us to identify two important quantities needed to determine the dynamics of linear systems. Suppose we first plot a complex number $\alpha + \beta i$, and then connect it to the origin using a straight line segment. There is a certain angle θ that this line segment makes with the positive Real Axis, and a unique distance r along this line from the origin to the point (see Fig. 4.16). These two values θ and r are called the **polar coordinates** of the point.

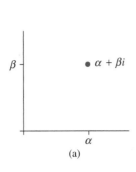

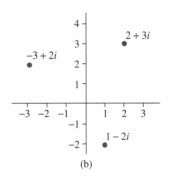

(a)   (b)

FIGURE 4.15

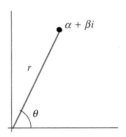

FIGURE 4.16

The distance formula allows r to be computed unambiguously by $r = \sqrt{\alpha^2 + \beta^2}$, but the designation of the angle θ is not unique owing to the periodic nature of angles around a circle. So, to avoid confusion, we will simply assume that $\theta = \theta \pm 2k\pi$ for all integers k.

EXAMPLE 5

Find the polar coordinates of (a) $2 - 2i$; (b) $3 + 8i$; (c) $-1 + i\sqrt{3}$.

Solution: (a) The line connecting the origin with $2-2i$ makes an angle of $7\pi/4$ (or perhaps $-\pi/4$) with the positive Real Axis. The distance between the two points is $\sqrt{(-2)^2 + 2^2} = \sqrt{8}$. So the polar coordinates are $\theta = 7\pi/4$ and $r = \sqrt{8}$.

(b) Since $\alpha = 3$ and $\beta = 8$, then $\theta = \tan^{-1} 8/3 \approx 1.212$. Also, $r = \sqrt{3^2 + 8^2} = \sqrt{73}$.

(c) The line connecting the origin with $-1 + i\sqrt{3}$ makes angle of $\pi/3$ with the *negative* Real Axis, and so $\theta = \pi - \pi/3 = 2\pi/3$. The value of r is $\sqrt{(-1)^2 + (\sqrt{3})^2} = 2$.   ■

When $\beta = 0$ the value of $r = \sqrt{\alpha^2 + \beta^2}$ is equal to $|\alpha|$. Because of this connection to absolute value when the number is real, for any complex number $\alpha + \beta i$ one often writes

$$|\alpha + \beta i| = r = \sqrt{\alpha^2 + \beta^2}$$

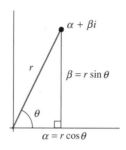

FIGURE 4.17

although this is now called the **modulus** rather than the absolute value. It is easily seen that the moduli of complex conjugates are equal:

$$|\alpha + \beta i| = |\alpha - \beta i|$$

EXAMPLE 6

Compute (a) $|1 - i|$; (b) $|-2 - 2i\sqrt{3}|$.

Solution: (a) $|1 - i| = \sqrt{1^2 + (-1)^2} = \sqrt{2}$.
(b) $|-2 - 2i\sqrt{3}| = \sqrt{(-2)^2 + (-2\sqrt{3})^2} = \sqrt{16} = 4$.   ∎

Using trigonometry, α and β can be expressed in terms of the polar coordinates of $\alpha + \beta i$:

$$\alpha = r \cos\theta \quad \text{and} \quad \beta = r \sin\theta$$

(see Fig. 4.17). This implies

$$\alpha + \beta i = r \cos\theta + r i \sin\theta = r(\cos\theta + i \sin\theta)$$

which means also

$$(\alpha + \beta i)^n = r^n(\cos\theta + i \sin\theta)^n$$

But because of (2) we have

$$(\alpha + \beta i)^n = r^n(\cos n\theta + i \sin n\theta) \tag{3}$$

This provides a convenient way to compute powers of any complex number $z = \alpha + \beta i$. A similar argument gives the corresponding result for the complex conjugate:

$$(\alpha - \beta i)^n = r^n(\cos n\theta - i \sin n\theta) \tag{4}$$

EXAMPLE 7

(a) Find the complex number z whose polar coordinates are $\theta = 2$ and $r = 3$. (b) Simplify $(1 + i)^{10}$.

Solution: (a) $\alpha = 3 \cos 2 \approx 3(-0.416) = -1.248$ and $\beta = 3 \sin 2 \approx 3(0.909) = 2.728$, which implies $z \approx -1.248 + 2.728i$.

(b) The polar coordinates of $(1 + i)$ are $\theta = \pi/4$ and $r = \sqrt{2}$, and so

$$(1 + i)^{10} = (\sqrt{2})^{10}(\cos(10\pi/4) + i \sin(10\pi/4))$$

from (3). But since

$$(\sqrt{2})^{10} = 32, \quad \cos(10\pi/4) = \cos(\pi/2) = 0, \quad \text{and} \quad \sin(10\pi/4) = \sin(\pi/2) = 1$$

then $(1 + i)^{10} = 32i$. ■

Although we have chosen to present only this small sampling of the unusual and fascinating relationships that emerge in the Complex Plane, it will suffice for our needs in the next chapter. Some additional features are developed in the exercises.

Exercises 4.6

Compute and simplify each of the expressions in Exercises 1–8.

1. $(3 - 7i) + (-4 - 2i)$

2. $(5 - i) - (i + 8)$

3. $\dfrac{i}{2} \cdot (5 + 2i)$

4. $(8 + 6i) \cdot (1 - i)$

5. $(3 - 4i)^2$

6. $\dfrac{8 + i}{3 + i}$

7. $\dfrac{-5i}{4 + 3i}$

8. $\dfrac{15}{2i - 6}$

9. If $z = 1 + 2i$ graph each of the following in the Complex Plane: z, $2z$, $3z$, $\frac{1}{2}z$.
10. Repeat the previous exercise for $z = -2 + 3i$.
11. Compute (a) i^2; (b) i^3; (c) i^4; (d) i^5.
12. Compute (a) $(-i)^2$; (b) $(-i)^3$; (c) $(-i)^4$; (d) $(-i)^5$.

In Exercises 13–20 find all real and complex solutions of the given equation.

13. $2z^2 + 6 = 0$

14. $z^2 - 2z + 10 = 0$

15. $z^2 + z + 1 = 0$

16. $3z + \dfrac{2}{z} + 1 = 0$

17. $z^3 + 2z^2 + 3z = 0$

18. $z^3 - 4z^2 + 40z = 0$

19. $z^4 = 1$

20. $z^4 = 4$

In Exercises 21 and 22 write each expression in the form $\alpha + \beta i$.

21. $8e^{i\pi/6}$

22. $3e^{-i\pi/4}$

In Exercises 23–28 write each expression in polar form $re^{i\theta}$.

23. $-3 + 3i$

24. $-2i/5$

25. $\dfrac{1 - i}{\sqrt{2}}$

26. $\sqrt{27} + 3i$

27. $\sqrt{3} + 3i$

28. $\dfrac{-5(1 + i)}{2}$

In Exercises 29–32 write each expression in the form $\alpha + \beta i$.

29. $(1 - i)^8$

31. $(2\cos(\pi/16) - 2i\sin(\pi/16))^8$

30. $(i - 1)^5$

32. $(-1 + i\sqrt{3})^7$

33. Prove that for any complex number z:

(a) $\dfrac{z + \bar{z}}{2}$ equals the real part of z.

(b) $\dfrac{z - \bar{z}}{2i}$ equals the imaginary part of z.

(c) $\sqrt{z\bar{z}} = |z|$.

34. Prove that $e^{i\theta} = \cos\theta + i\sin\theta$ by the following steps:

(a) Compute the Taylor (Maclaurin) Series for $\cos\theta$ and $\sin\theta$.

(b) Compute the Taylor (Maclaurin) Series for e^z and then substitute $z = i\theta$.

(c) Simplify your answer to part (b) and separate into real and imaginary parts.

(d) Show that the real part of $e^{i\theta}$ is $\cos\theta$, and the imaginary part is $\sin\theta$.

▬▬▬▬▬ Computer Projects 4.6

1. The complex solution of the linear equation $z_{n+1} = w\, z_n$, where $z_0 = 1$ and w is any complex number, is $z_n = w^n$. For each of the following compute z_n for $n = 0, \ldots, 10$ and graph each trajectory in the Complex Plane: (a) $w = i$; (b) $w = 2i$; (c) $w = i/2$.

   Note: If $z_n = x_n + iy_n$ and $w = a + bi$, then iterating $z_{n+1} = w\, z_n$ is equivalent to

   iterating the system $\begin{cases} x_{n+1} = a\, x_n - b\, y_n \\ y_{n+1} = b\, x_n + a\, y_n \end{cases}$.

2. Repeat the previous project for (a) $w = (1 + i)/\sqrt{2}$; (b) $w = 1 + i$; (c) $w = (1 + i)/2$.

3. Repeat Project 1 for (a) $w = (3 - 4i)/5$; (b) $w = 4 + 3i$; (c) $w = (-3 + 4i)/10$.

4. Based on your observations in the preceding projects:

(a) Given any w can you predict the trajectory of z_n?

(b) Establish criteria for $\lim_{n\to\infty} z_n = 0$.

(c) Test the validity of your answers to (a) and (b) by computing z_n for some additional choices of w.

SECTION
4.7   Complex Eigenvalues

With the groundwork in complex numbers and their arithmetic provided in the previous section, we are finally in a position to complete our analysis of homogeneous linear systems. We address here the last remaining eigenvalue scenario that may arise for a coefficient matrix — the case in which they are complex valued.

Complex Conjugate Eigenvalues

First recall from Section 4.4 that finding the eigenvalues of the matrix $A = \begin{bmatrix} a & b \\ c & d \end{bmatrix}$ involves finding the roots of a certain quadratic equation. Since we have already covered the cases of two distinct real roots and one repeated real root for that equation, the only case remaining is that of complex roots. So it is appropriate to now use the complex variable z when writing that quadratic equation:

$$z^2 - (a+b)z + ad - bc = 0 \tag{1}$$

But as we learned in the previous section, if the equation has complex roots, then they must be complex conjugates. This means that there must exist two complex conjugate eigenvalues $z = \alpha + \beta i$ and $\bar{z} = \alpha - \beta i$ that satisfy (1).

Also recall from Section 4.5 that the first component x_n of the vector $\mathbf{x}_n = \begin{bmatrix} x_n \\ y_n \end{bmatrix}$ that solves $\mathbf{x}_{n+1} = A\mathbf{x}_n$ satisfies the second-order equation

$$x_{n+2} - (a+d)x_{n+1} + (ad - bc)x_n = 0 \tag{2}$$

A similar equation is satisfied by y_n. Since, as we saw earlier, all solutions of (2) when two distinct real eigenvalues r and s exist are linear combinations of the form

$$x_n = v_1 r^n + u_1 s^n$$

it is reasonable to assume now that all solutions have the form

$$x_n = v_1' z^n + u_1' \bar{z}^n$$

since z and $\bar{z}$ are the two distinct eigenvalues this time. However, since these eigenvalues are complex, it is very likely that so too are the constants v_1' and u_1'.

It is easy to verify that $x_n = v_1' z^n + u_1' \bar{z}^n$, where v_1' and u_1' are any complex numbers, does indeed solve (2), provided of course we use the algebra of complex numbers. To accomplish this, it will suffice to show that $x_n = z^n$ and $x_n = \bar{z}^n$ are both solutions of (2), since in that case, as we established earlier, all linear combinations of them $x_n = v_1' z^n + u_1' \bar{z}^n$ would also have to be solutions.

Substituting $x_n = z^n$, $x_{n+1} = z^{n+1}$ and $x_{n+2} = z^{n+2}$ into (2) yields

$$z^{n+2} - (a+d)z^{n+1} + (ad - bc)z^n = 0$$

and dividing by z^n this becomes

$$z^2 - (a+d)z + (ad - bc) = 0$$

This is obviously true since z satisfies (1). A similar argument can be made for $\bar{z}$. The conclusion is therefore that $x_n = v_1' z^n + u_1' \bar{z}^n$ solves (2). Although we shall not do so here, it can further be shown that *every* solution of (2) can be written this way.

The form that we have found for the solution x_n is not very convenient though, since it involves raising the two complex conjugate numbers $z = \alpha + \beta i$ and $\bar{z} = \alpha - \beta i$ to the power n. Fortunately, there is a way to simplify things. Using equations (3) and (4) from Section 4.6, which say respectively that

$$(\alpha + \beta i)^n = r^n(\cos n\theta + i \sin n\theta) \quad \text{and} \quad (\alpha - \beta i)^n = r^n(\cos n\theta - i \sin n\theta)$$

where r and $\pm\theta$ are the polar coordinates of $\alpha \pm \beta i$, our solution

$$x_n = v_1' \, z^n + u_1' \, \bar{z}^n = v_1' \, (\alpha + \beta i)^n + u_1' \, (\alpha - \beta i)^n$$

can instead be written as

$$x_n = v_1' \, r^n(\cos n\theta + i \, \sin n\theta) + u_1' \, r^n(\cos n\theta - i \, \sin n\theta) \tag{3}$$

We thus obtain a simple expression for x_n that does not involve raising complex numbers to the power n. The problem now is that this solution is not necessarily real. To observe only those dynamics of the system that actually occur in the *real world*, we need to strip off in some way the imaginary parts. This can be accomplished with one final step.

Suppose we make the cleverly chosen substitutions

$$v_1' = \tfrac{1}{2} v_1 - \tfrac{1}{2} u_1 \, i \quad \text{and} \quad u_1' = \tfrac{1}{2} v_1 + \tfrac{1}{2} u_1 \, i$$

where v_1 and u_1 are any *real* numbers. Substituting these into (3) gives

$$x_n = \left(\tfrac{1}{2} v_1 - \tfrac{1}{2} u_1 \, i \right) r^n(\cos n\theta + i \, \sin n\theta) + \left(\tfrac{1}{2} v_1 + \tfrac{1}{2} u_1 \, i \right) r^n \, (\cos n\theta - i \, \sin n\theta)$$

The imaginary part of this expression (that is, the coefficient of i) is

$$r^n \left(\tfrac{1}{2} v_1 \sin n\theta - \tfrac{1}{2} u_1 \cos n\theta - \tfrac{1}{2} v_1 \sin n\theta + \tfrac{1}{2} u_1 \cos n\theta \right) = 0$$

This means that the resulting expression

$$x_n = r^n \left(\tfrac{1}{2} v_1 \cos n\theta + \tfrac{1}{2} u_1 \sin n\theta + \tfrac{1}{2} v_1 \cos n\theta + \tfrac{1}{2} u_1 \sin n\theta \right)$$

must be real. In fact, it's not only real, but after combining terms, it's also surprisingly simple:

$$x_n = v_1 \, r^n \cos n\theta + u_1 \, r^n \sin n\theta$$

This is now a linear combination of $r^n \cos n\theta$ and $r^n \sin n\theta$ that involves only real valued quantities. So in the end, we finally arrive at a form for x_n that doesn't involve imaginary numbers. Our trip through the *make-believe* world of the Complex Plane was temporary, but well worth making since it provided sufficient insight into the nature of solutions to yield a real and usable result. That is often the case with the complex number system.

It can be shown that the second component y_n of the vector $\mathbf{x}_n = \begin{bmatrix} x_n \\ y_n \end{bmatrix}$ that solves $\mathbf{x}_{n+1} = A\mathbf{x}_n$ in this complex eigenvalue case has a form similar to that of x_n:

$$y_n = v_2 \, r^n \cos n\theta + u_2 \, r^n \sin n\theta$$

This means that

$$\mathbf{x}_n = \begin{bmatrix} x_n \\ y_n \end{bmatrix} = \begin{bmatrix} v_1 \, r^n \cos n\theta + u_1 \, r^n \sin n\theta \\ v_2 \, r^n \cos n\theta + u_2 \, r^n \sin n\theta \end{bmatrix} = \begin{bmatrix} v_1 \\ v_2 \end{bmatrix} r^n \cos n\theta + \begin{bmatrix} u_1 \\ u_2 \end{bmatrix} r^n \sin n\theta.$$

If we call $\mathbf{v} = \begin{bmatrix} v_1 \\ v_2 \end{bmatrix}$ and $\mathbf{u} = \begin{bmatrix} u_1 \\ u_2 \end{bmatrix}$ then we can say more concisely that when A has complex conjugate eigenvalues, all solutions of $\mathbf{x}_{n+1} = A\mathbf{x}_n$ have the form

$$\mathbf{x}_n = \mathbf{v} \, r^n \cos n\theta + \mathbf{u} \, r^n \sin n\theta$$

where r and $\pm\theta$ are the polar coordinates of the eigenvalues. Additionally, although we will not prove so here, both $\mathbf{v}$ and $\mathbf{u}$ must be non-zero vectors (unless all $\mathbf{x}_n = \mathbf{0}$).

EXAMPLE   1

(a) Find the general form of all real solutions of the system

$$x_{n+1} = 2x_n - y_n$$
$$y_{n+1} = 4x_n$$

(b) Find the unique solution for $x_0 = 1$ and $y_0 = 2$.

Solution: (a) The eigenvalues are found to be $1 \pm i\sqrt{3}$, and their polar coordinates are $\pm\theta = \pm\pi/3$ and $r = 2$. This means that all solutions have the form

$$x_n = v_1\, 2^n \cos(n\pi/3) + u_1\, 2^n \sin(n\pi/3) \quad \text{and} \quad y_n = v_2\, 2^n \cos(n\pi/3) + u_2\, 2^n \sin(n\pi/3)$$

where v_1, v_2, u_1, u_2 are constants.

  (b) We must find the specific values of the four constants v_1, u_1, v_2, u_2 that make $x_0 = 1$ and $y_0 = 2$. Iterating the system for $n = 0$ gives

$$x_1 = 2x_0 - y_0 = 2 \cdot 1 - 2 = 0 \quad \text{and} \quad y_1 = 4x_0 = 4 \cdot 1 = 4$$

Using the form for x_n found in part (a) with $n = 0$ gives

$$x_0 = v_1\, 2^0 \cos(0\pi/3) + u_1\, 2^0 \sin(0\pi/3) = v_1$$

Since we were given $x_0 = 1$, then we must have $v_1 = 1$, which also makes

$$x_1 = v_1\, 2\cos(\pi/3) + u_1\, 2\sin(\pi/3) = 2\cos(\pi/3) + u_1\, 2\sin(\pi/3) = 1 + u_1\sqrt{3}$$

Also since $x_1 = 0$, then $x_1 = 1 + u_1\sqrt{3}$ becomes $0 = 1 + u_1\sqrt{3}$ or $u_1 = -1/\sqrt{3}$. Having found that $v_1 = 1$ and $u_1 = -1/\sqrt{3}$, then x_n from part (a) therefore becomes

$$x_n = v_1\, 2^n \cos(n\pi/3) + u_1\, 2^n \sin(n\pi/3) = 2^n \cos(n\pi/3) - \frac{2^n \sin(n\pi/3)}{\sqrt{3}}$$

A similar set of calculations shows that $v_2 = 2$ and $u_2 = 2/\sqrt{3}$, and so

$$y_n = v_2\, 2^n \cos(n\pi/3) + u_2\, 2^n \sin(n\pi/3) = 2^{n+1} \cos(n\pi/3) + \frac{2^{n+1} \sin(n\pi/3)}{\sqrt{3}} \qquad \blacksquare$$

Rotation and Complex Eigenvalues

It is now possible to determine the dynamics of all solutions of $\mathbf{x}_{n+1} = A\mathbf{x}_n$ when A has complex conjugate eigenvalues. Before taking up stability issues, let us first see what

$$\mathbf{x}_n = \mathbf{v}\,r^n \cos n\theta + \mathbf{u}\,r^n \sin n\theta = r^n(\mathbf{v}\cos n\theta + \mathbf{u}\sin n\theta)$$

tells us about the *rotation* of the solution vector $\mathbf{x}_n$. Suppose we call

$$\mathbf{w}_n = \mathbf{v}\cos n\theta + \mathbf{u}\sin n\theta$$

and pretend for a moment that n is a continuous rather than discrete variable. If that were the case, then we would have

$$
\begin{aligned}
\mathbf{w}_{n+2\pi/\theta} &= \mathbf{v}\cos[(n+2\pi/\theta)\theta] + \mathbf{u}\sin[(n+2\pi/\theta)\theta] \\
&= \mathbf{v}\cos(n\theta + 2\pi) + \mathbf{u}\sin(n\theta + 2\pi) \\
&= \mathbf{v}\cos n\theta + \mathbf{u}\sin n\theta \\
&= \mathbf{w}_n
\end{aligned}
$$

That is, if n were continuous, then $\mathbf{w}_n$ would be a periodic function of n with period $2\pi/\theta$. The parameterized graph in the plane corresponding to $\mathbf{w}_n$ would therefore repeat itself after every interval of size $2\pi/\theta$ for n.

Not only would this graph repeat itself, but it would also trace out a very special type of geometric shape. For any non-zero vectors $\mathbf{v}$ and $\mathbf{u}$ the graph of $\mathbf{w}_n = \mathbf{v}\cos n\theta + \mathbf{u}\sin n\theta$ always describes an ellipse or perfect circle with center at the origin. The following example demonstrates this.

EXAMPLE 2

Draw the graph of $\mathbf{w}_n = \mathbf{v}\cos n\theta + \mathbf{u}\sin n\theta$ for all continuous $n \geq 0$ if (a) $\mathbf{v} = \begin{bmatrix} 3 \\ 0 \end{bmatrix}$, $\mathbf{u} = \begin{bmatrix} 0 \\ 3 \end{bmatrix}$ and $\theta = 1$; (b) $\mathbf{v} = \begin{bmatrix} -3 \\ 5 \end{bmatrix}$, $\mathbf{u} = \begin{bmatrix} 10 \\ 0 \end{bmatrix}$ and $\theta = \pi/2$

Solution: The graphs for parts (a) and (b) are shown in Figures 4.18(a) and 4.18(b) respectively. In (a) since $\mathbf{v}$ and $\mathbf{u}$ are perpendicular and their lengths both equal 3, the graph is a circle of radius 3. In (b) the graph traces out an ellipse.   ∎

Since n is an integer rather than a continuous variable, however, the discrete points described by $\mathbf{w}_n$ for $n = 0, 1, 2, \ldots$, which all lie somewhere on the aforementioned ellipse or circle, will repeat themselves only under certain (rare) conditions. In general, if θ is a rational multiple of π (such as $\theta = \pi/4$ or $\theta = 5\pi/9$), then the points $\mathbf{w}_n$ are periodic with period m, where m equals the smallest whole number that makes $\dfrac{m\theta}{2\pi}$ an integer. If θ is an irrational multiple of π, as it is more likely to be (such as $\theta = 2$ or $\theta = \tan^{-1} 4/5$),

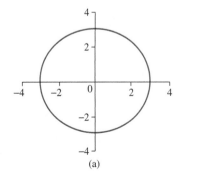

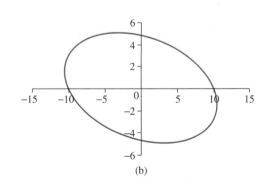

(a)   (b)

FIGURE 4.18

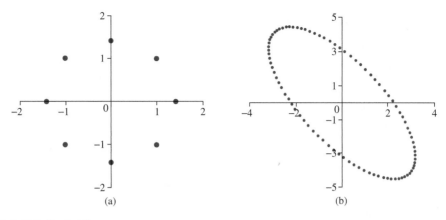

FIGURE 4.19

then the points $\mathbf{w}_n$ will never quite repeat themselves exactly. In this case we might say that $\mathbf{w}_n$ is **irrationally periodic.**

EXAMPLE 3

Graph the set of points in the plane given by $\mathbf{v} \cos n\theta + \mathbf{u} \sin n\theta$ for $n = 0, 1, \ldots, 75$ if

(a) $v = \begin{bmatrix} 1 \\ -1 \end{bmatrix}$, $u = \begin{bmatrix} 1 \\ 1 \end{bmatrix}$, $\theta = \pi/4$; (b) $v = \begin{bmatrix} 1 \\ 2 \end{bmatrix}$, $u = \begin{bmatrix} 3 \\ -4 \end{bmatrix}$, $\theta = 0.5$

Solution: The graphs are given in Figures 4.19(a) and 4.19(b) respectively. The set of points in (a) all fall on a circle around the origin, and in (b) on an ellipse. However the points in (a) repeat themselves after every 8 iterations since $(m\pi/4)/(2\pi) = m/8$, and so $m = 8$ is the smallest whole number that makes $m\theta/(2\pi)$ an integer. But in (b) $0.5/2\pi = 1/(4\pi)$ is not rational, and so the points never repeat exactly. Instead they are irrationally periodic, and if continued would trace out the entire ellipse. ■

Since $\mathbf{x}_n = r^n \mathbf{w}_n$ where r^n is a scalar, then $\mathbf{x}_n$ is always a scalar multiple of $\mathbf{w}_n$. So the vectors $\mathbf{w}_n$ and $\mathbf{x}_n$ always make the same angle with respect to the positive x-axis. This implies that similar to $\mathbf{w}_n$, the solution vector $\mathbf{x}_n$ rotates around the origin with an angle that is either a periodic or an irrationally periodic function of n, although unlike $\mathbf{w}_n$ its length either grows or shrinks with n (provided $r \neq 1$). We therefore state the following.

THEOREM *Rotation: Complex Conjugate Eigenvalues*

Suppose the coefficient matrix of a homogeneous linear system has complex conjugate eigenvalues with polar coordinates r and $\pm\theta$. Then solutions rotate around the fixed point $(0, 0)$ of the system with an angle that is either periodic if θ is a rational multiple of π, or irrationally periodic if θ is an irrational multiple of π. In the former case, the period m is the smallest whole number that makes $\dfrac{m\theta}{2\pi}$ an integer. ■

Stability and Complex Eigenvalues

With this, we can now derive stability criteria for the equilibrium of any homogeneous linear system when the coefficient matrix has complex conjugate eigenvalues. We remind the reader that in the following $|\alpha \pm \beta i| = \sqrt{\alpha^2 + \beta^2}$ represents the polar coordinate r.

THEOREM   *Stability: Complex Conjugate Eigenvalues*

Suppose the coefficient matrix of a homogeneous linear system has complex conjugate eigenvalues $\alpha \pm \beta i$. Then the fixed point $(0, 0)$ of the system is

(i) a sink if $|\alpha \pm \beta i| < 1$; or
(ii) a source if $|\alpha \pm \beta i| > 1$.

Proof: As we saw above, the points $\mathbf{w}_n = \mathbf{v} \cos n\theta + \mathbf{u} \sin n\theta$ for any non-zero vectors $\mathbf{v}$ and $\mathbf{u}$ always lie on the graph of a non-zero circle or ellipse centered at the origin. This means that $\mathbf{x}_n = r^n \mathbf{w}_n$ always falls on a circle or ellipse whose *size* is directly proportional to r^n. That is, the radius of the circle or the major and minor axes of the ellipse that $\mathbf{x}_n$ lies on has r^n as a factor. In part *(i)* where $r = |\alpha \pm \beta i| < 1$, solutions $\mathbf{x}_n$ therefore decay to $\mathbf{0}$, since $r^n \to 0$ as $n \to \infty$. This makes the fixed point $\mathbf{0}$ a sink (see Fig. 4.20(a)). On the other hand, in part *(ii)* where $r = |\alpha \pm \beta i| > 1$, the length of $\mathbf{x}_n$ becomes infinite, since now $r^n \to \infty$ as $n \to \infty$. This means that $\mathbf{0}$ is a source (see Fig. 4.20(b)). ∎ ■

EXAMPLE 4

Determine the stability of the system

$$x_{n+1} = 2x_n - 4y_n$$
$$y_{n+1} = x_n - y_n$$

Solution: The eigenvalues are found to be $(1 \pm i\sqrt{7})/2$. Since $r = |(1 \pm i\sqrt{7})/2| = \sqrt{2}$, then, according to the stability theorem, all solutions diverge to ∞, indicating that the origin is a source. ■

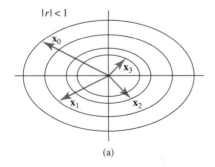
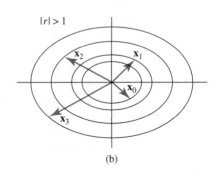

(a)   (b)

FIGURE 4.20

EXAMPLE 5

Determine the stability of prey–predator model

$$P_{n+1} = 0.4P_n - 0.2Q_n$$
$$Q_{n+1} = 0.8P_n + 0.5Q_n$$

Solution: Since the eigenvalues $0.45 \pm i\sqrt{0.1575}$ satisfy $r = |0.45 \pm i\sqrt{0.1575}| = 0.6$, then the origin is a sink, and so all solutions decay to 0. However, we should not conclude that the predator and prey populations both really die out. As before, the model breaks down as soon as either population becomes negative.   ■

If we combine both types of motion predicted by the form we have derived for the solution vector $\mathbf{x}_n = r^n \mathbf{w}_n$, i.e., a periodic or irrationally periodic rotation caused by $\mathbf{w}_n$ and an exponential stretching or shrinking caused by r^n, we can now explain the spiral trajectories we saw for some systems in Section 4.2. When complex conjugate eigenvalues exist, trajectories resemble spirals that are inward-directed for $r < 1$, or outward-directed for $r > 1$. When r has the borderline value $r = 1$, the amplitude stays fixed, and so all trajectories travel in ellipses or perfect circles around the origin, which is why neither a sink nor a source exist for this case.

EXAMPLE 6

Determine the nature of the trajectories in solution space for the system

$$x_{n+1} = 0.80x_n - 0.44y_n$$
$$y_{n+1} = x_n + 0.70y_n$$

Solution: The eigenvalues are $(3 \pm i\sqrt{7})/4$, which yields $r = |(3 \pm i\sqrt{7})/4| = 1$ and $\theta = \tan^{-1}(\sqrt{7}/3)$, which is an irrational multiple of π. All trajectories are elliptical around the origin and irrationally periodic, and so the points (x_n, y_n) never quite repeat themselves. Instead, they gradually trace out an entire ellipse as $n \to \infty$ (see Fig. 4.21).   ■

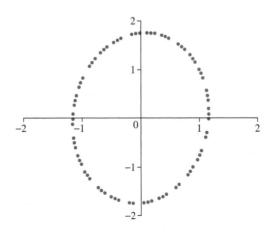

FIGURE 4.21

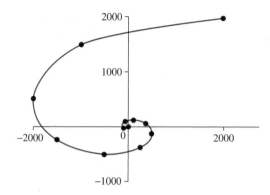

FIGURE 4.22

▬▬▬▬ **EXAMPLE 7**

Determine the nature of trajectories in solution space for the prey–predator model

$$P_{n+1} = 0.5P_n - Q_n$$
$$Q_{n+1} = 0.25P_n + 0.5Q_n$$

Solution: This time the eigenvalues $(1 \pm i)/2$ satisfy $r = |(1 \pm i)/2| = 1/\sqrt{2}$ and $\theta = \pi/4$. The origin must be a sink and trajectories must spiral inward toward it. The angle of the solution vector is periodic with period $m = 8$, since $8(\pi/4)/(2\pi) = 1$ is an integer (see Fig. 4.22). We remark once again that the system ceases to model the true populations when either becomes negative.   ■

Exercises 4.7 _____

In Exercises 1–4 find the eigenvalues of the given matrix.

1. $\begin{bmatrix} 2 & -5 \\ 1/2 & -1 \end{bmatrix}$ 3. $\begin{bmatrix} 1 & -3 \\ 1 & 1 \end{bmatrix}$

2. $\begin{bmatrix} -1/3 & -1 \\ 1/3 & -1/3 \end{bmatrix}$ 4. $\begin{bmatrix} 1 & 2 \\ -1 & 3 \end{bmatrix}$

For each of the matrices A given in Exercises 5–12 determine whether the origin is a sink or a source of $x_{n+1} = Ax_n$.

5. The matrix A from Exercise 1.

6. The matrix A from Exercise 2.

7. The matrix A from Exercise 3.

8. The matrix A from Exercise 4.

9. $A = \begin{bmatrix} 0.5 & -0.5 \\ 0.4 & 0.1 \end{bmatrix}$ 10. $A = \begin{bmatrix} 0 & -2 \\ 1/6 & -1 \end{bmatrix}$

11. $A = \begin{bmatrix} 4/3 & -5 \\ 1/3 & -1 \end{bmatrix}$

12. $A = \begin{bmatrix} 5 & 4 \\ -5 & 1 \end{bmatrix}$

13. Determine all real values of the constant t for which the origin is a sink for the system:
$$\begin{cases} x_{n+1} = tx_n - 0.5y_n \\ y_{n+1} = 0.5x_n + ty_n \end{cases}.$$

14. Repeat the previous exercise for the system: $\begin{cases} x_{n+1} = tx_n - 0.18y_n \\ y_{n+1} = 2x_n + ty_n \end{cases}.$

15. Show by direct substitution that each of the following are complex solutions of the second-order equation $x_{n+2} - 2x_{n+1} + 2x_n = 0$:

    (a) $x_n = (1+i)^n$

    (b) $x_n = (1-i)^n$

16. Repeat the previous exercise for $x_{n+2} - 6x_{n+1} + 58x_n = 0$ and

    (a) $x_n = (3+7i)^n$

    (b) $x_n = (3-7i)^n$

17. Show by direct substitution that each of the following are solutions of the second-order equation $x_{n+2} + 9x_n = 0$:

    (a) $x_n = 3^n \cos(n\pi/2)$

    (b) $x_n = 3^n \sin(n\pi/2)$

18. Repeat the previous exercise for $x_{n+2} - \sqrt{2}x_{n+1} + x_n = 0$ and

    (a) $x_n = \cos(n\pi/4)$

    (b) $x_n = \sin(n\pi/4)$

For each of the systems given in Exercises 19–22: (a) Find the general form of all solutions; (b) Find the unique solution that satisfies the given initial values; (c) Compute its limit as $n \to \infty$ *if it exists.*

19. $\begin{aligned} x_{n+1} &= 2y_n \\ y_{n+1} &= -2x_n \end{aligned}$; $x_0 = 1, \ y_0 = 2$

20. $\begin{aligned} P_{n+1} &= P_n - Q_n \\ Q_{n+1} &= P_n + Q_n \end{aligned}$; $P_0 = 8000, \ Q_0 = 1000$

21. $\begin{aligned} P_{n+1} &= 0.4P_n - 0.6Q_n \\ Q_{n+1} &= 0.8P_n + 0.4Q_n \end{aligned}$; $P_0 = 5000, \ Q_0 = 2000$

22. $\begin{aligned} P_{n+1} &= 2P_n - Q_n \\ Q_{n+1} &= P_n + Q_n \end{aligned}$; $P_0 = 1000, \ Q_0 = 1000$

23. Let $x_n = v_1 \cos n\theta$ and $y_n = u_1 \sin n\theta$ for any constants v_1, u_1 and θ. Prove that the set of points (x_n, y_n) for all n lie on the same ellipse in the (x, y) plane by showing that $\dfrac{x_n^2}{v_1^2} + \dfrac{y_n^2}{u_1^2} = 1$. What are the lengths of the major and minor axes?

24. Let $x_n = \cos n\theta - \sin n\theta$ and $y_n = \cos n\theta + \sin n\theta$ for any constant θ. Prove that the set of points (x_n, y_n) for all n lie on the same circle in the (x, y) plane by showing that $x_n^2 + y_n^2 = r^2$ for some r. What is the radius r?

In Exercises 25–32 determine the period of the given function w_n *where* n *is an integer, i.e., find the smallest positive integer* m *that makes* $w_{n+m} = w_n$ *for all* n ≥ 0.

25. $w_n = 2\sin(3n\pi/4)$

26. $w_n = -1.5\cos(4n\pi/6)$

27. $w_n = \dfrac{\sin(7n\pi/5)}{1000}$

28. $w_n = 2\cos(13n\pi/14) + 5\sin(13n\pi/14)$

29. $w_n = \sin(12n\pi/15) - \cos(12n\pi/15)$

30. $w_n = 5\sin(1.8n\pi)$

31. $w_n = \cos(0.35n\pi)$

32. $w_n = \sin(1.75n\pi/15)$

Computer Projects 4.7

Each of the matrices A in Projects 1–4 has complex conjugate eigenvalues with modulus equal to 1. Graph the solutions of $x_{n+1} = Ax_n$ for several choices of initial points, and verify that each trajectory lies on an ellipse. Identify which trajectories comprise a finite number of periodic points that keep repeating, and which never repeat.

1. $A = \begin{bmatrix} 1/\sqrt{2} & 1/\sqrt{2} \\ -1/\sqrt{2} & 1/\sqrt{2} \end{bmatrix}$

2. $A = \begin{bmatrix} 2 & 3 \\ -1 & -1 \end{bmatrix}$

3. $A = \begin{bmatrix} 2/9 & -1 \\ 2/3 & 3/2 \end{bmatrix}$

4. $A = \begin{bmatrix} -1 & 1/2 \\ -1 & -1/2 \end{bmatrix}$

For each of the matrices A given in Projects 5–8 verify that iteration using either A, $B = 2A$ or $C = \frac{1}{2}A$ rotates vectors by the same angle θ. Choose a common non-zero initial vector $x_0 = y_0 = z_0$ and iterate $x_{n+1} = Ax_n$, $y_{n+1} = By_n$ and $z_{n+1} = Cz_n$ for $n = 0, \ldots, 20$. Graph these vectors and verify that x_n, y_n and z_n all lie along the same line.

5. The matrix A from Project 1.

6. The matrix A from Project 2.

7. The matrix A from Project 3.

8. The matrix A from Project 4.

SECTION

4.8   Non-Homogeneous Systems

The stability and rotation theorems developed in the previous sections were directed primarily toward homogeneous linear systems and the fixed point (0, 0). Many of the models constructed in Section 4.1, however, are non-homogeneous and have non-zero fixed points. Does this mean that in order to establish their dynamics we have to start over? Fortunately, the answer is *no*. Nearly all the arguments made with regard to homogeneous systems carry over immediately to non-homogeneous ones as well. In this section we explain why this must be so.

We also revisit Markov processes, the simplest types of which were first introduced in Chapter 2. In light of the stability criteria derived in this chapter, Markov processes of greater complexity can now be addressed.

Dynamics of Non-Homogeneous Systems

Recall that a non-homogeneous linear systems is one having the form

$$x_{n+1} = a\,x_n + b\,y_n + h$$
$$y_{n+1} = c\,x_n + d\,y_n + k$$

where h and k are not both 0. Using matrix notation any such system can now be written first as

$$\begin{bmatrix} x_{n+1} \\ y_{n+1} \end{bmatrix} = \begin{bmatrix} a & b \\ c & d \end{bmatrix} \begin{bmatrix} x_n \\ y_n \end{bmatrix} + \begin{bmatrix} h \\ k \end{bmatrix}$$

and then also as

$$\mathbf{x}_{n+1} = A\mathbf{x}_n + \mathbf{b}$$

where $A = \begin{bmatrix} a & b \\ c & d \end{bmatrix}$, $\mathbf{x}_n = \begin{bmatrix} x_n \\ y_n \end{bmatrix}$ and $\mathbf{b} = \begin{bmatrix} h \\ k \end{bmatrix}$.

The trick to proving instantly that all the dynamics uncovered in previous sections for $\mathbf{x}_{n+1} = A\mathbf{x}_n$ are shared by $\mathbf{x}_{n+1} = A\mathbf{x}_n + \mathbf{b}$ is to convert the latter to the former by a change of variable analogous to those of Chapter 2. This can be done for any non-homogeneous system that has a unique fixed point, which is almost always the case, especially for systems that model naturally evolving processes.

So let us suppose then that a fixed point $\mathbf{p} = \begin{bmatrix} p \\ q \end{bmatrix}$ exists for the system. We have already seen that any such $\mathbf{p}$ must satisfy equation (8) of Section 4.2, which says

$$p = a\,p + b\,q + h$$
$$q = c\,p + d\,q + k$$

Using vectors and matrices, this can be written

$$\mathbf{p} = A\mathbf{p} + \mathbf{b} \quad \text{or} \quad A\mathbf{p} - \mathbf{p} + \mathbf{b} = \mathbf{0}$$

If we make the change of variable $\mathbf{x}_n = \mathbf{y}_n + \mathbf{p}$, which of course implies $\mathbf{x}_{n+1} = \mathbf{y}_{n+1} + \mathbf{p}$, then

$$\mathbf{x}_{n+1} = A\mathbf{x}_n + \mathbf{b} \quad \text{becomes} \quad \mathbf{y}_{n+1} + \mathbf{p} = A(\mathbf{y}_n + \mathbf{p}) + \mathbf{b}$$
$$\text{or} \quad \mathbf{y}_{n+1} = A\mathbf{y}_n + A\mathbf{p} - \mathbf{p} + \mathbf{b}$$

Because $A\mathbf{p} - \mathbf{p} + \mathbf{b} = \mathbf{0}$, the previous equation reduces to

$$\mathbf{y}_{n+1} = A\mathbf{y}_n$$

which is a homogeneous system.

Not only does this show that any non-homogeneous system with a unique fixed point $\mathbf{p}$ can be converted into a homogeneous system, but several other important implications may also be discerned. First, since $\mathbf{x}_n = \mathbf{y}_n + \mathbf{p}$, then the only difference between the solutions of $\mathbf{x}_{n+1} = A\mathbf{x}_n + \mathbf{b}$ and those of $\mathbf{y}_{n+1} = A\mathbf{y}_n$ is a *translation* by the vector $\mathbf{p}$. This means that the existence of a sink, source or saddle of one implies the same for the other, but located at a different point in solution space. Any rotation around sinks and sources must also be the same. Finally, since the coefficient matrix A is identical for both, its eigenvalues can be used in the same way to determine the dynamics of either type of system. That is, we have proven the following.

THEOREM   *Stability and Rotation*

All the stability and rotation theorems of this chapter concerning homogeneous linear systems $\mathbf{x}_{n+1} = A\mathbf{x}_n$ and the fixed point $(0, 0)$ apply to non-homogeneous linear systems $\mathbf{x}_{n+1} = A\mathbf{x}_n + \mathbf{b}$ and their fixed points (p, q).   ∎

EXAMPLE 1

Determine the dynamics of solutions around the equilibrium point $(p, q) = (2500, 25{,}000)$ for the population model

$$P_{n+1} = 0.8P_n - 0.3Q_n + 8000$$
$$Q_{n+1} = 0.2P_n + 0.9Q_n + 2000$$

previously investigated in Examples 4 and 8 of Section 4.2.

Solution: Here, the coefficient matrix has the complex eigenvalues $0.85 \pm i\sqrt{0.0575}$. Since $|0.85 \pm i\sqrt{0.0575}| \approx 0.883 < 1$, the equilibrium point $(p, q) = (2500, 25{,}000)$ must be a sink and solutions spiral to it, as previously found numerically.   ∎

EXAMPLE 2

Determine the dynamics of solutions around the equilibrium point $(p, q) = (50/3, 50)$ for the price–demand model

$$P_{n+1} = P_n + 0.4D_n - 20$$
$$D_{n+1} = -0.3P_n + D_n + 5$$

previously investigated in Examples 5 and 7 of Section 4.2.

Solution: The eigenvalues of the coefficient matrix can be shown to be $1 \pm i\sqrt{0.12}$. Since $|1 \pm i\sqrt{0.12}| \approx 1.058 > 1$ the equilibrium point $(p, q) = (50/3, 50)$ is a source. Also, solutions must spiral outward from that point to ∞. This confirms previous numerical findings. Again, however, the model breaks down when the price or demand becomes negative.   ∎

Three-State Markov Processes

In addition to these population and economic models, the mathematical tools developed in this chapter can be applied as well to an important class of problems first introduced in Chapter 2, i.e., Markov processes. Recall, for example, the two-state Markov process described there involving the distribution of urban vs. suburban dwellers. There we assumed that each decade 10% of those living in the city move to the suburbs, and 15% of those in the suburbs move to the city, while the rest remain where they are. This process was summarized in the transition diagram shown in Figure 4.23.

With C_n and S_n representing the respective populations of city and suburbs n decades after the process begins, we previously saw that the next city population C_{n+1} consists of those who stay there from the nth decade $0.90C_n$ plus those who move there from the suburbs $0.15S_n$ during this time. Similarly, S_{n+1} is made up of those who move to the suburbs from

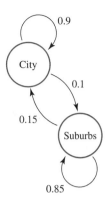

FIGURE 4.23

the city $0.10C_n$ plus those who remain in the suburbs during the previous decade $0.85S_n$. These dynamics can now be expressed in the form

$$\begin{aligned} C_{n+1} &= 0.90C_n + 0.15S_n \\ S_{n+1} &= 0.10C_n + 0.85S_n \end{aligned} \quad \text{or} \quad \begin{bmatrix} C_{n+1} \\ S_{n+1} \end{bmatrix} = \begin{bmatrix} 0.90 & 0.15 \\ 0.10 & 0.85 \end{bmatrix} \begin{bmatrix} C_n \\ S_n \end{bmatrix}$$

This is a homogeneous linear system of the form $\mathbf{x}_{n+1} = T\mathbf{x}_n$ where the **state vector** is $\mathbf{x}_n = \begin{bmatrix} C_n \\ S_n \end{bmatrix}$ and the **transition matrix** is $T = \begin{bmatrix} 0.90 & 0.15 \\ 0.10 & 0.85 \end{bmatrix}$. However, as we quickly discovered back in Chapter 2, the *dimension* of this problem can be reduced by one, since for a constant total population of 100,000 we must always have $S_n = 100,000 - C_n$. This led to the uncoupled non-homogeneous linear equation

$$C_{n+1} = 0.90C_n + 0.15(100,000 - C_n) \quad \text{or} \quad C_{n+1} = 0.75C_n + 15,000$$

which has the stable fixed point or **steady-state** of 60,000 as the city population and $100,000 - 60,000 = 40,000$ as the suburban population.

This is always the case with Markov processes of this type. One of the variables can always be eliminated to reduce the dimension of the system by one. Because of this, it is now possible to investigate three-state homogeneous Markov processes by similarly eliminating one of the variables of the state vector to produce a non-homogeneous system involving two variables. To demonstrate the technique, we develop a more elaborate version of the city/suburbs problem:

> Suppose that each decade 20% of city dwellers move to the suburbs, and an additional 10% move to rural areas; 15% of suburban residents move to the city and another 10% move to rural areas; 5% of those living in rural areas move to the city and another 10% to the suburbs; the remainder of the total population of 100,000 remain where they are.

This process can be summarized in the transition diagram shown in Figure 4.24. For decade n, if we call C_n the number of city dwellers, S_n the number of suburban dwellers and R_n the number of rural dwellers, then to determine the number of residents in each region in step $n + 1$ we follow the same logic as before. For example, C_{n+1} will be the total of the number of those who remain in the city during the previous decade $0.70C_n$, plus the

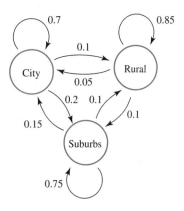

FIGURE 4.24

number who move there from the suburbs $0.15 S_n$, plus those who move there from rural areas $0.05 R_n$. This means that

$$C_{n+1} = 0.70 C_n + 0.15 S_n + 0.05 R_n$$

The other two can be computed similarly:

$$S_{n+1} = 0.20 C_n + 0.75 S_n + 0.10 R_n \quad \text{and} \quad R_{n+1} = 0.10 C_n + 0.10 S_n + 0.85 R_n.$$

Using vector and matrix notation, the problem may be written

$$\begin{bmatrix} C_{n+1} \\ S_{n+1} \\ R_{n+1} \end{bmatrix} = \begin{bmatrix} 0.70 & 0.15 & 0.05 \\ 0.20 & 0.75 & 0.10 \\ 0.10 & 0.10 & 0.85 \end{bmatrix} \begin{bmatrix} C_n \\ S_n \\ R_n \end{bmatrix}$$

This system involves multiplying the three-dimensional state vector representing the distribution in decade n:

$$\begin{bmatrix} C_n \\ S_n \\ R_n \end{bmatrix}, \quad \text{by the 3 by 3 transition matrix:} \quad T = \begin{bmatrix} 0.70 & 0.15 & 0.05 \\ 0.20 & 0.75 & 0.10 \\ 0.10 & 0.10 & 0.85 \end{bmatrix}$$

to arrive at the distribution in the decade that follows. The multiplication here is done in a manner analogous to how 2 by 2 matrices and two-dimensional vectors are multiplied.

Using the trick of eliminating one of the variables, however, this system can be simplified. Assuming the total population is still 100,000, then we know that

$$R_n = 100{,}000 - C_n - S_n$$

This means we can write

$$C_{n+1} = 0.70 C_n + 0.15 S_n + 0.05 \,(100{,}000 - C_n - S_n) = 0.65 C_n + 0.10 S_n + 5000$$

and

$$S_{n+1} = 0.20 C_n + 0.75 S_n + 0.10 \,(100{,}000 - C_n - S_n) = 0.10 C_n + 0.65 S_n + 10{,}000$$

We thus arrive at the non-homogeneous system

$$C_{n+1} = 0.65C_n + 0.10S_n + 5000$$
$$S_{n+1} = 0.10C_n + 0.65S_n + 10{,}000$$

Using the theorem developed earlier in this section, we are now able to determine easily the dynamics of this simplified system. It is found quickly that the fixed point is approximately (24,444, 35,556) and the coefficient matrix $A = \begin{bmatrix} 0.65 & 0.10 \\ 0.10 & 0.65 \end{bmatrix}$ has two real eigenvalues 0.75 and 0.55, whose magnitudes indicate that this fixed point or steady-state is stable. So the process is in equilibrium when the city population is 24,444, the suburban population is 35,556 and the rural population is

$$100{,}000 - 24{,}444 - 35{,}556 = 40{,}000$$

Further, this is a stable state. That is, any other starting distribution of populations will converge to this distribution as $n \to \infty$.

In general, any three-state Markov process

$$\begin{bmatrix} x_{n+1} \\ y_{n+1} \\ z_{n+1} \end{bmatrix} = \begin{bmatrix} a_1 & a_2 & a_3 \\ b_1 & b_2 & b_3 \\ c_1 & c_2 & c_3 \end{bmatrix} \begin{bmatrix} x_n \\ y_n \\ z_n \end{bmatrix}$$

which is a homogeneous system of the form $\mathbf{x}_{n+1} = T\mathbf{x}_n$ where the state vector and transition matrix are

$$\mathbf{x}_n = \begin{bmatrix} x_n \\ y_n \\ z_n \end{bmatrix} \quad \text{and} \quad T = \begin{bmatrix} a_1 & a_2 & a_3 \\ b_1 & b_2 & b_3 \\ c_1 & c_2 & c_3 \end{bmatrix}$$

respectively, can be converted into a non-homogeneous system for any two of its variables. Assuming that N is the total population of the three states, then we could choose to eliminate z_n from the system by first writing

$$z_n = N - x_n - y_n$$

for all n, and then substituting this for z_n in the first two equations. Dropping the now unnecessary third equation from the system yields

$$x_{n+1} = a_1 x_n + a_2 y_n + a_3(N - x_n - y_n)$$
$$y_{n+1} = b_1 x_n + b_2 y_n + b_3(N - x_n - y_n)$$

which can be written more simply as

$$x_{n+1} = (a_1 - a_3)x_n + (a_2 - a_3)y_n + a_3 N$$
$$y_{n+1} = (b_1 - b_3)x_n + (b_2 - b_3)y_n + b_3 N \tag{1}$$

This allows the steady-state vector $\mathbf{x} = \begin{bmatrix} x \\ y \\ z \end{bmatrix}$ of the Markov process $\mathbf{x}_{n+1} = T\mathbf{x}_n$ to be computed by first solving

$$x = (a_1 - a_3)x + (a_2 - a_3)y + a_3 N$$
$$y = (b_1 - b_3)x + (b_2 - b_3)y + b_3 N$$

which gives the fixed point (x, y) of the simplified system, and then combining this with $z = N - x - y$ to obtain the steady-state vector $\mathbf{x}$. Finally, since the stability or instability of one system implies the same for the other, these can be determined by computing the eigenvalues of

$$A = \begin{bmatrix} a_1 - a_3 & a_2 - a_3 \\ b_1 - b_3 & b_2 - b_3 \end{bmatrix}$$

which is the coefficient matrix of the simplified system.

EXAMPLE 3

Suppose a certain chronic illness has three distinct stages of severity. If someone is in Stage I, the most benign state, there is a 90% chance of remaining there, and a 10% chance of progressing to the intermediate Stage II. If someone is in Stage II there is an 80% chance of staying there, a 10% chance of going on to the most severe Stage III, but another 10% chance of returning to Stage I. If someone is in Stage III there is an 80% chance of remaining there, and a 20% chance of returning to Stage II. If 5000 patients are studied over an extended period of time how many should we expect to see in each of the three stages of the illness at any particular time?

Solution: The process is summarized in Figure 4.25. If we let F_n, S_n and T_n be the numbers of those at time n who are in the first, second and third stages respectively, then the dynamics

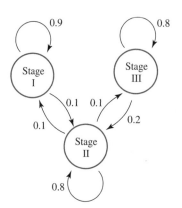

FIGURE 4.25

may be described by the system

$$
\begin{bmatrix} F_{n+1} \\ S_{n+1} \\ T_{n+1} \end{bmatrix} = \begin{bmatrix} 0.9 & 0.1 & 0.0 \\ 0.1 & 0.8 & 0.2 \\ 0.0 & 0.1 & 0.8 \end{bmatrix} \begin{bmatrix} F_n \\ S_n \\ T_n \end{bmatrix}
$$

Using (1), this can be simplified to

$$
F_{n+1} = 0.9F_n + 0.1S_n
$$
$$
S_{n+1} = -0.1F_n + 0.6S_n + 1000
$$

where $T_n = 5000 - F_n - S_n$ for all n. The fixed point of this reduced system is (2000, 2000), which makes the steady-state vector of the Markov process equal to $\begin{bmatrix} 2000 \\ 2000 \\ 1000 \end{bmatrix}$. The eigenvalues of $A = \begin{bmatrix} 0.9 & 0.1 \\ -0.1 & 0.6 \end{bmatrix}$ are approximately 0.86 and 0.64, which means the steady-state is stable. So at any given time we should expect approximately 2000 patients out of the 5000 in the study to be in Stage I, another 2000 in Stage II and 1000 in Stage III. ∎

Sometimes, rather than having available the actual initial population sizes for all the states of a Markov process, the *fraction* of the total is instead known for each. This means that with each iteration, the model generates the next fraction of the total shared by each state. In such a case, the only practical difference is that we use $N = 1$ in our computations.

EXAMPLE 4

Suppose current socioeconomic data reveals that 50% of children whose parents were low-income wage earners will also be low-income earners, 40% of these children will instead be middle-income earners and another 10% will reach high-income earnings. For middle-income parents, 80% of their children will also be middle-income earners, but 10% will be low-income earners and another 10% will be high-income earners. For high-income parents, 70% of their children will also be high-income earners, but 25% will be middle-income earners and 5% will be low-income earners. If these trends continue, what fraction of the population would we expect to consistently see in each of the three categories of wage earners?

Solution: The dynamics are shown in Figure 4.26, and may be written

$$
\begin{bmatrix} L_{n+1} \\ M_{n+1} \\ H_{n+1} \end{bmatrix} = \begin{bmatrix} 0.50 & 0.10 & 0.05 \\ 0.40 & 0.80 & 0.25 \\ 0.10 & 0.10 & 0.70 \end{bmatrix} \begin{bmatrix} L_n \\ M_n \\ H_n \end{bmatrix}
$$

where L_n, M_n and H_n respectively represent the fractions of low-, middle- and high-income earners in the population n generations after the process begins. Since we are dealing with fractions of the total population, then we can use equation (1) with $N = 1$ to reduce the problem to

$$
L_{n+1} = 0.45L_n + 0.05M_n + 0.05
$$
$$
M_{n+1} = 0.15L_n + 0.55M_n + 0.25
$$

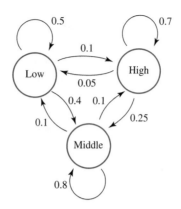

FIGURE 4.26

where $H_n = 1 - L_n - M_n$. The fixed point of this system works out to be approximately $(0.15, 0.60)$, and the eigenvalues of $A = \begin{bmatrix} 0.45 & 0.05 \\ 0.15 & 0.55 \end{bmatrix}$ are exactly 0.4 and 0.6. The conclusion therefore is that if current trends continue, then over time we should expect that approximately 15% of the population will be low-income earners, 60% will be middle-income earners and 25% $= 1 - 15\% - 60\%$ will be high-income earners. The distribution will gradually stabilize at these levels regardless of the initial distribution. ■

Exercises 4.8

In Exercises 1–10 find the fixed point of the given system and determine whether it is a sink, source or saddle.

1. $P_{n+1} = P_n + 0.4D_n - 10$
   $D_{n+1} = -0.5P_n + D_n + 2$

2. $P_{n+1} = \frac{2}{3}P_n - \frac{1}{3}Q_n + 4000$
   $Q_{n+1} = \frac{1}{6}P_n + \frac{1}{6}Q_n + 5000$

3. $P_{n+1} = 0.75P_n - Q_n + 1000$
   $Q_{n+1} = -0.5P_n + 0.25Q_n + 1500$

4. $P_{n+1} = P_n + 1.5D_n - 20$
   $D_{n+1} = -0.75P_n + D_n + 6$

5. $x_{n+1} = -\frac{1}{8}x_n + \frac{1}{4}y_n + 1$
   $y_{n+1} = -\frac{1}{2}x_n + \frac{1}{4}y_n + 2$

6. $x_{n+1} = 3x_n + 2y_n + 2$
   $y_{n+1} = -x_n + y_n - 2$

7. $P_{n+1} = P_n + 0.25D_n + 3$
   $D_{n+1} = -0.5P_n + D_n + 5$

8. $P_{n+1} = 0.25P_n - Q_n + 2200$
   $Q_{n+1} = -P_n + 0.25Q_n + 2000$

9. $P_{n+1} = \frac{1}{10}P_n - \frac{1}{5}Q_n + 1000$
   $Q_{n+1} = \frac{1}{5}P_n + \frac{1}{2}Q_n + 2500$

10. $P_{n+1} = 0.25P_n - 0.25Q_n + 1000$
    $Q_{n+1} = -0.5P_n + 0.25Q_n + 1250$

In Exercises 11 and 12 find the fixed point of the given second-order equation and determine its stability by converting it to a system.

11. $9x_{n+2} - 12x_{n+1} + 4x_n = 8$

12. $5x_{n+2} - 8x_{n+1} + 8x_n = 20$

13. Show that solutions never rotate around the equilibrium point of Linear Competition Model I/II with positive parameters r_1, r_2, s_1, s_2. *Hint:* Show that the eigenvalues are real.

14. Show that solutions always rotate around the equilibrium point of Linear Prey–Predator Model I/II with positive parameters: r_1, r_2, s_1, s_2, if $r_1 = r_2$.

For the Markov process corresponding to each of the 3 by 3 transition matrices T given in Exercises 15–18: (a) Convert into a 2 by 2 non-homogeneous system; (b) Find the steady-state (assuming a 100% total) and determine its stability.

15. $T = \begin{bmatrix} 1 & 1/2 & 1/3 \\ 0 & 1/2 & 1/3 \\ 0 & 0 & 1/3 \end{bmatrix}$

17. $T = \begin{bmatrix} 0 & 1 & 1/3 \\ 1 & 0 & 1/3 \\ 0 & 0 & 1/3 \end{bmatrix}$

16. $T = \begin{bmatrix} 1 & 0 & 0 \\ 0 & 1/2 & 1/2 \\ 0 & 1/2 & 1/2 \end{bmatrix}$

18. $T = \begin{bmatrix} 0 & 0 & 1/2 \\ 1/2 & 0 & 1/2 \\ 1/2 & 1 & 0 \end{bmatrix}$

For each of the Markov processes given in Exercises 19–22: (a) Find the 3 by 3 transition matrix; (b) Convert into a 2 by 2 non-homogeneous system; (c) Find the steady-state (assuming a 100% total) and determine its stability.

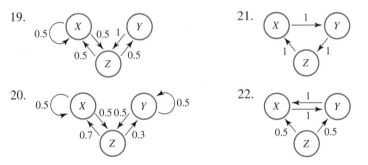

19.

21.

20.

22.

23. In the city/suburbs/rural dwellers problem discussed in this section, suppose instead that each decade 10% of city dwellers move to the suburbs and an additional 5% to rural areas; 10% of suburban dwellers move to the city and another 10% to rural areas; 5% of rural dwellers move to the city and another 5% to the suburbs; the remainder of the total population of 100,000 remain where they are. Find the steady-state in this case and determine its stability.

24. In Example 4 of this section suppose instead that of the children of low-income earners 50% will be low-, 40% will be middle- and 10% will be high-income earners. Of the children of middle-income earners 10% will be low-, 70% will be middle- and 20% will be high-income earners. Of the children of high-income earners 5% will be low-, 15% will be middle- and 80% will be high-income earners. Find the steady-state distribution of income earners and determine its stability.

25. Suppose that the next choice of vehicle for those who currently drive a small car is as follows: 60% will again choose a small car, 20% a large car, and the rest an SUV. Of those who currently drive a large car 10% will next choose a small car, 50% another large car, and the rest an SUV. Of SUV drivers 10% will next choose a small car, 10% a

large car, and the rest another SUV. What percent of drivers will eventually drive small cars, what percent large cars and what percent SUV's?

26. Suppose trainers of guide dogs classify their subjects as being either untrained, partially trained or fully trained. Each month 50% of untrained dogs become partially trained, but the rest remain untrained; 20% of partially trained dogs remain partially trained, another 70% become fully trained, but 10% revert to untrained status; 80% of fully trained dogs remain fully trained, but 20% revert to partially trained status. What percent of these dogs will eventually be in each of these three classifications?

27. For the general non-homogeneous system $\mathbf{x}_{n+1} = A\mathbf{x}_n + \mathbf{b}$ derive the following formula for the exact solution: $\mathbf{x}_n = A^n\mathbf{x}_0 + (I + A + A^2 + A^3 + \cdots + A^{n-1})\mathbf{b}$.

28. Derive an alternate formula for the exact solution of $\mathbf{x}_{n+1} = A\mathbf{x}_n + \mathbf{b}$ as follows:

(a) For any A show that $(A - I)(A^{n-1} + A^{n-2} + \cdots + A^2 + A + I) = A^n - I$.

(b) Show that if the matrix $A - I$ has an inverse (which occurs when there is a unique fixed point) then

$$A^{n-1} + A^{n-2} + \cdots + A^2 + A + I = (A - I)^{-1}(A^n - I)$$

(c) Use the result of the previous exercise to conclude that

$$\mathbf{x}_n = A^n\mathbf{x}_0 + (A - I)^{-1}(A^n - I)\mathbf{b}$$

Compare this with the formula for the exact solution of $x_{n+1} = ax_n + b$.

■■■■■■ Computer Projects 4.8

Note: To compute the powers of a matrix, see Exercise 35 of Section 4.3.

For a regular Markov process, i.e., one that has a stable steady-state, powers of the transition matrix satisfy a certain property in the limit. Determine which of the transition matrices T *given in Projects 1–4 are regular, and then compute:* T, T^2, T^3, ..., T^{10}. *What happens to those that are regular?*

1. $T = \begin{bmatrix} 1/2 & 1/3 \\ 1/2 & 2/3 \end{bmatrix}$ 3. The matrix T from Exercise 15.

2. $T = \begin{bmatrix} 0 & 1 \\ 1 & 0 \end{bmatrix}$ 4. The matrix T from Exercise 17.

If $A^n \to O$ *(the matrix of all 0's) as* $n \to \infty$, *then Exercise 28(b) implies that* $I + A + A^2 + A^3 + \cdots = (I - A)^{-1}$. *For each of the matrices* A *given in Projects 5–8, find* $(I - A)^{-1}$ *and then compute* $I + A + A^2 + A^3 + \cdots A^n$ *for* n = 5, 10, 15, 20. *Does the infinite sum appear to converge to* $(I - A)^{-1}$?

5. $A = \begin{bmatrix} 0.5 & 0.3 \\ 0.4 & 0.7 \end{bmatrix}$ 7. $A = \begin{bmatrix} 1/2 & 2/3 \\ -1/3 & -1/2 \end{bmatrix}$

6. $A = \begin{bmatrix} 4/5 & -1 \\ 1/4 & 4/5 \end{bmatrix}$ 8. $A = \begin{bmatrix} 1 & -1/2 \\ 1/2 & 0 \end{bmatrix}$

5

Modeling with Nonlinear Systems

Not all interacting processes that may be of interest in the natural and social sciences can be adequately modeled using systems of iterative equations having linear forms. For many, the equations that most accurately represent their evolution over time are inherently nonlinear, which implies that the theory developed in the previous chapter for the linear case might not apply.

In this final chapter we investigate the dynamics of discrete processes that are best modeled using systems of nonlinear equations. In addition to their being the most sophisticated type of iterative process we shall consider, they also pose for us the greatest challenge. Due to their complexity, the dynamics of nonlinear systems are far from being fully understood, and a comprehensive theory still awaits development. For this reason, the discussion presented in an introductory text such as this must be understandably brief. Here, even more so than in the preceding chapters, we must rely substantially upon the computer analysis of such problems, while only touching upon the few known elementary mathematical techniques ordinarily used for their investigation.

5.1   Nonlinear Systems and Their Dynamics

Recall that several of the nonlinear models constructed in Chapter 3 were refinements of simpler linear models introduced in Chapter 2. Analogously, nonlinear systems may result by similarly refining models based upon systems of linear equations, such as those just discussed in Chapter 4. In this section we adopt that approach, revisiting some previously described areas of application of discrete systems from population biology and from economics. We also take the first few steps in our attempt to uncover the dynamics of nonlinear systems, which is the primary goal of this chapter.

Some Nonlinear Systems Models

In our construction of the linear prey–predator models of the previous chapter, it was assumed that, independent of the other, each population possesses a constant intrinsic growth rate: r_1 for the prey and r_2 for the predator. Also, since the predator consumes the prey, the next prey population P_{n+1} is decreased by an amount directly proportional to the size of the present predator population Q_n, while the next predator population Q_{n+1} is increased by an amount directly proportional to the present prey population P_n.

However, common sense might tell us that it would be more fitting to assume instead that the next prey population is diminished to a degree that is directly proportional to the *number of contacts* between prey and predator during the previous time step, and that the next predator population is enhanced by a similar quantity. As seen in Chapter 3, the number of contacts between two groups is often considered to be directly proportional to the product of their sizes, in this case: P_n and Q_n. Consequently, a quantity such as $s_1 P_n Q_n$ should be subtracted from the next prey population, and a quantity such as $s_2 P_n Q_n$ added to the next predator population. This yields an alternate and perhaps more accurate model for the population growth of a prey and its predator.

NONLINEAR PREY–PREDATOR MODEL

The nonlinear system

$$P_{n+1} = r_1 P_n - s_1 P_n Q_n$$

$$Q_{n+1} = r_2 Q_n + s_2 P_n Q_n$$

where all parameters are non-negative, models a prey–predator relationship between two interacting species. P_n represents the prey and Q_n the predator.

We say that this system is *nonlinear* since at least one of the equations being iterated corresponds to a nonlinear function of its two primary variables, and consequently has a non-planar graph. In this case, the graphs of both $z = r_1 x - s_1 xy$ and $z = r_2 y + s_2 xy$ describe curved surfaces in three-dimensional (x, y, z) space, provided of course that s_1 and s_2 are nonzero. This system is also referred to as *autonomous* since neither equation depends explicitly upon the variable n. It should be apparent that, although the equations involved are nonlinear, given values for P_0 and Q_0, they can be iterated just like those of an autonomous linear system.

We remark that any number of different nonlinear models might be constructed that describe the interplay between a prey and its predator, or equivalently a host and its parasite. Each would differ primarily with regard to the forms of the nonlinear terms used to measure the effects that contact between the two species has on their respective populations. This, as well as the issues of immigration, migration and harvesting, will be taken up further in the exercises at the end of this section. Similar remarks might be made with regard to each of the population models considered below.

With a small modification, this nonlinear prey–predator system can be converted into one that models the population dynamics of certain types of competing species. Suppose that two species with intrinsic growth rates of r_1 and r_2 are each diminished in size to an extent that is directly proportional to the number of contacts between them during any time period. This may be due to the resulting loss of members of at least one of these species after contact with a competitor. If the number of contacts is once again assumed to be directly proportional to the product of the present population sizes, then the following model describes this interaction.

NONLINEAR COMPETITION MODEL

The nonlinear system

$$P_{n+1} = r_1 P_n - s_1 P_n Q_n$$

$$Q_{n+1} = r_2 Q_n - s_2 P_n Q_n$$

where all parameters are non-negative, models the populations of two competing species.

If density dependence is taken into consideration, the overlapping-generations models constructed in Chapter 4 will also be transformed into nonlinear systems. As before, suppose that for some species the populations of the present and previous generations, P_n and P_{n-1} respectively, both contribute to that of the next generation P_{n+1}, each with its own growth rate. Now, however, rather than these growth rates being constant, due to density dependence, we assume instead that they are decreasing functions of the total population size $P_n + P_{n-1}$. Further, these functions equal 0 if the total population ever reaches the carrying capacity C of the environment.

The simplest growth rate functions that satisfy these criteria are linear ones having the forms

$$r\left(1 - \frac{P_n + P_{n-1}}{C}\right) \quad \text{and} \quad s\left(1 - \frac{P_n + P_{n-1}}{C}\right)$$

respectively, where r and s are growth rate parameters. This implies that the model may be written in the form

$$P_{n+1} = r\left(1 - \frac{P_n + P_{n-1}}{C}\right) P_n + s\left(1 - \frac{P_n + P_{n-1}}{C}\right) P_{n-1}$$

and after factoring also as

$$P_{n+1} = (r\, P_n + s\, P_{n-1}) \left(1 - \frac{P_n + P_{n-1}}{C} \right) \tag{1}$$

Finally, if we convert this second-order equation to a system as before, by letting $Q_n = P_{n-1}$, which implies $Q_{n+1} = P_n$, the following model results.

NONLINEAR OVERLAPPING GENERATIONS MODEL

The nonlinear system

$$P_{n+1} = (r\, P_n + s\, Q_n) \left(1 - \frac{P_n + Q_n}{C} \right)$$

$$Q_{n+1} = P_n$$

for $r,\ s \geq 0$ and $C > 0$ is a density-dependent population model for a species with overlapping generations. The system is equivalent to the second-order equation (1) with $Q_n = P_{n-1}$.

The spread of a contagious disease may also need to be modeled using a nonlinear system. In constructing such models in Chapter 3, it was assumed that the fraction r of infected individuals, where $0 \leq r \leq 1$, recover at each step, and immediately after recovery an individual is again susceptible to the disease. Suppose now, however, that a temporary immunity is conferred after infection and recovery. This creates a new group of recovered and immune individuals, whose population R_n changes along with the number of infected I_n. Since immunity is temporary, let us assume that the fraction t of the recovered population, where $0 \leq t \leq 1$, lose immunity at each step. Since $t\, R_n$ leave the recovered and immune group each time, while $r\, I_n$ enter it, we must have

$$R_{n+1} = R_n - t\, R_n + r\, I_n \tag{2}$$

Note that, because of this new group, the number of individuals susceptible to the disease at any time n must now equal $N - I_n - R_n$, where N is the size of the total population.

In Chapter 3, it was observed that the next number of infected individuals I_{n+1} should be equal to the number currently infected I_n, minus the number recently recovered $r\, I_n$, plus the number of new cases. That is,

$$I_{n+1} = I_n - r\, I_n + New\ Cases \tag{3}$$

It was also assumed there that the spread of the disease is directly proportional to the number of contacts between the infected and the susceptible groups, or equivalently, the product of the population sizes of the two groups. But since these sizes are now

$$I_n \quad \text{and} \quad N - I_n - R_n$$

respectively, then the number of new cases must therefore take the form

$$New\ Cases = k\, I_n\, (N - I_n - R_n) = k\, N\, I_n \left(1 - \frac{I_n + R_n}{N} \right)$$

Letting $s = kN$ in this expression, and combining it with (2) and (3), provides us with the following model.

NONLINEAR INFECTION-RECOVERY MODEL

The nonlinear system

$$I_{n+1} = I_n - r\, I_n + s\, I_n \left(1 - \frac{I_n + R_n}{N} \right)$$

$$R_{n+1} = R_n - t\, R_n + r\, I_n$$

where $s \geq 0$ and $0 \leq r, t \leq 1$, models the spread of a contagious disease in a population of size $N > 0$. I_n represents the number infected and R_n the number who have recovered and are temporarily immune.

As a final application of nonlinear systems, consider the economic models constructed in the last chapter. There it was assumed that the next price of a product P_{n+1} is related to the present price P_n, as well as to the present supply S_n and demand D_n, by an equation of the form

$$P_{n+1} = P_n + a_1\, D_n - a_2\, S_n + a_3 \tag{4}$$

where $a_1, a_2 \geq 0$. Clearly, this iterative equation is linear, as are the associated ones used in Section 4.1 to describe the evolution over time of S_n and D_n. Since there is little reason to believe that all these relationships are necessarily linear, here we will instead use nonlinear equations for the latter two.

To satisfy the demand principle cited in Chapter 4, i.e., that a high price causes the demand to decrease from its present level, and a low price causes the demand to increase from its present level, let us now assume

$$D_{n+1} = D_n + \frac{c_1}{P_n} - k_1 \tag{5}$$

where $c_1, k_1 \geq 0$, which is one of many nonlinear equations satisfying that economic principle. Further, since supply is an increasing (or at least non-decreasing) function of price, as discussed in Chapters 2 and 3, suppose we now write that function as

$$S_n = c_2 P_n^2 + d_2$$

where $c_2, d_2 \geq 0$. To avoid a system of three iterative equations, we have assumed here that supply either remains constant or reacts immediately to any price changes. Substituting this expression for S_n into (4) gives

$$P_{n+1} = P_n + a_1\, D_n - b_1\, P_n^2 + h_1$$

where $b_1 = a_2 c_2$ and $h_1 = a_3 - a_2 d_2$. Combining this nonlinear equation with (5) yields the following model.

NONLINEAR PRICE-DEMAND MODEL

The nonlinear system

$$P_{n+1} = P_n + a_1 \, D_n - b_1 \, P_n^2 + h_1$$

$$D_{n+1} = D_n + \frac{c_1}{P_n} - k_1$$

where all parameters (except perhaps h_1) are non-negative, describes the interaction between the price P_n and the demand D_n.

Dynamics of Nonlinear Systems

It should be apparent that all of the nonlinear models just introduced fall under the following classification scheme.

DEFINITION

The general form of an **autonomous nonlinear system** is

$$x_{n+1} = f(x_n, y_n)$$
$$y_{n+1} = g(x_n, y_n) \tag{6}$$

where at least one of the real-valued functions $f(x, y)$ and $g(x, y)$ is nonlinear. This is equivalent to

$$(x_{n+1}, y_{n+1}) = F(x_n, y_n) \quad \text{where} \quad F(x, y) = (f(x, y), g(x, y))$$

When initial values are provided for x_0 and y_0, a unique solution exists for x_n and y_n for all $n \geq 0$.

Similar to the situation in Chapter 3, we must generally require that the range of the function $F(x, y)$ be either contained within or exactly equal to its domain in the (x, y) plane. Otherwise, a solution (x_n, y_n) may not exist for all $n \geq 0$. We will also assume that $f(x, y)$ and $g(x, y)$, and consequently $F(x, y)$, are at least continuous, and in most cases differentiable for all (x, y) in the domain of interest. This is true for all the models we have been discussing.

It should come as no surprise that fixed points play a significant role in the dynamics we are investigating. So we also define the following.

DEFINITION

A point (p, q) is a **fixed point** of (6), or equivalently a fixed point of $F(x, y)$, if it satisfies

$$p = f(p, q)$$
$$q = g(p, q)$$

or equivalently $F(p, q) = (p, q)$. Letting $(x_0, y_0) = (p, q)$ will make the $(x_n, y_n) = (p, q)$ for all $n \geq 0$, and so (p, q) is an **equilibrium point** of the model.

Like those of linear systems, fixed points of nonlinear systems are also classified as **sinks, sources** or **saddles,** depending upon the geometry that solutions around them exhibit when graphed in solution space. However, there is a difference. Unlike the linear case, these terms refer now only to the *local* behavior of solutions near the fixed point, rather than to the *global* dynamics. The situation is analogous to local stability of nonlinear equations investigated in Chapter 3, where a fixed point was considered stable if it attracted all *nearby* solutions, or unstable if it repelled them. What happened to solutions elsewhere had no bearing on whether a fixed point was classified as an attractor or a repeller. In the present setting we say that a fixed point (p, q) is **locally stable** if it is a sink for the nonlinear system, or **unstable** otherwise. The word *locally* is usually omitted here, since we seldom refer to global stability.

Since the pair of simultaneous equations

$$x = f(x, y) \quad \text{and} \quad y = g(x, y)$$

that must be solved to find the fixed points $(p, q) = (x, y)$ of (6) potentially yields any number of solutions, also as in Chapter 3, there may exist multiple fixed points, or perhaps none at all. Complicating the matter further, the algebra involved in solving that pair of equations may prohibit these fixed points from ever being computed exactly.

The following examples demonstrate some of these ideas.

EXAMPLE 1

Find all fixed points, if any, of the nonlinear system:

$$x_{n+1} = x_n^2 - y_n + 5$$
$$y_{n+1} = 2x_n - y_n + 4$$

Solution: Any fixed point of this system must satisfy

$$x = x^2 - y + 5$$
$$y = 2x - y + 4$$

Solving the second equation for y gives $y = x + 2$, and substituting this into the first equation and simplifying yields the quadratic equation

$$x^2 - 2x + 3 = 0.$$

Since the quadratic formula tells us that $x = 1 \pm \sqrt{-2}$, there are no (real) fixed points of the system. ∎

EXAMPLE 2

(a) Find all equilibrium points of the prey–predator model:

$$P_{n+1} = 5P_n - 0.02 P_n Q_n$$
$$Q_{n+1} = 0.1 Q_n + 0.01 P_n Q_n$$

(b) Determine whether each is a sink, a source or a saddle.

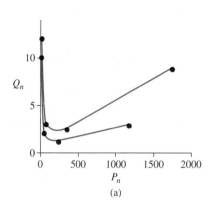

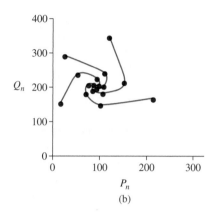

FIGURE 5.1

Solution: (a) Here we must find all solutions of

$$x = 5x - 0.02xy$$
$$y = 0.1y + 0.01xy$$

By inspection, it should be apparent that one possible solution is $(x, y) = (0, 0)$. Dividing the first equation by x and the second by y yields

$$1 = 5 - 0.02y \quad \text{and} \quad 1 = 0.1 + 0.01x$$

respectively, from which we find another solution $(x, y) = (90, 200)$. So, $(0, 0)$ and $(90, 200)$ are the equilibrium points of the model.

(b) Since we have not yet developed any mathematical tools for distinguishing sinks, sources and saddles for a nonlinear system, we have no alternative other than to iterate numerically and observe the dynamics in solution space. Figure 5.1(a) shows the first few iterates of two trajectories with initial points chosen from the first quadrant close to the y-axis. Each appears to follow a hyperbolic path, indicating that $(0, 0)$ is a saddle. Figure 5.1(b) shows the first few iterates of several trajectories originating near $(90, 200)$, each of which spirals outward from that point. Based on this, $(90, 200)$ must be a source. It is worth noting, however, that the dynamics described here are local, i.e., none of the trajectories maintain their hyperbolic or spiral shapes farther away from the fixed points.   ∎

Not only might several different sinks, sources and saddles coexist for the same nonlinear system, as the previous example implies, but other forms of asymptotic dynamics that we have not witnessed before may be present as well.

EXAMPLE 3

Investigate the dynamics of the Nonlinear Overlapping-Generations Model with carrying capacity $C = 1000$ and growth rates (a) $r = 2.75$, $s = 0.85$; (b) $r = 1.1$, $s = 4.4$.

Solution: (a) We first find the equilibrium points by solving

$$x = (2.75x + 0.85y)(1 - (x + y)/1000)$$
$$y = x$$

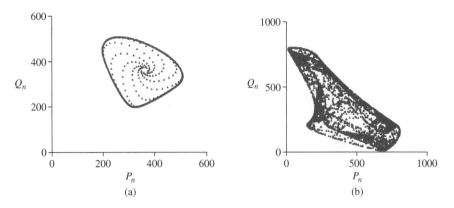

FIGURE 5.2

Substituting x in place of y in the first equation yields

$$x = 3.6x(1 - x/500)$$

This quadratic equation has two solutions: $x = 0$ and $x = 3250/9 \approx 361$. Since we must have $y = x$, then the equilibrium points are $(0, 0)$ and $(361, 361)$ (after rounding off to whole numbers due to the population model interpretation).

Not only would solution-space graphs show that $(0, 0)$ is a saddle and $(361, 361)$ is a source, but this time they also indicate some additional interesting dynamics: Virtually any trajectory that remains bounded approaches the same *continuous curve* in solution space that loops completely around the fixed point $(361, 361)$. This curve is *invariant,* i.e., any trajectory that hits the curve remains on it thereafter. Figure 5.2(a) shows points of a trajectory that originates close to $(361, 361)$, but spirals outward toward that invariant curve. This type of behavior cannot exist for autonomous linear systems.

(b) In a similar fashion, the equilibrium points are found in this case to be $(0, 0)$ and $(4500/11, 4500/11) \approx (409, 409)$. Although solution-space graphs once again indicate that the former is a saddle and the latter a source, trajectories this time do not approach a simple invariant curve. Instead, the points of nearly every bounded trajectory wander in a random-like manner, appearing to ultimately fill up an entire two-dimensional region in solution space. Figure 5.2(b) shows the first 5000 points of a typical trajectory. Again, this does not occur for autonomous linear systems. ∎

In the following sections, we will investigate these ideas more carefully, and as in previous chapters attempt to explain once again mathematically the dynamics we are seeing.

Exercises 5.1

1. For the Nonlinear Competition Model with $r_1 = 8$, $r_2 = 4$, $s_1 = 0.005$, $s_2 = 0.002$, $P_0 = 1510$ and $Q_0 = 1420$, compute (a) P_1 and Q_1 (b) P_2 and Q_2

2. Construct the Nonlinear Prey–Predator Model that satisfies

   (a) The prey's growth rate is 3 and the predator's is 0.5, and if $P_0 = 100$ and $Q_0 = 50$, then $P_1 = 250$ and $Q_1 = 425$.

   (b) $P_0 = 100$, $Q_0 = 100$, $P_1 = 200$, $Q_1 = 300$, $P_2 = 150$ and $Q_2 = 1200$.

3. For the disease model $\begin{cases} I_{n+1} = 0.85I_n + 1.75I_n(1 - (I_n + R_n)/10^5) \\ R_{n+1} = 0.95R_n + 0.15I_n \end{cases}$

   (a) Identify the constants r, s, t and N.
   (b) If $I_0 = 4000$ and $R_0 = 500$, compute I_1 and R_1.

4. Construct the Nonlinear Infection–Recovery Model for which the total population is 10,000, each month 40% of those infected recover and 10% of those recovered lose their immunity, and

   (a) if 50 are currently infected and 350 are currently recovered and immune, then there are six new cases the following month.
   (b) if $I_0 = 100$ and $R_0 = 400$, then $I_1 = 136$.

5. For the Nonlinear Competition Model with $r_1, r_2, s_1 \, s_2, P_0$ and Q_0 as given in Exercise 1, suppose additionally that the first species migrates at a constant rate of 120 per step, and the second undergoes immigration at a constant rate of 50 per step. (a) Construct this new model. (b) Compute P_1 and Q_1 for this new model.

6. (a) Modify the second-order, overlapping-generations equation (1) of this section to account for immigration, migration or harvesting at a constant positive or negative rate of k per step. (b) Convert this new equation into a first-order system.

7. (a) Modify the second-order, overlapping-generations equation (1) of this section by assuming instead that the growth rates for the present and previous generations are $re^{-(P_n+P_{n-1})/N}$ and $se^{-(P_n+P_{n-1})/N}$ respectively. (b) Convert this new equation into a first-order system.

8. Modify the Nonlinear Prey–Predator Model by replacing the constant growth rates r_1 and r_2 with the density dependent growth rates

   (a) $r_1(1 - P_n/C_1)$ and $r_2(1 - Q_n/C_2)$ respectively.
   (b) $\dfrac{r_1}{1 + P_n^2/N_1^2}$ and $\dfrac{r_2}{1 + Q_n^2/N_2^2}$ respectively.

9. Modify the Nonlinear Competition Model by assuming instead that at every step each population is diminished by an amount that is directly proportional to

   (a) the product of the squares of the two population sizes.
   (b) the product of the square roots of the two population sizes.

10. Modify the Nonlinear Infection–Recovery Model by assuming instead that at each step the number of new cases is directly proportional to

   (a) $I_n^2(1 - (I_n + R_n)/N)$ (b) $I_n^2(1 - (I_n + R_n)/N)^2$

11. Modify the Nonlinear Price–Demand Model by assuming instead that supply always satisfies

   (a) $S_n = c_2\sqrt{P_n} + d_2$ (b) $S_n = c_2 e^{P_n} + d_2$

12. Construct a nonlinear price–supply model by assuming instead that demand is either constant or reacts immediately to price $D_n = c_1/P_n + d_1$, and that supply satisfies $S_{n+1} = S_n + c_2 P_n^2 - k_2$. Combine these with equation (4).

In Exercises 13–20 find all fixed points (if any) of each system.

13. $P_{n+1} = P_n + 4D_n - 2P_n^2 + 7$ 14. $P_{n+1} = 2P_n e^{-Q_n/100}$
    $D_{n+1} = D_n + 3/P_n - 1.5$ $Q_{n+1} = 0.6Q_n e^{P_n/500}$

15. $I_{n+1} = 3I_n(1 - (I_n + R_n)/2700)$
     $R_{n+1} = 0.875R_n + I_n$

18. $P_{n+1} = \dfrac{1.8P_n + 2.7Q_n}{1 + (P_n + Q_n)^2/10^8}$
     $Q_{n+1} = P_n$

16. $I_{n+1} = 0.8I_n + 2I_n(1 - (I_n + R_n)/10^5)$
     $R_{n+1} = 0.5R_n + 0.2I_n$

19. $x_{n+1} = x_n y_n + 1$
     $y_{n+1} = y_n/x_n + 4$

17. $P_{n+1} = (3P_n + Q_n)e^{-(P_n+Q_n)/3000}$
     $Q_{n+1} = P_n$

20. $x_{n+1} = x_n^2 - y_n^2 + 1$
     $y_{n+1} = 2x_n y_n$

21. For the general Nonlinear Overlapping-Generations Model find the non-zero fixed point in terms of the parameters r, s and C.

22. For the general Nonlinear Price–Demand Model find the non-zero fixed point in terms of its parameters.

23. Find a formula for the exact solution of the nonlinear system $\begin{cases} x_{n+1} = x_n^a \\ y_{n+1} = b\,y_n \end{cases}$ for any constants a and b.

24. Find a formula for the exact solution of the nonlinear system $\begin{cases} x_{n+1} = y_n^a \\ y_{n+1} = x_n^a \end{cases}$ for any constant a.

25. Find a formula for the exact solution of the nonlinear system $\begin{cases} x_{n+1} = a/y_n \\ y_{n+1} = a/x_n \end{cases}$ for any constant a.

26. Several of the models constructed in this section can be simplified somewhat by making a change of variables that eliminates one of its parameters. (a) Use the substitutions $P_n = Cx_n$ and $Q_n = Cy_n$ to convert the Nonlinear Overlapping-Generations Model into a nonlinear system for x_n and y_n that does not depend upon the parameter C. (b) Use the substitutions $I_n = Nx_n$ and $R_n = Ny_n$ to convert the Nonlinear Infection–Recovery Model into a nonlinear system for x_n and y_n that does not depend upon the parameter N.

27. (a) Use the substitutions $x_n = 1/u_n$ and $y_n = 1/v_n$ to convert the nonlinear system $\begin{cases} x_{n+1} = x_n y_n/(8x_n + 2y_n) \\ y_{n+1} = x_n \end{cases}$ into a linear one for u_n and v_n.
    (b) Determine the asymptotic behavior of all solutions (u_n, v_n) of this linear system as $n \to \infty$.
    (c) What does this say about the asymptotic behavior of all solutions (x_n, y_n) of the original nonlinear system as $n \to \infty$

28. Another use of discrete nonlinear systems is to find approximate solutions of nonlinear systems of differential equations, such as the autonomous system $\begin{cases} x'(t) = f(x(t), y(t)) \\ y'(t) = g(x(t), y(t)) \end{cases}$ where $x(0)$ and $y(0)$ are known.
    (a) Use the method of Section 3.4 to derive a two-dimensional Euler Method for this system with a step size of $h > 0$, where (x_n, y_n) approximates $(x(nh), y(nh))$ for $n = 0, 1, 2, \ldots$. *Hint:* First let $x'(t) \approx (x(t + h) - x(t))/h$ and $y'(t) \approx (y(t + h) - y(t))/h$, and then replace t by nh.
    (b) What conditions must a fixed-point solution (p, q) satisfy in this case?

Computer Projects 5.1

In Projects 1 and 2 find all fixed points exactly and determine whether each is a sink, source or saddle by graphing the first few iterates of several nearby trajectories. Also, graph at least the first 5000 points of several trajectories (that remain bounded) and describe what other types of dynamics you observe.

1. The Nonlinear Infection–Recovery Model for $N = 1000$ and
   - (a) $r = 0.75, s = 2, t = 0.5$
   - (b) $r = 0.295, s = 5, t = 0.5$

2. The Nonlinear Price–Demand Model for $a_1 = 2.5, h_1 = -1, k_1 = 1$ and
   - (a) $b_1 = 0.11, c_1 = 2.6$
   - (b) $b_1 = 0.125, c_1 = 2.34$

3. The system $\begin{cases} I_{n+1} = \dfrac{s I_n^2}{6000}\left(1 - \dfrac{I_n + R_n}{6000}\right) \\ R_{n+1} = I_n \end{cases}$ may be considered a nonlinear disease

   model in which $N = 6000, r = t = 1$ and the number of new cases is now directly proportional to $\dfrac{I_n^2}{N}\left(1 - \dfrac{I_n + R_n}{N}\right)$. (a) For $s = 9$ first find all fixed points exactly, and then use numerical experimentation to show that two of them are stable. (b) Use numerical experimentation to show that this model has a threshold for all $s \geq 0$. That is, regardless of the value of s, small initial values I_0 and R_0 always lead to $I_n \to 0$ and $R_n \to 0$, or eradication of the disease. But for larger initial populations this does not necessarily occur.

In Projects 4–7 use numerical experimentation to determine the validity of the given extinction assertion. Note that extinction occurs to the first species whose population becomes 0 or negative.

4. For any Nonlinear Prey–Predator model with $r_1 > 1$ and any $P_0, Q_0 > 0$, the prey always becomes extinct.

5. For any Nonlinear Competition Model with $r_1, r_2 > 1, s_1 = s_2 > 0$ and $P_0 = Q_0 > 0$, the species with the smaller growth rate r_1 or r_2 always becomes extinct.

6. For any Nonlinear Competition Model with $r_1 = r_2 > 1$ and $P_0 = Q_0 > 0$, the species with the larger constant of proportionality s_1 or s_2 always becomes extinct.

7. For any Nonlinear Competition Model with $r_1 = r_2 > 1$ and $s_1 = s_2 > 0$, the species with the smaller initial population P_0 or Q_0 always becomes extinct.

S E C T I O N
5.2   Linearization and Local Dynamics

To some extent the qualitative behavior of nonlinear systems may be characterized as a mixture of the dynamics of both nonlinear equations from Chapter 3 and systems of linear equations from Chapter 4. Nonlinear systems have much in common with each of those previously investigated types of problems, as one may have already gathered from our brief analysis thus far.

In this section we make many more connections to the dynamics encountered in those previous chapters, as well as to the tools of analysis developed there. We begin with an important technique for classifying fixed points that makes use of both derivatives, as in Chapter 3, and eigenvalues, as in Chapter 4. We later extend those ideas to periodic points.

Linearization

As we have seen repeatedly throughout this text, the general behavior of a dynamical system is often linked to the type of equilibrium point(s) it has. Although we were able to classify each of the fixed points encountered in the previous section as a sink, source or saddle through numerical experimentation, deciding which one it is without extensive calculation would once again prove much more convenient. Fortunately, there is a way to accomplish this for a differentiable system, provided the coordinates of the equilibrium point can actually be computed.

First, recall from Chapter 3 that the local dynamics of a nonlinear equation of the form $x_{n+1} = f(x_n)$ near a fixed point p can be determined by linearizing the function $f(x)$ at that point. This means approximating the curve $y = f(x)$ by the straight line $y = ax + b$ that is tangent to it at $x = p$. One then determines the local dynamics of $x_{n+1} = f(x_n)$ through an analysis of the simpler linear equation $x_{n+1} = ax_n + b$. Among the local dynamical features that must be the same for both equations are stability or instability of p, and any oscillation that occurs around p.

By similarly linearizing a nonlinear system, one can determine whether an equilibrium point is locally a sink, source or saddle, and whether or not rotation occurs locally around it. Linearizing the system

$$x_{n+1} = f(x_n, y_n)$$
$$y_{n+1} = g(x_n, y_n)$$

(1)

involves approximating the two surfaces $z = f(x, y)$ and $z = g(x, y)$ in three-dimensional (x, y, z) space by the unique planes

$$z = ax + by + h \quad \text{and} \quad z = cx + dy + k$$

(2)

that are respectively tangent to them at the fixed point (p, q). Figure 5.3 illustrates the idea for $z = f(x, y)$. The geometry is similar for $z = g(x, y)$.

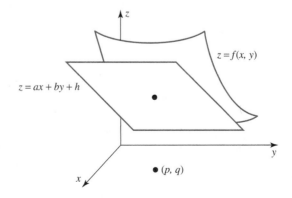

FIGURE 5.3

To construct these tangent planes, we use

$$f(x, y) \approx f(p, q) + f_x(p, q)(x - p) + f_y(p, q)(y - q)$$
$$g(x, y) \approx g(p, q) + g_x(p, q)(x - p) + g_y(p, q)(y - q) \tag{3}$$

Although we shall not go into their derivation here, these equations come from multivariate calculus. Here, the four quantities

$$f_x(p, q), \quad f_y(p, q), \quad g_x(p, q), \quad g_y(p, q)$$

represent the partial derivatives of the functions $f(x, y)$ and $g(x, y)$, evaluated at the fixed point (p, q). For those unfamiliar with the process of finding partial derivatives, examples were given in Section 2.7. Additional examples appear below.

Next, since (p, q) is a fixed point of (1) then we must have $f(p, q) = p$ and $g(p, q) = q$. Substituting these into (3), using the distributive law and rearranging some terms afterward gives

$$f(x, y) \approx f_x(p, q) x + f_y(p, q) y + p - f_x(p, q) p - f_y(p, q) q$$
$$g(x, y) \approx g_x(p, q) x + g_y(p, q) y + q - g_x(p, q) p - g_y(p, q) q$$

Although it is rather difficult to see, these approximations have simple linear forms:

$$f(x, y) \approx a x + b y + h$$
$$g(x, y) \approx c x + d y + k \tag{4}$$

where the following are constants:

$$a = f_x(p, q), \quad b = f_y(p, q), \quad c = g_x(p, q), \quad d = g_y(p, q)$$
$$h = p - f_x(p, q) p - f_y(p, q) q, \quad k = q - g_x(p, q) p - g_y(p, q) q$$

These are the unique values that make the planes given in (2) respectively tangent to $z = f(x, y)$ and $z = g(x, y)$ at (p, q).

Letting $x = x_n$ and $y = y_n$ in (4) implies that (1) may be approximated by the system

$$x_{n+1} = a x_n + b y_n + h$$
$$y_{n+1} = c x_n + d y_n + k$$

This non-homogeneous linear system, which is called the **linearization** of (1) at (p, q), may also be written in vector/matrix form as $\mathbf{x}_{n+1} = D\mathbf{x}_n + \mathbf{b}$ where

$$\mathbf{x}_n = \begin{bmatrix} x_n \\ y_n \end{bmatrix}, \quad D = \begin{bmatrix} a & b \\ c & d \end{bmatrix} \quad \text{and} \quad \mathbf{b} = \begin{bmatrix} h \\ k \end{bmatrix}$$

Note that, rather than using A to designate the coefficient matrix as in Chapter 4, here D is used since it represents a **matrix of partial derivatives.** D is often referred to simply as the **derivative** of the function $F(x, y) = (f(x, y), g(x, y))$ at the point (p, q).

A stability and rotation theorem can now be formulated for (1) that is based upon an eigenvalue analysis of D. Rather than stating it as we have those for linear systems, we take this opportunity to rephrase in a more unified way some of those earlier findings.

If a real eigenvalue is viewed as a complex number with imaginary part equal to 0, its modulus equals its absolute value. The criteria developed in Chapter 4 for stability or instability of a fixed point may therefore be stated more concisely as follows: A fixed point of a linear system is a sink if each eigenvalue of the coefficient matrix has modulus less than 1, a source if each has modulus greater than 1 or a saddle if one has modulus less than 1 while the other has modulus greater than 1. This one simple rule applies whether or not the eigenvalues are real and distinct, real and repeated or complex conjugates.

Since $\mathbf{x}_{n+1} = D\mathbf{x}_n + \mathbf{b}$ represents the linearization of (1) at (p, q), this rule can be used to determine not only the dynamics of that linear system, but also the local dynamics of the nonlinear system (1) near (p, q). Hence, we state the following.

THEOREM   *Stability and Rotation*

Suppose (p, q) is a fixed point of the nonlinear system (1), and D is the corresponding matrix of partial derivatives given by

$$D = \begin{bmatrix} f_x(p, q) & f_y(p, q) \\ g_x(p, q) & g_y(p, q) \end{bmatrix}$$

Then locally (p, q) is a sink if each eigenvalue of D has modulus less than 1, a source if each has modulus greater than 1 or a saddle if one has modulus less than 1 while the other has modulus greater than 1. Also, if the eigenvalues of D are complex conjugates, then solutions rotate locally around (p, q).   ∎

Note that we make no attempt here to characterize further the nature of the rotation when complex eigenvalues exist. Unlike the linear case, this is not so straightforward for nonlinear systems.

EXAMPLE 1

Find the equilibrium point of the price–demand model

$$P_{n+1} = P_n + 2.5D_n - 0.1P_n^2 - 1$$
$$D_{n+1} = D_n + \frac{5}{P_n} - 1$$

and determine whether it is a sink, source or saddle.

Solution: To find the fixed point we solve the system of simultaneous equations

$$x = x + 2.5y - 0.1x^2 - 1 \quad \text{and} \quad y = y + \frac{5}{x} - 1$$

The second equation yields $x = 5$. Substituting this into the first then gives $y = 1.4$. So, the equilibrium point is $(5, 1.4)$.

Determining whether this point is a sink, source or saddle requires first computing the partial derivatives of the functions $f(x, y)$ and $g(x, y)$ corresponding to the system. Since $f(x, y) = x + 2.5y - 0.1x^2 - 1$, then

$$f_x(x, y) = 1 - 0.2x \quad \text{and} \quad f_y(x, y) = 2.5$$

Recall here that, when taking the partial derivative with respect to either variable, the other is treated as a constant. Similarly, since $g(x, y) = y + \dfrac{5}{x} - 1$, then

$$g_x(x, y) = -\frac{5}{x^2} \quad \text{and} \quad g_y(x, y) = 1$$

We next evaluate these four partial derivatives at the point $(5, 1.4)$, which yields

$$f_x(5, 1.4) = 0, \quad f_y(5, 1.4) = 2.5, \quad g_x(5, 1.4) = -0.2 \quad \text{and} \quad g_y(5, 1.4) = 1$$

This means that the matrix associated with the linearization of the system at $(5, 1.4)$ is $D = \begin{bmatrix} 0 & 2.5 \\ -0.2 & 1 \end{bmatrix}$. Using the techniques of Chapter 4, we find its eigenvalues to be $(1 \pm i)/2$. Since

$$|(1 \pm i\sqrt{7})/4| = 1/\sqrt{2} < 1$$

then, according to our new theorem, the point $(5, 1.4)$ must be a sink. This implies that for any initial price P_0 sufficiently close to 5 and demand D_0 sufficiently close to 1.4, the price P_n and demand D_n will eventually stabilize at those levels. ∎

EXAMPLE 2

Use linearization to classify each of the fixed points $(0, 0)$ and $(90, 200)$, previously found in Example 2 of Section 5.1 for the prey–predator system

$$P_{n+1} = 5P_n - 0.02 P_n Q_n$$
$$Q_{n+1} = 0.1 Q_n + 0.01 P_n Q_n$$

Solution: To determine these dynamics, we first compute the four partial derivatives that will be needed. Since $f(x, y) = 5x - 0.02xy$ and $g(x, y) = 0.1y + 0.01xy$, then

$$f_x(x, y) = 5 - 0.02y, \quad f_y(x, y) = -0.02x, \quad g_x(x, y) = 0.01y \quad \text{and} \quad g_y(x, y) = 0.1 + 0.01x$$

Evaluating these at $(0, 0)$ yields

$$f_x(0, 0) = 5, \quad f_y(0, 0) = 0, \quad g_x(0, 0) = 0 \quad \text{and} \quad g_y(0, 0) = 0.1$$

which makes $D = \begin{bmatrix} 5 & 0 \\ 0 & 0.1 \end{bmatrix}$. The eigenvalues of D are easily found to be 5 and 0.1, which indicates that $(0, 0)$ is a saddle. Consequently, nearby trajectories must be hyperbolic around it. This verifies numerical observations described in the previous example.

For the fixed point $(90, 200)$ we find that

$$f_x(90, 200) = 1, \quad f_y(90, 200) = -1.8, \quad g_x(90, 200) = 2 \quad \text{and} \quad g_y(90, 200) = 1$$

which means $D = \begin{bmatrix} 1 & -1.8 \\ 2 & 1 \end{bmatrix}$. The eigenvalues are now $1 \pm i\sqrt{3.6}$, which satisfy

$$|1 \pm i\sqrt{3.6}| = \sqrt{4.6} \approx 2.145 > 1$$

Not only does this imply that $(90, 200)$ is a source, but also, due to complex eigenvalues, solutions rotate locally around this source. The combination of these two dynamics makes nearby solutions spiral away from (90, 200), again confirming previous numerical findings.   ■

The reader may wish to similarly verify that the two fixed points of the overlapping-generations model considered in Example 3 of the previous section also constitute a saddle and a source as claimed there.

EXAMPLE 3

For the Nonlinear Overlapping-Generations Model of the previous section, find all non-negative equilibrium points and determine their stability for $N = 3000$ and (a) $r = 0.5$, $s = 0.25$; (b) $r = 0.75$, $s = 0.5$. (c) Discuss the implications of these results for the model.

Solution: (a) The fixed points of the system

$$P_{n+1} = (0.5P_n + 0.25Q_n)(1 - (P_n + Q_n)/3000)$$
$$Q_{n+1} = P_n$$

are found to be (0, 0) and (−500, −500). The origin is therefore the only non-negative equilibrium point. To determine its stability, we evaluate the matrix of partial derivatives of

$$f(x, y) = (0.5x + 0.25y)(1 - (x + y)/3000) \quad \text{and} \quad g(x, y) = x$$

at (0, 0), which yields $D = \begin{bmatrix} 0.5 & 0.25 \\ 1 & 0 \end{bmatrix}$. Since the eigenvalues of this matrix are $(1 \pm \sqrt{5})/4 \approx 0.81, -0.31$, then (0, 0) must be stable.

(b) This time there are two non-negative fixed points of the system

$$P_{n+1} = (0.75P_n + 0.5Q_n)(1 - (P_n + Q_n)/3000)$$
$$Q_{n+1} = P_n$$

namely (0, 0) and (300, 300). Since

$$f(x, y) = (0.75x + 0.5y)(1 - (x + y)/3000) \quad \text{and} \quad g(x, y) = x$$

then the matrix of partial derivatives evaluated at (0, 0) is $D = \begin{bmatrix} 0.75 & 0.5 \\ 1 & 0 \end{bmatrix}$, whose eigenvalues are $(3 \pm \sqrt{41})/8 \approx 1.18, -0.43$. This indicates that the origin is now unstable. On the other hand, for the point (300, 300) this matrix becomes $D = \begin{bmatrix} 0.475 & 0.275 \\ 1 & 0 \end{bmatrix}$ with eigenvalues $(19 \pm \sqrt{2121})/80 \approx 0.81, -0.34$, which means (300, 300) is a stable equilibrium.

(c) These results suggest that with growth rates of $r = 0.5$ and $s = 0.25$ as in part (a) (and perhaps for all growth rates below a certain level), all populations P_n and Q_n that are initially small will approach 0 over time, since (0, 0) is a local attractor. In fact, numerical results confirm that this actually occurs for all populations that remain non-negative for all n. Such behavior implies eventual extinction of the species.

However, for the growth rates $r = 0.75$ and $s = 0.5$ of (b) (and perhaps for all larger growth rates), solutions diverge from (0, 0), and there is instead a locally stable positive equilibrium (300, 300) that attracts them. Not only are populations initially close to 300 attracted to that value, but numerical experimentation would show that all populations that remain non-negative for all n approach 300. This means that the species will not only survive, but will stabilize over time at a constant level.   ■

Periodic Points and Cycles

The techniques just developed for determining stability and instability of fixed points can be adapted to cover the case of **periodic points** as well. Like those of nonlinear equations investigated in Chapter 3, periodic points of nonlinear systems repeat after a certain number of iterations. Such points may exist, as the following example demonstrates.

EXAMPLE 4

Show that a periodic point exists for the system

$$x_{n+1} = y_n^2$$
$$y_{n+1} = x_n^2$$

Solution: To show this we simply verify that $(0, 1)$ is one. Letting $n = 0$ in the system with $(x_0, y_0) = (0, 1)$, we find that

$$x_1 = y_0^2 = 1^2 = 1 \quad \text{and} \quad y_1 = x_0^2 = 0^2 = 0$$

Also, since we now have $(x_1, y_1) = (1, 0)$ then for $n = 1$

$$x_2 = y_1^2 = 0^2 = 0 \quad \text{and} \quad y_2 = x_1^2 = 1^2 = 1$$

Since (x_0, y_0) and (x_2, y_2) both equal $(0, 1)$, but (x_1, y_1) does not, then $(0, 1)$ must be a point of period 2.   ■

In this example it is easy to see that $(1, 0)$ is also a point of period 2. The pair of points $(0, 1)$ and $(1, 0)$ together form what is again called a **2-cycle** of the system. More generally, a 2-cycle of (1) consists of two distinct points (p_1, q_1) and (p_2, q_2) that satisfy

$$(p_2, q_2) = (f(p_1, q_1), g(p_1, q_1)) \quad \text{and} \quad (p_1, q_1) = (f(p_2, q_2), g(p_2, q_2))$$

If we again define the function $F(x, y) = (f(x, y), g(x, y))$, then we can say more simply that a 2-cycle satisfies

$$F(p_1, q_1) = (p_2, q_2) \quad \text{and} \quad F(p_2, q_2) = (p_1, q_1)$$

From these relationships one may see that

$$F(F(p_1, q_1)) = F(p_2, q_2) = (p_1, q_1)$$

which means (p_1, q_1) is a fixed point of the composition $F(F(x, y))$. The same is true for (p_2, q_2). Although this implies that all points of period 2 are fixed points of $F(F(x, y))$, solving the system of simultaneous equations $F(F(x, y)) = (x, y)$ to find them is not likely an easy task. To obtain the 2-cycle in the above example, guesswork was used instead.

Extending these ideas, we define the following for any $m \geq 1$.

DEFINITION

A set of m distinct points $(p_1, q_1), (p_2, q_2), \ldots, (p_m, q_m)$ satisfying

$$F(p_1, q_1) = (p_2, q_2), \quad F(p_2, q_2) = (p_3, q_3), \quad \ldots$$

$$\ldots, \quad F(p_{m-1}, q_{m-1}) = (p_m, q_m), \quad F(p_m, q_m) = (p_1, q_1)$$

where $F(x, y) = (f(x, y), g(x, y))$, is called an **m-cycle** of (1). Each point (p_i, q_i) of an m-cycle is called a **point of period m** for $F(x, y)$.

Similar to the notation introduced in Chapter 3, if we let $F^n(x, y)$ for $n \geq 2$ represent the composition of $F(x, y)$ with itself $n - 1$ times, then it can easily be verified that each of these points of period m satisfies $F^m(p_i, q_i) = (p_i, q_i)$. That is, any point of period m is a fixed point of $F^m(x, y)$. Because of this, we can use the theorem developed earlier in this section to establish stability criteria for periodic points. Assuming that an m-cycle can actually be found, this will enable us to determine its stability or instability.

To accomplish this, we first need the matrix of partial derivatives of the function $F^m(x, y)$ evaluated at a point of the cycle. For simplicity we choose (p_1, q_1), which, by renumbering if necessary, may actually refer to any point of the cycle. To find this matrix, we use the following result from multivariate calculus: Let D_i be the matrix of partial derivatives of $F(x, y)$ evaluated at $(x, y) = (p_i, q_i)$ for $i = 1, 2, \ldots, m$. By repeated use of the chain rule for functions of several variables, the matrix of partial derivatives of the composition $F^m(x, y)$ evaluated at $(x, y) = (p_1, q_1)$ can be shown to equal the matrix product

$$\Pi = D_m \cdot D_{m-1} \cdot \cdots \cdot D_2 \cdot D_1 \tag{5}$$

Note that the order of multiplication is important here since matrix multiplication is not in general commutative, as we learned in Section 4.3.

The eigenvalues of the resulting matrix Π can now be used to determine whether (p_1, q_1) is a sink, source or saddle of $F^m(x, y)$. It can be additionally shown that the eigenvalues of the matrix of partial derivatives of $F^m(x, y)$ evaluated at any other point of the m-cycle must be identical to these eigenvalues, which means that the entire m-cycle may be classified at once as either **locally stable,** if these points are sinks under $F^m(x, y)$, or **unstable** otherwise. Again, the word *locally* is generally omitted.

We have therefore established the following.

THEOREM *Stability of m-Cycles*

An m-cycle $(p_1, q_1), (p_2, q_2), \ldots, (p_m, q_m)$ of (1) is locally stable if each eigenvalue of the matrix Π given by (5) has modulus less than 1, or unstable otherwise. ∎

We remark that being more specific with regard to the instability of a periodic cycle, i.e., saddle or source, is generally unnecessary since the particular geometry of trajectories around the points of an m-cycle is seldom of use. Knowing whether the cycle is stable or unstable is usually sufficient.

EXAMPLE 5

Determine whether the 2-cycle $(0, 1)$ and $(1, 0)$ of the system from Example 4 is stable or unstable.

Solution: Since $F(x, y) = (f(x, y), g(x, y)) = (y^2, x^2)$, the general matrix of partial derivatives is $D = \begin{bmatrix} 0 & 2y \\ 2x & 0 \end{bmatrix}$. Evaluating D first at $(x, y) = (0, 1)$ and then at $(x, y) = (1, 0)$ give, respectively,

$$D_1 = \begin{bmatrix} 0 & 2 \\ 0 & 0 \end{bmatrix} \quad \text{and} \quad D_2 = \begin{bmatrix} 0 & 0 \\ 2 & 0 \end{bmatrix}$$

The product of these matrices is $\Pi = D_2 \cdot D_1 = \begin{bmatrix} 4 & 0 \\ 0 & 0 \end{bmatrix}$, whose eigenvalues are quickly found to be 4 and 0. Consequently, $(0, 1)$ is a saddle under the composition $F(F(x, y))$, and the 2-cycle is unstable. ∎

The reader may wish to check in this example that the same eigenvalues result if D is evaluated first at $(1, 0)$ and then at $(0, 1)$, indicating that $(1, 0)$ is also a saddle and confirming that the entire 2-cycle is indeed unstable.

EXAMPLE 6

The point $(5000, 5000)$ is a periodic point of the Nonlinear Overlapping-Generations Model with $r = s = 2$ and $C = 10,000$, i.e.,

$$P_{n+1} = (2P_n + 2Q_n)(1 - (P_n + Q_n)/10{,}000)$$

$$Q_{n+1} = P_n$$

(a) Find the other points and period of its cycle. (b) Determine whether or not the cycle is stable.

Solution: (a) Given one point of an m-cycle, to find the others and their period we need only iterate the system. With $(P_0, Q_0) = (5000, 5000)$, we find

$$(P_1, Q_1) = (0, 5000), \quad (P_2, Q_2) = (5000, 0) \quad \text{and} \quad (P_3, Q_3) = (5000, 5000).$$

From this we see that the other points of the cycle are $(0, 5000)$ and $(5000, 0)$, and that the period is 3.

(b) To determine whether this 3-cycle is stable, we first compute the general matrix of partial derivatives D for the system. Since

$$f(x, y) = (2x + 2y)(1 - (x + y)/10{,}000) \quad \text{and} \quad g(x, y) = x$$

then

$$D = \begin{bmatrix} 2 - (x + y)/2500 & 2 - (x + y)/2500 \\ 1 & 0 \end{bmatrix}$$

Evaluating D at each of the points $(5000, 5000)$, $(0, 5000)$ and $(5000, 0)$ yields, respectively,

$$D_1 = \begin{bmatrix} -2 & -2 \\ 1 & 0 \end{bmatrix}, \quad D_2 = \begin{bmatrix} 0 & 0 \\ 1 & 0 \end{bmatrix} \quad \text{and} \quad D_3 = \begin{bmatrix} 0 & 0 \\ 1 & 0 \end{bmatrix}$$

The product of these (in reverse) is

$$\Pi = D_3 \cdot D_2 \cdot D_1 = \begin{bmatrix} 0 & 0 \\ 1 & 0 \end{bmatrix} \cdot \begin{bmatrix} 0 & 0 \\ 1 & 0 \end{bmatrix} \cdot \begin{bmatrix} -2 & -2 \\ 1 & 0 \end{bmatrix} = \begin{bmatrix} 0 & 0 \\ 0 & 0 \end{bmatrix}$$

Since both eigenvalues of $\Pi = \begin{bmatrix} 0 & 0 \\ 0 & 0 \end{bmatrix}$ are 0, the 3-cycle must be stable.   ∎

Exercises 5.2 _____

In Exercises 1–4 find the matrix of partial derivatives D.

1. $f(x, y) = x^2 - y^2$
   $g(x, y) = 2xy$

2. $f(x, y) = x/(x + y)$
   $g(x, y) = y(x - y)$

3. $f(x, y) = x \ln x + y \ln y$
   $g(x, y) = xe^y - ye^x$

4. $f(x, y) = \sin(xy)$
   $g(x, y) = \cos(x + y)$

In Exercises 5–8 determine whether the fixed point (p, q) *is a sink, source or saddle of the given nonlinear system, and whether or not solutions rotate around it.*

5. $x_{n+1} = 2x_n(x_n + y_n)$
   $y_{n+1} = (x_n - y_n)/2$, $(p, q) = (0, 0)$

6. $x_{n+1} = 0.66 - \sqrt{x_n + y_n}$
   $y_{n+1} = 0.79 - \sqrt{x_n} - \sqrt{y_n}$, $(p, q) = (0.16, 0.09)$

7. $x_{n+1} = 2x_n \ln(y_n) - 1$
   $y_{n+1} = x_n + y_n + 1$, $(p, q) = (-1, 1)$

8. $x_{n+1} = 2e^{4-y_n}$
   $y_{n+1} = 4e^{2-x_n}$, $(p, q) = (2, 4)$

In Exercises 9–16 find all fixed points and determine whether each is a sink, source or saddle.

9. $P_{n+1} = 0.8P_n - 0.05P_nQ_n$
   $Q_{n+1} = 0.6Q_n + 0.04P_nQ_n$

10. $P_{n+1} = 2P_ne^{-Q_n/1500}$
    $Q_{n+1} = 0.25Q_ne^{P_n/1000}$

11. $P_{n+1} = (1.75P_n + 2.25Q_n)(1 - (P_n + Q_n)/8000)$
    $Q_{n+1} = P_n$

12. $P_{n+1} = (2.5P_n + 1.5Q_n)(1 - (P_n + Q_n)/4000)$
    $Q_{n+1} = P_n$

13. $P_{n+1} = P_n + 2.5D_n - 0.1P_n^2 - 1$
    $D_{n+1} = D_n + 5/P_n - 1$

14. $P_{n+1} = P_n + 2D_n - 0.1P_n^2 - 2$
    $D_{n+1} = D_n + 4/P_n - 2$

15. $x_{n+1} = 2x_n^2(1 - x_n - y_n)$
$y_{n+1} = x_n$

16. $P_{n+1} = (4x_n + y_n)/(1 + x_n^2 + y_n^2)$
$Q_{n+1} = P_n$

17. If $F(x, y) = (x/y, y/x)$, compute $F^2(x, y)$, $F^3(x, y)$ and $F^4(x, y)$. Can you find a general formula for $F^n(x, y)$ for $n = 1, 2, 3, \ldots$?

18. Repeat the previous exercise for $F(x, y) = (xy, x/y)$.

19. If $F(x, y) = (y^2, x)$ find all solutions (x, y) of $F(F(x, y)) = (x, y)$, and decide which of these are fixed points and which are points of period 2. Also, determine the stability of each of these fixed points and periodic cycles.

20. Repeat the previous exercise for $F(x, y) = (y^2, -x^2)$

In Exercises 21–26 a nonlinear system and one point (p_1, q_1) *of an* m-*cycle is given. Find all other points of that cycle, and determine its period and stability.*

21. $x_{n+1} = 2 - 0.5x_n^2$, $y_{n+1} = 0.5y_n$, $(p_1, q_1) = (0, 0)$

22. $x_{n+1} = 3 - x_n^2$, $y_{n+1} = 2y_n$, $(p_1, q_1) = (2, 0)$

23. $x_{n+1} = 3 - y_n^2$, $y_{n+1} = x_n$, $(p_1, q_1) = (2, 2)$

24. $x_{n+1} = 1 - 0.5(x_n + y_n)^2$, $y_{n+1} = x_n$, $(p_1, q_1) = (1, 1)$

25. $x_{n+1} = 1 - (x_n + y_n)^2$, $y_{n+1} = x_n$, $(p_1, q_1) = (0, 0)$

26. $x_{n+1} = 1 - y_n^2$, $y_{n+1} = x_n$, $(p_1, q_1) = (0, 0)$

Computer Projects 5.2

Each of the nonlinear systems given in Projects 1–4 has a stable m-*cycle for the given parameter values. Verify this through numerical experimentation and determine the period of each cycle.*

1. The Nonlinear Infection–Recovery Model for $N = 5000$, $r = 0.25$, $s = 5$ and
   (a) $t = 0.32$ (b) $t = 0.3475$
2. The Nonlinear Overlapping-Generations Model for $C = 1000$ and
   (a) $r = 2.9, s = 1$ (b) $r = 2.8, s = 1.2$
3. The Nonlinear Price–Demand Model for $a_1 = 2.5$, $h_1 = -1$, $k_1 = 1$ and
   (a) $b_1 = 0.06, c_1 = 3.6$ (b) $b_1 = 0.05, c_1 = 3.82$
4. The nonlinear system $\begin{cases} x_{n+1} = x_n^2 - y_n^2 + a \\ y_{n+1} = 2x_n y_n + b \end{cases}$, which is known as the *Mandelbrot map,* for
   (a) $a = -0.1, b = 0.8$ (b) $a = 0.33, b = 0.41$

5. Verify mathematically that $(0, 0)$ is a locally stable fixed point of the nonlinear system
$$\begin{cases} x_{n+1} = 0.5(x_n + y_n)(1 - x_n - y_n) \\ y_{n+1} = x_n \end{cases}$$. Then use numerical experimentation to see if
this fixed point is globally stable (or appears to be) for all initial points satisfying $0 \leq x_0, y_0 \leq 0.25$.

6. Repeat the previous project for the system $\begin{cases} x_{n+1} = 0.5(x_n + y_n)e^{2-x_n-y_n} \\ y_{n+1} = x_n \end{cases}$, the fixed
point $(1, 1)$ and all initial points satisfying $x_0, y_0 > 0$.

SECTION
5.3   Bifurcation and Chaos

Contrary to the emphasis of the previous section on the orderly dynamics associated with their stable fixed and periodic points, nonlinear systems as a whole exhibit an even wider variety of complex behaviors than do the nonlinear equations of Chapter 3. As we shall now see, this includes bifurcation into stable periodic or *irrationally periodic* cycles, *period-multiplying* cascades, and of course chaos. As before, all of this is most conveniently investigated through an analysis of parameterized families and the transitions in the dynamics as their parameters are varied.

Parameterized Systems

Each of the models constructed in Section 5.1 depends upon at least two parameters. Rather than providing values for all parameters of a model before undertaking its investigation as we have been doing, a certain advantage may be gained by leaving one or more of the parameters unspecified. As we learned in Chapter 3, this allows the determination of the dynamics over a whole range of parameter values at once.

Computing equilibrium points of parameterized nonlinear systems and determining their local dynamics can still be accomplished using the techniques of the previous sections. However, the calculations involved may be a bit tedious, as the following example suggests.

EXAMPLE 1

(a) For the Nonlinear Infection–Recovery Model with $r = 0.5$, $t = 0.5$ and $N = 10,000$, find all non-negative equilibrium points and determine their stability for all $s \geq 0$. (b) Discuss the implications of these results for the model.

Solution: (a) The fixed points of the parameterized system in question,

$$I_{n+1} = 0.5I_n + sI_n(1 - (I_n + R_n)/10{,}000)$$
$$R_{n+1} = 0.5I_n + 0.5R_n$$

may be found by solving simultaneously

$$x = 0.5x + sx(1 - (x + y)/10{,}000) \quad \text{and} \quad y = 0.5x + 0.5y$$

From the second equation we must have $y = x$. Substituting this into the first then gives

$$x = 0.5x + sx(1 - x/5000) \quad \text{or after simplifying} \quad sx^2 + (2500 - 5000s)x = 0$$

The two solutions of this quadratic equation are $x = 0$ and $x = 5000 - 2500/s$. Combining these with $y = x$ gives the two fixed points $(0, 0)$ and (p_s, p_s), where $p_s = 5000 - 2500/s$.

Since $f(x, y) = 0.5x + sx(1 - (x+y)/10{,}000)$ and $g(x, y) = 0.5x + 0.5y$, to determine stability, we first compute

$$f_x(x, y) = 0.5 + s(1 - (2x + y)/10{,}000), \quad f_y(x, y) = -sx/10{,}000,$$

$$g_x(x, y) = 0.5 \quad \text{and} \quad g_y(x, y) = 0.5$$

At $(0, 0)$ this makes

$$D = \begin{bmatrix} s + 0.5 & 0 \\ 0.5 & 0.5 \end{bmatrix}$$

whose eigenvalues are quickly found to be 0.5 and $s + 0.5$. For $-1.5 < s < 0.5$ both eigenvalues lie between -1 and 1, but since we are interested in only $s \geq 0$, we will say that $(0, 0)$ is stable for $0 \leq s < 0.5$.

Evaluating and then simplifying the partial derivatives for $x = y = p_s$, we find that

$$D = \begin{bmatrix} 1.25 - 0.5s & 0.25 - 0.5s \\ 0.5 & 0.5 \end{bmatrix}$$

for the fixed point (p_s, p_s). The eigenvalues of this matrix can be computed, and using some algebra, then written as

$$\frac{7 - 2s \pm \sqrt{4s^2 - 28s + 17}}{8}$$

With some further analysis on these eigenvalues (that we shall not go into but the reader may wish to verify), we find that both are real and lie between -1 and 1 for

$$0.5 < s \leq 3.5 - 2\sqrt{2} \quad \text{and for} \quad 3.5 + 2\sqrt{2} \leq s < 6.5$$

Additionally, they are instead complex conjugates with a modulus of $1/\sqrt{2} < 1$ for

$$3.5 - 2\sqrt{2} < s < 3.5 + 2\sqrt{2}$$

This means that (p_s, p_s) is stable for $0.5 < s < 6.5$.

(b) From this analysis one can infer that for a sufficiently small infection-rate parameter s, both the number infected I_n and the number recovered and immune R_n will likely decay to 0 over time. This happens for $0 \leq s < 0.5$. But for $s > 0.5$ the disease does not necessarily die out on its own. There is a positive equilibrium that is locally stable for $0.5 < s < 6.5$, indicating instead that the disease will likely never be completely eradicated from the population. For any such s the number of infected I_n will generally stabilize at the level $p_s = 5000 - 2500/s$, which is a function of s. The equilibrium p_s increases from 0, for $s = 0$, to approximately 4615, for $s = 6.5$, implying that the infected population will stabilize at higher levels the larger the infection rate is (see Fig. 5.4). Since p_s is also the stable equilibrium for R_n, the same occurs for the recovered and immune population. ∎

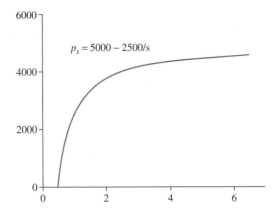

FIGURE 5.4

Borrowing some terminology from Chapter 3, the intervals $s \geq 0$ and $s > 0.5$ in the previous example, for which the non-negative fixed points $(0, 0)$ and (p_s, p_s) respectively exist, may again be referred to as **intervals of existence.** Similarly, the intervals $0 \leq s < 0.5$ and $0.5 < s < 6.5$, for which they are respectively stable, are **intervals of stability.**

When a nonlinear system is parameterized by a single parameter as the one in this example is, fixed points usually have such intervals. However, if two or more unspecified parameters appear in a system, these intervals turn into **regions** of existence and stability in a **parameter space,** whose dimension matches the number of such parameters. The following example demonstrates this idea.

EXAMPLE 2

Find all values of a and b for which the origin is a sink for the nonlinear system

$$x_{n+1} = ax_n - by_n + y_n^2$$
$$y_{n+1} = ay_n + bx_n - x_n^2$$

Solution: It is immediately evident that $(0, 0)$ is a fixed point for all a and b. This means that the region of existence consists of the entire (a, b) plane. After computing partial derivatives at $(0, 0)$ we find that $D = \begin{bmatrix} a & -b \\ b & a \end{bmatrix}$, which has complex conjugate eigenvalues equal to $a \pm bi$. To determine stability, we compute $|a \pm bi| = \sqrt{a^2 + b^2}$ and compare it to 1. This gives

$$\sqrt{a^2 + b^2} < 1 \quad \text{which is equivalent to} \quad a^2 + b^2 < 1$$

The region of stability in the (a, b) plane must therefore satisfy $a^2 + b^2 < 1$. This describes the region inside the unit circle $a^2 + b^2 = 1$ (see Fig. 5.5). ■

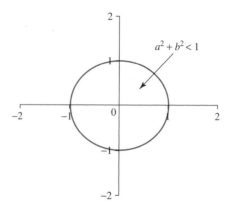

FIGURE 5.5

Bifurcation

It was no coincidence in Example 1 that the intervals of existence and stability of the nonzero fixed point (p_s, p_s) were found to begin at $s = 0.5$, where stability ends for the fixed point $(0, 0)$ and where $(p_s, p_s) = (0, 0)$. This s value corresponds to one kind of bifurcation that a fixed point of a parameterized nonlinear system $(x_{n+1}, y_{n+1}) = F_s(x_n, y_n)$ may undergo, in which an eigenvalue of the linearization matrix D passes through 1, and consequently a new stable fixed point emerges directly from it at that parameter value. This closely parallels the first type of bifurcation seen in Chapter 3 for parameterized nonlinear equations of the form $x_{n+1} = f_r(x_n)$, in which a fixed point p_r gives birth to a new stable fixed point when $f_r'(p_r)$ passes through 1 for some r value.

Also reminiscent of Chapter 3, period-doubling bifurcations may occur in the present setting as well, as the following example shows.

EXAMPLE 3

Show that the Nonlinear Infection–Recovery Model of Example 1 undergoes a period-doubling cascade for $s \geq 0.5$.

Solution: From Example 1 we know that the fixed point (p_s, p_s) is stable for $0 < s < 6.5$. But since attempting to determine the dynamics for $s > 6.5$ using precise mathematical methods is likely to prove impossible, we instead choose numerical iteration of the system. For s slightly larger than 6.5 we do indeed find a stable 2-cycle whose points begin at and diverge from the fixed point as s increases. This continues until s reaches approximately 7.2535, at which point the 2-cycle bifurcates into a stable 4-cycle. These events mark the beginning of an infinite period-doubling cascade in which we additionally find

$$\begin{aligned}
\text{stable 4-cycles for} \quad &7.2535 < s < 7.3684 \\
\text{stable 8-cycles for} \quad &7.3684 < s < 7.3941 \\
\text{stable 16-cycles for} \quad &7.3941 < s < 7.3988
\end{aligned}$$

One may uncover the mechanism behind these period-doublings by first noting that the eigenvalues of linearization matrix D at (p_s, p_s), which were found earlier to be

$$\frac{7 - 2s \pm \sqrt{4s^2 - 28s + 17}}{8}$$

equal -0.5 and -1 at $s = 6.5$. That is, one of these eigenvalues passes through -1 at $s = 6.5$, and consequently the fixed point bifurcates into a stable 2-cycle. This scenario is similar to the one observed in Chapter 3 for parameterized nonlinear equations, but with an eigenvalue replacing the derivative. Eigenvalues passing through -1 account as well for all the period-doubling bifurcations that follow. However, for those cases the eigenvalues belong to the linearization matrix product Π associated with the points of the 2^n-cycle. ∎

Based upon the findings of Examples 1 and 3, it appears that every type of bifurcation encountered in Chapter 3 for parameterized families of nonlinear equations, as well as their period-doubling cascades, have counterparts in the present context of parameterized non-linear systems.

However, while these scenarios exhaust the bifurcation possibilities that are available for the former, this is not so for the latter. There is another common means through which a fixed or periodic point of a nonlinear system loses its stability, besides having an eigenvalue of the linearization matrix pass through either 1 or -1. Frequently, this instability is instead concurrent with the *modulus* of complex conjugate eigenvalues passing through 1. When this occurs and a stable cycle emerges, an event called a **Hopf bifurcation,** the period neither remains the same nor doubles. The following example offers some possibilities of what just might happen in this case.

EXAMPLE 4

Determine the nature of the stable cycle that emerges from the non-zero fixed point of the Nonlinear Overlapping-Generations Model with $N = 1000$ at each of the bifurcation points (a) $(r, s) = (2, 2)$; (b) $(r, s) = (22/9, 11/9)$.

Solution: (a) The non-zero fixed point of the parameterized system

$$P_{n+1} = (r P_n + s Q_n)(1 - (P_n + Q_n)/1000)$$
$$Q_{n+1} = P_n$$

may be written as (p, p) where

$$p = p(r, s) = \frac{500(r + s - 1)}{r + s}$$

and for $(r, s) = (2, 2)$ this point becomes $(375, 375)$. Since the eigenvalues of the corresponding linearization matrix $D = \begin{bmatrix} -1 & -1 \\ 1 & 0 \end{bmatrix}$ are $(-1 \pm i\sqrt{3})/2$ with modulus equal to 1, we see that $(2, 2)$ must indeed sit on the border of this fixed point's region of stability in the (r, s) plane. Just to one side of this border the modulus of these eigenvalues is less than 1, implying stability of (p, p). On the other side they exceed 1, implying a loss of stability. Iterating numerically for parameter values just outside the stability region, a stable 3-cycle

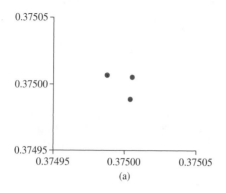
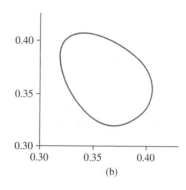

FIGURE 5.6

can be found whose points emerge from that fixed point at $(r, s) = (2, 2)$. Figure 5.6(a) shows this 3-cycle for $(r, s) = (2.00001, 2)$. Since it did not exist for parameter values within the stability region, this 3-cycle must bifurcate from (p, p) at $(r, s) = (2, 2)$.

(b) For $(r, s) = (22/9, 11/9)$ the fixed point (p, p) becomes $(4000/11, 4000/11)$ and $D = \begin{bmatrix} -2/3 & -1 \\ 1 & 0 \end{bmatrix}$. As in (a), since the eigenvalues $(-1 \pm i\sqrt{8})/3$ have modulus equal to 1, these parameter values again lie on the border of the stability region of (p, p). Numerical iteration indicates this time that a stable invariant curve surrounding (p, p) emerges from that fixed point at $(r, s) = (22/9, 11/9)$. Figure 5.6(b) shows the curve for $(r, s) = (2.45, 1.23)$. All nearby solutions are attracted to this curve, whose points constitute a trajectory that cycles around itself in a regular and orderly fashion. But the cycle, if it may be called that, is not periodic in the usual sense since no point ever quite repeats itself exactly. Instead, any trajectory that begins on the cycle eventually traces out the entire continuous curve. One may notice the similarity here to what was found in Example 3(b) in Section 5.1. ■

From this example we now see that a third type of bifurcation scheme is possible for parameterized nonlinear systems, the Hopf bifurcation, which has no equivalent for the parameterized nonlinear equations of Chapter 3.

It is interesting to note in this example that the period of the cycle that emerges from the Hopf bifurcation in each case appears to be related to the polar coordinate θ of the complex conjugate eigenvalues of the matrix D. Since $\theta = 2\pi/3$ for the eigenvalues $(-1 \pm i\sqrt{3})/2$ of part (a), the Rotation Theorem of Chapter 4 tells us that solutions of the linearized system rotate around the fixed point with an angle that is periodic. Further, since $m = 3$ is the smallest whole number that makes $m(2\pi/3)/(2\pi)$ an integer, the period of this rotation is 3. This is also the period of the cycle that arises in this case. For (b), where the eigenvalues are $(-1 \pm i\sqrt{8})/3$, we find that $\theta = \tan^{-1}(-\sqrt{8})$ is an irrational multiple of π, and so the Rotation Theorem says this time that the angular rotation of solutions of the linearized system is irrationally periodic. Since this is essentially what we found for the invariant curve that emerged in part (b), we henceforth describe this curve as an **irrationally periodic cycle.** And since these observations concerning θ are generally valid, they provide a simple way to predict the nature of any periodic cycle that emerges from a fixed point in a Hopf bifurcation.

EXAMPLE 5

For the system of Example 2 determine the period of the stable cycle that bifurcates from the fixed point $(0, 0)$ at each of the points in parameter space: (a) $(a, b) = (0, 1)$; (b) $(a, b) = (2/5, \sqrt{21}/5)$.

Solution: (a) Since the eigenvalues of the linearization matrix at $(0, 0)$ were found earlier to be $a \pm bi$, for $(a, b) = (0, 1)$ these become $\pm i$. This makes $\theta = \pi/2$, and the period of the cycle that bifurcates from $(0, 0)$ equal to 4, since $m = 4$ is the smallest whole number making $m(\pi/2)/(2\pi)$ an integer. So, assuming that a bifurcation actually occurs, it will generate a 4-cycle.

(b) In this case the eigenvalues are $(2 \pm i\sqrt{21})/5$, making $\theta = \tan^{-1}(\sqrt{21}/2)$, which is an irrational multiple of π. So, if a new stable cycle emerges, it will be an irrationally periodic one. ∎

It should be noted that, along with any stable periodic cycle that arises in a Hopf bifurcation, an unstable one of the same period always accompanies it. The points of the two cycles generally alternate along a closed curve. Also, although only one type of stable cycle may be produced at any particular bifurcation point, it is not uncommon for several different stable cycles to simultaneously coexist afterward, each with its own basin of attraction. For example, Figure 5.7 shows a 4-cycle and an irrationally periodic cycle for the Nonlinear Overlapping-Generations Model with $N = 1000$ and $(r, s) = (3, 0.6)$ that are both locally stable.

Mandelbrot Map

The three types of bifurcations we have just observed are quite common for parameterized nonlinear systems. In fact, each of the models constructed in Section 5.1 undergoes at least one such bifurcation, and most undergo several. A few even have **period-multiplying** cascades associated with them, in which an infinite sequence of bifurcations occurs as certain parameters are varied, and each time the period of the resulting stable cycle is a

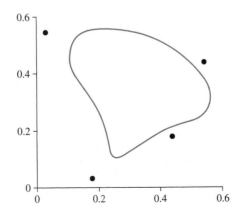

FIGURE 5.7

multiple of its predecessor's period. While repeated period-doubling is the most common of these, numerous other types of bifurcation sequences are possible.

The nonlinear system that has unquestionably become the paradigm for the bifurcation process in recent decades is

$$
\begin{aligned}
x_{n+1} &= x_n^2 - y_n^2 + a \\
y_{n+1} &= 2x_n y_n + b
\end{aligned}
\tag{1}
$$

first investigated by B. Mandelbrot in the 1970's. This famous system actually corresponds to iteration of the single nonlinear equation $z_{n+1} = z_n^2 + c$, called the **Mandelbrot map,** where $z_n = x_n + i y_n$ is a complex-valued sequence of iterates and $c = a + bi$ is a complex parameter. If broken apart into separate iterative equations for the real and imaginary parts of z_n, which are x_n and y_n respectively, the system (1) would result.

What is most significant about the Mandelbrot map is that it possesses virtually *every* type of period-multiplying cascade imaginable! That is, for (a, b) close to $(0, 0)$ a stable fixed point exists that bifurcates into a stable m-cycle for any positive integer m, depending upon the path that (a, b) takes in its two-dimensional parameter space. And, again depending upon that path, each point of these m-cycles similarly bifurcates into a stable k-cycle (of the composite function involved), which yields a stable cycle of (1) having period mk. As this continues, virtually all period-multiplying cascades can be created, and stable cycles of any period can be found if one looks in the right location in parameter space.

This remarkable process is best understood with the aid of the **Mandelbrot set,** depicted in the center of Figure 5.8, which shows the entire set of (a, b)-values for which solutions (x_n, y_n) of (1) beginning at $(x_0, y_0) = (0, 0)$ stay bounded for all n. The large *cardioid* shape constitutes the region of stability of the fixed point, and each of the infinite number of smaller *circular* nodes sitting on its border, the region of stability of an m-cycle that bifurcates from this fixed point. Parameter values chosen from these nodes yield stable

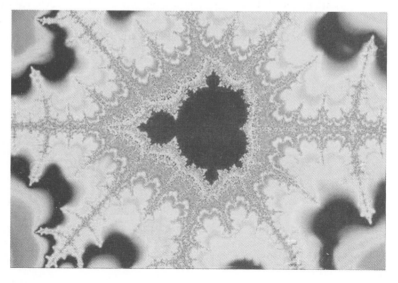

FIGURE 5.8

cycles of every positive integer period, for example:

$$\text{a stable 11-cycle for} \quad (a, b) = (-0.3, 0.635)$$
$$\text{a stable 18-cycle for} \quad (a, b) = (-0.425, 0.57)$$
$$\text{a stable 21-cycle for} \quad (a, b) = (-0.385, 0.59)$$

One might be able to make out from Figure 5.8 that an infinite number of smaller secondary nodes grows out from each of these nodes, and an infinite number of even smaller ones from these secondary nodes, etc. This accounts for the infinite collection of bifurcation sequences mentioned earlier, and for the Mandelbrot set's *fractal* boundary, of which more will be mentioned in the next section. Surrounding this are other fractal sets associated with the Mandelbrot map, as Figure 5.8 also indicates.

Chaos

Like the analogous situation in Chapter 3, the final stages of a period-multiplying bifurcation cascade include smaller and smaller intervals and/or regions of stability involving cycles having increasingly lengthier periods. And, predictably, at the end of such a cascade the dynamics once again resemble **chaos.** However, this time the onset of chaos need not be viewed by way of an increasing multitude of periodic cycles only, but also by following the gradual deformation of the stable invariant curves also generated by the bifurcations. Figure 5.9 shows just such a deformation process for the Nonlinear Price–Demand Model with $a_1 = 2.5$, $b_1 = 0.1$, $h_1 = -1$, $k_1 = 1$ and (a) $c_1 = 3$, (b) $c_1 = 2.7$, (c) $c_1 = 2.6265$, (d) $c_1 = 2.59$. Interestingly, the dynamics become more complex as c_1 *decreases.*

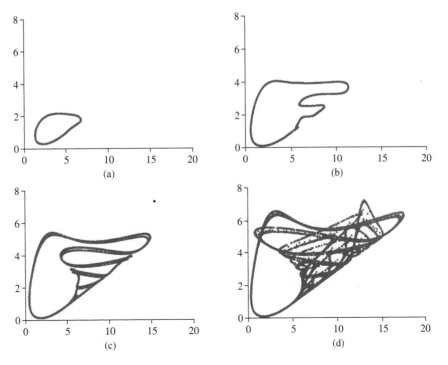

FIGURE 5.9

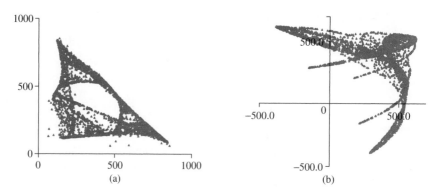

(a)   (b)

FIGURE 5.10

Chaos for nonlinear systems includes all of the complexity previously described in Chapter 3 for nonlinear equations: a scrambled set, an infinite number of periodic points of different periods, dense orbits, sensitive dependence on initial conditions, etc. Due to their two-dimensional solution spaces, chaotic nonlinear systems also display a wide variety of possible geometries with regard to their scrambled sets, which are referred to in this setting as **strange attractors** since they frequently take on rather bizarre shapes and attract all nearby trajectories. Besides those shown in Figure 5.2(b) (of Section 5.1) and Figure 5.9(d), other possibilities include the strange attractor of the Nonlinear Overlapping-Generations Model for $N = 1000$ with $(r, s) = (2.44, 2.15)$, and with $(r, s) = (1.5, 3.75)$, shown in Figures 5.10(a) and 5.10(b) respectively.

As before, it is useful to have some mathematical indicators of chaos that would allow one to determine its presence without needing to perform extensive iteration. Unfortunately, the most famous rule in the field of chaos theory, which says that *period 3 implies chaos,* applies to functions of one variable only. This idea cannot be to extended to higher dimensions since many systems have 3-cycles but not chaos.

However, one indicator of chaos that does successfully generalize is the horseshoe idea from Chapter 3. Recall that if a continuous function $f(x)$ satisfies

$$I \cup J \subset f(I) \cap f(J)$$

for two intervals I and J that are disjoint except for perhaps a common end-point, then $x_{n+1} = f(x_n)$ is chaotic. Suppose now that A and B are two closed rectangles in the (x, y) plane that do not intersect, except perhaps at their boundaries. For some system $(x_{n+1}, y_{n+1}) = F(x_n, y_n)$, if we let

$$F(A) = \{F(x, y) : (x, y) \in A\} \quad \text{and} \quad F(B) = \{F(x, y) : (x, y) \in B\}$$

then a sufficient condition for chaos is that $F(A)$ and $F(B)$ each *cover* both A and B (see Fig. 5.11). As before, this geometric property, which is once again referred to as a **horseshoe,** is written more precisely as

$$A \cup B \subset F(A) \cap F(B) \tag{2}$$

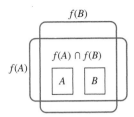

FIGURE 5.11

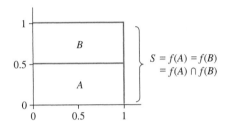

FIGURE 5.12

EXAMPLE 6

Show that the system

$$x_{n+1} = 4y_n(1 - y_n)$$
$$y_{n+1} = x_n$$

is chaotic by finding a horseshoe.

Solution: We begin by letting A be the rectangle in the (x, y) plane described by $0 \leq x \leq 1, 0 \leq y \leq 1/2$, and letting B be the rectangle described by $0 \leq x \leq 1, 1/2 \leq y \leq 1$. Note that $A \cup B$ equals the square S in the (x, y) plane given by $0 \leq x \leq 1, 0 \leq y \leq 1$. Since the function corresponding to the system is $F(x, y) = (4y(1 - y), x)$, it can be checked that $F(A) = S$ and $F(B) = S$, which means that $F(A) \cap F(B) = S$ (see Fig. 5.12). Since both $A \cup B = S$ and $F(A) \cap F(B) = S$, then (2) is satisfied. This implies that the system is chaotic. ∎

Although this situation does not really resemble a horseshoe, it is related to another indicator of chaos that does. Suppose this time that a closed rectangle R in the (x, y) plane, with corner points p_1, p_2, p_3 and p_4, is sent by the function $F(x, y)$ onto the *truly* horseshoe-shaped region $F(R)$, with corner points q_1, q_2, q_3 and q_4, as shown in Figure 5.13. Here, we assume that

$$F(p_1) = q_1, \quad F(p_2) = q_2, \quad F(p_3) = q_3 \quad \text{and} \quad F(p_4) = q_4$$

Not only is this also referred to as a **horseshoe,** but this is the geometric property from which the name originated. In the 1960's S. Smale first described this kind of horseshoe and proved that if one exists, then the nonlinear system $(x_{n+1}, y_{n+1}) = F(x_n, y_n)$ is chaotic. Since then, this discovery has become one of the classic results in the field of chaos theory.

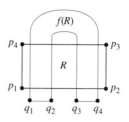

F I G U R E   5.13

In the following theorem we summarize our findings concerning these two types of horseshoes, but omit its proof since it goes well beyond the scope of this book.

T H E O R E M   *Chaos: Horseshoe*

If $F(x, y)$ has a horseshoe then $(x_{n+1}, y_{n+1}) = F(x_n, y_n)$ is chaotic.   ∎

Although it is slightly easier to find the first type of horseshoe described above than it is the second, for most functions $F(x, y)$, observing either type directly would be difficult, if not impossible. But the existence of a horseshoe, and therefore chaos, can be inferred by other means. Recall from Chapter 3 the criteria for chaos that involved a solution of $x_{n+1} = f(x_n)$ that begins and ends at an unstable fixed point p, but with not all $x_n = p$. This was called a *homoclinic orbit,* and the point p a *snap-back repeller,* since the homoclinic orbit is at first repelled by p but then snaps back to hit it. Analogous solutions may exist in the present setting as well, and when they do, chaos often results as before.

To see this, first suppose that in addition to computing the *forward* trajectory of an initial point (x_0, y_0) by using $(x_{n+1}, y_{n+1}) = F(x_n, y_n)$ with $n = 0, 1, 2, \ldots$, we also write the system as

$$(x_n, y_n) = F^{-1}(x_{n+1}, y_{n+1})$$

and then iterate *backwards* by letting $n = -1, -2, -3, \ldots$. Not only would this generate a trajectory consisting of a *doubly-infinite* sequence of iterates

$$\ldots, (x_{-2}, y_{-2}), (x_{-1}, y_{-1}), (x_0, y_0), (x_1, y_1), (x_2, y_2), \ldots$$

but also perhaps a multitude of such trajectories, since the *inverse* $F^{-1}(x, y)$ of a nonlinear function $F(x, y)$ is often multivalued.

Next, suppose that (p, q) is an unstable fixed point of $F(x, y)$ and that for some initial point $(x_0, y_0) \neq (p, q)$ the resulting trajectory (x_n, y_n) begins and ends at (p, q), i.e.,

$$\lim_{n \to -\infty} (x_n, y_n) = (p, q) \quad \text{and} \quad \lim_{n \to +\infty} (x_n, y_n) = (p, q)$$

Such a trajectory is again referred to as a **homoclinic orbit,** and if (p, q) is a source then it is sometimes called a **snap-back repeller** for the same reason as before.

Although we shall not prove so here, the existence of a homoclinic orbit generally implies that $F(x, y)$ has a horseshoe: the first type of horseshoe if (p, q) is a source, or the second type if it is a saddle. Since the existence of a horseshoe is associated with chaos, this gives us the following.

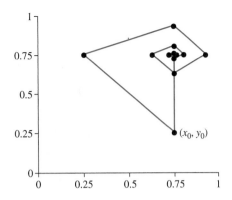

FIGURE 5.14

THEOREM   *Chaos: Homoclinic Orbit*

If $F(x, y)$ has a homoclinic orbit then $(x_{n+1}, y_{n+1}) = F(x_n, y_n)$ is almost always chaotic. ∎

Note that the phrase *almost always* is used here. For technical reasons one additional condition, called *transversality,* must be met before chaos can be absolutely concluded. Since this condition is nearly always satisfied, we shall not go into it. For our purposes, it suffices to say *almost always*.

EXAMPLE 7

Show that the system from Example 6 is chaotic by finding a homoclinic orbit.

Solution: Since the system may be written as $(x_{n+1}, y_{n+1}) = F(x_n, y_n)$ where $F(x, y) = (4y(1 - y), x)$, the non-zero fixed point is easily found to be $(3/4, 3/4)$. If we let $(x_0, y_0) = (3/4, 1/4)$, then we also find that

$$(x_1, y_1) = F(x_0, y_0) = F(3/4, 1/4) = (3/4, 3/4)$$

Since $(3/4, 3/4)$ is a fixed point and $(x_1, y_1) = (3/4, 3/4)$, then we must have $(x_n, y_n) = (3/4, 3/4)$ for all $n \geq 1$. This implies that $\lim_{n \to +\infty}(x_n, y_n) = (3/4, 3/4)$. To see that $\lim_{n \to -\infty}(x_n, y_n) = (3/4, 3/4)$ as well, we use direct numerical iteration of the inverse. Figure 5.14 shows the trajectory (x_n, y_n) for all $n < 0$, as well as for $n \geq 0$. It clearly shows a homoclinic orbit that begins and ends at $(3/4, 3/4)$. ∎

Exercises 5.3

In Exercises 1–8 find the interval of stability of the fixed point $(0, 0)$.

1. $x_{n+1} = x_n^2 - y_n^2$
    $y_{n+1} = y_n - a(x_n + y_n)$

2. $x_{n+1} = ax_n - y_n^2$
    $y_{n+1} = x_n^2 - ay_n$

3. $x_{n+1} = ax_n/(4 + 5y_n^2)$
    $y_{n+1} = x_n$

4. $x_{n+1} = ay_n - x_n^2$
    $y_{n+1} = x_n^2 + y_n^2$

5. $x_{n+1} = 2x_n - ax_n e^{4y_n}$

   $y_{n+1} = 3y_n - 3y_n e^{5x_n}$

6. $x_{n+1} = x_n^2 - ay_n$

   $y_{n+1} = x_n + y_n^2$

7. $x_{n+1} = ay_n/(2x_n + 5)$

   $y_{n+1} = -ax_n/(3y_n + 4)$

8. $x_{n+1} = ax_n - 3e^{y_n/4} + 3$

   $y_{n+1} = ay_n + 2e^{x_n/6} - 2$

9. Find the interval of stability of the fixed point $(0, 0)$ of the Nonlinear Overlapping-Generations Model:

   (a) for $r \geq 0$, if $s = 0$ and $N = 10{,}000$
   (b) for $s \geq 0$, if $r = 0$ and $N = 25{,}000$

In Exercises 10–13 find the region of stability in the (a, b) *plane of the fixed point* $(0, 0)$.

10. $x_{n+1} = ax_n^2 - y_n^2$

   $y_{n+1} = by_n^2 + x_n^2$

11. $x_{n+1} = ax_n - y_n^2$

   $y_{n+1} = by_n + x_n^2$

12. $x_{n+1} = ax_n + by_n/(x_n + 2)$

   $y_{n+1} = ay_n - bx_n/(y_n + 8)$

13. $x_{n+1} = ay_n - x_n^2$

   $y_{n+1} = bx_n + y_n^2$

14. Find the region of stability in the first quadrant of the (r_1, r_2) plane for the fixed point $(0, 0)$ of:

   (a) the general Nonlinear Prey–Predator Model.
   (b) the general Nonlinear Competition Model.

In Exercises 15 and 16: (a) Find the positive fixed point (p_a, p_a) *and its interval of existence; (b) Find the interval of stability of* $(0, 0)$ *and of* (p_a, p_a); *(c) Show that* (p_a, p_a) *bifurcates from* $(0, 0)$.

15. $x_{n+1} = ay_n(1 - y_n)$

   $y_{n+1} = x_n$

16. $x_{n+1} = ay_n - y_n^3$

   $y_{n+1} = x_n$

For the parameterized system considered in Example 2, the eigenvalues of the linearization matrix D *were found to be* $a \pm bi$. *Use this in Exercises 17–20 to find the period of the stable periodic cycle that bifurcates from* $(0, 0)$ *at each of the given bifurcation points in the* (a, b) *plane, or say whether the cycle is irrationally periodic.*

17. $(a, b) = (1/\sqrt{2}, 1/\sqrt{2})$

18. $(a, b) = (0, -1)$

19. $(a, b) = (-\sqrt{3}/2, 1/2)$

20. $(a, b) = (\sqrt{2}/3, \sqrt{7}/3)$

21. Show that the system $\begin{cases} x_{n+1} = 4x_n(1 - x_n) \\ y_{n+1} = 2y_n \end{cases}$ is chaotic by finding two rectangles A and B in the (x, y) plane that intersect only on one edge and satisfy the horseshoe principle $A \cup B \subset F(A) \cap F(B)$.

22. For the system $\begin{cases} x_{n+1} = 2|x_n| - 1 \\ y_{n+1} = 2y_n \end{cases}$ first verify that $(1, 0)$ is an unstable fixed point, and then prove that the sequence (x_n, y_n) defined by

$$(x_n, y_n) = (1 - 2^n, 0) \quad \text{for } n \leq 1 \quad \text{and} \quad (x_n, y_n) = (1, 0) \quad \text{for } n \geq 2$$

constitutes a homoclinic orbit of the system.

Computer Projects 5.3

1. Show that the Nonlinear Infection–Recovery Model with $N = 1000$, $r = 0.1$ and $s = 4.5$ has a period-doubling cascade as $t > 0$ increases, which ultimately leads to chaos.

2. Show that the Mandelbrot system with $b = 0$ has a period-doubling cascade as $a < 0$ decreases.

3. The system $\begin{cases} x_{n+1} = 1 + y_n - ax_n^2 \\ y_{n+1} = bx_n \end{cases}$ is known as the *Hénon map*. Show that for $b = 0.4$ this system has a period-doubling cascade as $a > 0$ increases, which ultimately leads to chaos.

4. When the two growth-rate parameters of the Nonlinear Overlapping-Generations Model with $N = 1000$ are equal, it may be written as

$$P_{n+1} = a(P_n + Q_n)(1 - (P_n + Q_n)/1000)$$
$$Q_{n+1} = P_n$$

   where $r = s = a$. As a increases, this system has a period-doubling cascade that begins with a stable 3-cycle instead of a stable fixed point. Show this through direct numerical iteration, by first finding the stable 3-cycle for $a = 2.05$, and then observing that as a increases, this 3-cycle bifurcates into a stable 6-cycle, this 6-cycle bifurcates into stable 12-cycle, this 12-cycle bifurcates into stable 24-cycle, etc., all of which eventually lead to chaos.

5. Through numerical experimentation, estimate the value of the parameter $c_1 > 0$ at which the fixed point of the Nonlinear Price–Demand Model with $a_1 = 2.5$, $b_1 = 0.1$, $h_1 = -1$ and $k_1 = 1$ bifurcates into a stable cycle. What kind of bifurcation does this appear to be?

6. Verify your conclusions regarding the bifurcations in Exercises 17–20 by numerically iterating each system for several choices of (a, b) just outside the unit circle $a^2 + b^2 = 1$, but close to the (a, b)-value given in each case.

7. Through numerical experimentation find as many stable periodic cycles as possible for the Mandelbrot system.

8. Observe the deformation of the stable invariant curve associated with the Nonlinear Overlapping-Generations Model with $r = 1.5$ and $N = 1000$ as s increases, by numerically iterating the system for $s = 3.50, 3.51, 3.52, \ldots, 3.60$. Perform at least 5000 iterations in each case, beginning with $P_0 = Q_0 = 0.3$, and skip the first 100 or so points when plotting each trajectory.

9. Compare the strange attractor of the Nonlinear Infection–Recovery Model for $r = 0.25$, $s = 4.9$ and $t = 0.4$ with that of the Nonlinear Overlapping-Generations Model for $r = 1$ and $s = 4.7$. Can you say which of these is likely to have the first type of horseshoe and which the second type?

10. Chaos cannot occur for systems of autonomous differential equations with fewer than three equations. But it does exist for the following model, which is related

to weather prediction:

$$x'(t) = 10y(t) - 10x(t)$$
$$y'(t) = 28x(t) - y(t) - x(t)z(t)$$
$$z'(t) = x(t)y(t) - 8z(t)/3$$

This system was made famous in the 1960's by E. Lorenz, one of the pioneers of chaos theory.

Use the techniques of Section 3.4 and Exercise 28 of Section 5.1 to derive a three-dimensional Euler approximation for this system. Then, using $(x_0, y_0, z_0) = (1, 1, 1)$ and a step size of $h = 0.01$, generate an approximate solution of the system for $n = 0, 1, \ldots, 4000$. If three-dimensional graphics software is available, plot the trajectories (x_n, y_n, z_n) in three-dimensional solution space. Otherwise, make three separate solution-space graphs, one each for (x_n, y_n), (x_n, z_n) and (y_n, z_n). Can you describe in words the shape of the strange attractor?

5.4   Fractals

No investigation of the dynamics of nonlinear systems would be complete without some mention of *fractals,* whose images are among the most interesting and bizarre ever generated by a computer. In this section we present a brief introduction to fractal geometry, first showing how fractals naturally arise in the study of chaotic dynamics, then how they are often generated using an *iterated-function system* and finally how their *fractional* dimensions may be computed.

Fractals on the Line

To understand the relationship between chaos and fractals, it is perhaps best to begin in the setting where chaos was first introduced in this text. Recall from Chapter 3 the period-doubling route to chaos experienced by many parameterized families of unimodal equations $x_{n+1} = f_r(x_n)$, in particular the logistic family where $f_r(x) = rx(1 - x)$. As explained in that chapter, the logistic family exhibits chaos for all $r > 3.57$, and this was verified numerically for several values of r between 3.57 and 4. In Section 3.8 it was further shown that for $r = 4$, not only do horseshoes and homoclinic orbits exist, both of which imply chaos, but the scrambled set is *dense* in the interval between 0 and 1.

But for reasons that should now become clear, no detailed investigations were ever undertaken with regard to the dynamics of the logistic family for $r > 4$ (unless Computer Project 9 of Section 3.8 was attempted). If one were to randomly select a large sample of initial points x_0 between 0 and 1, and iterate the logistic equation for any choice of $r > 4$, one would find that nearly all solutions x_n diverge to $\pm\infty$. That is, even though horseshoes, homoclinic orbits and hence chaos still exist, the scrambled sets seem to have virtually disappeared!

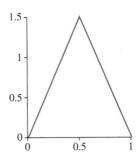

FIGURE 5.15

To help explain how this can happen, suppose we instead iterate the simpler piecewise-linear equation $x_{n+1} = f(x_n)$, where

$$f(x) = \begin{cases} 3x & \text{if } x \leq 1/2 \\ 3 - 3x & \text{if } x > 1/2 \end{cases}$$

As one may see from its graph in Figure 5.15, this function $f(x)$ has a unimodal shape similar to that of $f_r(x) = rx(1 - x)$ for any $r > 4$. With a few steps of cobwebbing, it can be checked that $f(x)$ has both a horseshoe and a homoclinic orbit, and therefore a chaotic scrambled set S must exist somewhere. Cobwebbing also indicates that for $x_0 < 0$ or $x_0 > 1$, solutions satisfy $x_n \to -\infty$ as $n \to \infty$, which implies that wherever S is, no part of it resides outside the closed interval $[0, 1]$.

To locate S within $[0, 1]$, first observe that for any x_0 chosen from the *middle third* of that interval, i.e., for $1/3 < x_0 < 2/3$, the very next iterate satisfies $x_1 > 1$, which again means that $x_n \to -\infty$. This implies that no part of S is included within the open interval $(1/3, 2/3)$, which therefore leaves only the two closed intervals

$$[0, 1/3] \quad \text{and} \quad [2/3, 1] \tag{1}$$

to be considered further.

Although it is less apparent, for any x_0 chosen from the middle thirds of those two intervals, i.e., for

$$1/9 < x_0 < 2/9 \quad \text{or} \quad 7/9 < x_0 < 8/9$$

we must have $1/3 < x_1 < 2/3$. Consequently, $x_2 > 1$, which once again leads to $x_n \to -\infty$, and so we must exclude from further consideration the two open intervals $(1/9, 2/9)$ and $(7/9, 8/9)$. This means that the set S must be contained within the four closed intervals

$$[0, 1/9], \quad [2/9, 1/3], \quad [2/3, 7/9] \quad \text{and} \quad [8/9, 1]. \tag{2}$$

For any x_0 chosen from the middle thirds of these four intervals, the next iterate must satisfy either

$$1/9 < x_1 < 2/9 \quad \text{or} \quad 7/9 < x_1 < 8/9$$

and so in either case $1/3 < x_2 < 2/3$, $x_3 > 1$ and $x_n \to -\infty$. And if this reasoning is continued, we see that the middle thirds of all remaining intervals must be successively excluded when searching for S. Figure 5.16 shows the first 4 steps of this process.

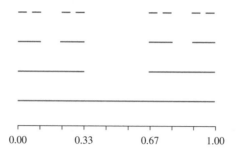

0.00   0.33   0.67   1.00

FIGURE 5.16

Gathering together all the open intervals removed from [0, 1] this way, we find

1   interval of length   1/3,

2   intervals of length   1/9,

4   intervals of length   1/27,

$$\vdots$$

and in general

2^n intervals of length $1/3^{n+1}$

Adding the lengths of all these intervals, while making use of the formula from Chapter 2 for the sum of an infinite geometric series, yields

$$\sum_{n=0}^{\infty} \frac{2^n}{3^{n+1}} = \frac{1}{3} \sum_{n=0}^{\infty} \left(\frac{2}{3}\right)^n = \frac{1}{3} \cdot \frac{1}{1 - 2/3} = 1$$

In other words, the lengths of all intervals that must be excluded from [0, 1] when searching for S add up to 1 exactly, which is the length of [0, 1] itself!

Although one might conclude from this that the entire interval [0, 1] must have been removed, that is not the case. There is in fact an uncountably infinite number of points that remain. This collection of points, which is called a **Cantor set,** constitutes the scrambled set S for this piecewise-linear equation. All the complex dynamics normally associated with chaos occur on this Cantor set, except for one significant feature. While an infinite number of periodic points of different periods and sensitive dependence on initial conditions do exist, S is not dense anywhere on the real line. In this case the set S is also a *repeller,* i.e., choosing x_0 not within S always leads to $x_n \to \pm\infty$. This explains why finding a point of S by random means is unlikely, not just for this function, but also for any member of the logistic family with $r > 4$, which has a scrambled set with similar properties.

However, in exchange for the density that is lost, such a scrambled set gains one additional characteristic. It is now a **fractal.** It has properties typical of fractal sets, in particular an infinite amount of detail. That is, no matter how much one *magnifies* the Cantor set, it never simplifies to an elementary geometric object, such as a point, line, rectangle, etc. In fact, its parts are comprised of an infinite number of copies of itself, and if one enlarges any small piece of it, the entire Cantor set can be recreated.

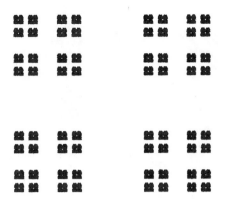

FIGURE 5.17

Fractals in the Plane

Not only are the scrambled sets of nonlinear systems also sometimes fractals, but the variety of fractal shapes that they may take on is much more diverse. Consider first the scrambled set of the system

$$x_{n+1} = \begin{cases} 3y_n & \text{if } y_n \leq 1/2 \\ 3 - 3y_n & \text{if } y_n > 1/2 \end{cases}$$

$$y_{n+1} = x_n$$

shown in Figure 5.17. This fractal may be considered a Cantor set in the plane, in which we begin with the square defined by $0 \leq x \leq 1, 0 \leq y \leq 1$, and remove at each step the *middle thirds,* along both the x-axis and the y-axis, of all squares that remain from the previous step. Like the Cantor set discussed earlier, this fractal is also a repeller, i.e., all solutions (x_n, y_n) become unbounded for virtually all (x_0, y_0) not within the fractal. This therefore makes the name, *strange attractor,* normally used to describe the scrambled set of a chaotic nonlinear system, rather inappropriate here. In recent years these sets have become known as **Julia sets.**

However, the previous example may be a bit misleading. The fractal sets that arise for most nonlinear systems are seldom as orderly and simple to describe as a Cantor set is. Figure 5.18 shows two examples of Julia sets for the Mandelbrot system: (a) for $a = 0.3$ and $b = 0.04$, and (b) for $a = -0.1$ and $b = 0.8$. These clearly possess much greater complexity than does a Cantor set.

In fact, the boundary of the Mandelbrot set itself, shown in Figure 5.8 (in the previous section), constitutes a fractal, perhaps the most famous one ever generated. By greatly magnifying this boundary and using a certain coloring algorithm, some fascinating images can be created, such as the one from Chapter 1 that is reproduced in Figure 5.19. The complexity, and perhaps beauty, of shapes such as these is more typical of what one finds for fractals associated with nonlinear systems. Numerous books and web sites exist that are devoted to the creation, exploration and/or aesthetic appreciation of fractal images such as these.

(a)   (b)

FIGURE 5.18

FIGURE 5.19

Generating Images of Julia Sets

Since Julia sets are generally repellers, one may be wondering how their images in solution space are obtained. After all, direct numerical iteration of the nonlinear system involved always results in trajectories diverging from these fractal sets, not converging to them as if they were strange attractors.

The trick is to transform the iterative process into one for which the Julia set is an attractor rather then a repeller. One way to accomplish this for a nonlinear system of the form

$$(x_{n+1}, y_{n+1}) = F(x_n, y_n)$$

is to first find the *inverse* function $F^{-1}(x, y)$ of $F(x, y)$, assuming that is possible, and then iterate the system

$$(x_{n+1}, y_{n+1}) = F^{-1}(x_n, y_n) \tag{3}$$

instead. Since a repeller of the former is an attractor of latter, trajectories of this inverse system should in theory converge to the Julia set.

However, there is one difficulty with this method: The inverse function $F^{-1}(x, y)$ is nearly always multivalued. This means that when iterating (3), given any initial point (x_0, y_0), there are likely to be several locations possible for (x_1, y_1), and for each of these points, several possible for (x_2, y_2), etc. To see the problem this creates, suppose there are always exactly two inverse points (x_{n+1}, y_{n+1}) associated with any iterate (x_n, y_n). This is a rather common occurrence since most nonlinear systems that have been investigated in recent decades, including the Mandelbrot system, are quadratic. Given any point (x_0, y_0), there must consequently be two points (x_1, y_1), four points (x_2, y_2), eight points (x_3, y_3), $\ldots$, and in general 2^n points (x_n, y_n). After only 100 or so iterations, there would be $2^{100} \approx 10^{30}$ points to keep track of! This is not possible for even the most powerful computers.

The following more practical approach is therefore commonly used. Suppose that each time (3) is iterated, *only one* of the two or more possible inverse points is chosen. This way there will always be a unique point (x_n, y_n) associated with any particular value of n. But how is one to decide which inverse point should be chosen each time? The simplest answer is to not even attempt to choose them *deterministically* at all. It is best if each point is selected *randomly* from the several available choices. The rationale is that a large random sample of iterates is more likely to spread out evenly over the entire Julia set.

To demonstrate this idea, consider the Mandelbrot system, for which

$$F(x, y) = (x^2 - y^2 + a, 2xy + b)$$

After some algebra (including the algebra of complex numbers), the multivalued inverse of this function can be written as

$$F^{-1}(x, y) = \left(\pm f(x, y), \pm \frac{y - b}{2f(x, y)} \right) \tag{4}$$

where

$$f(x, y) = \sqrt{\frac{x - a + \sqrt{(x - a)^2 + (y - b)^2}}{2}} \tag{5}$$

Note that each time (3) is iterated for this $F^{-1}(x, y)$, a $+$ or $-$ must be chosen in (4). Beginning with $(x_0, y_0) = (0, 0)$, if this sign is selected randomly each time, then after 30,000 or so iterations the images appearing in Figure 5.18 will appear. We should add, however, that to improve the quality of these images, a non-uniform distribution of signs was used. In Figure 5.18(a) approximately 3/4 are $+$ and 1/4 are $-$, and in Figure 5.18(b) approximately 1/4 are $+$ and 3/4 are $-$. Also, the first 100 or so iterates were skipped when plotting the solution-space graphs of (x_n, y_n).

Iterated-Function Systems

The type of iteration process just performed may be viewed in a slightly different way. Suppose we call

$$F_1(x, y) = \left(+f(x, y), +\frac{y - b}{2f(x, y)} \right) \quad \text{and} \quad F_2(x, y) = \left(-f(x, y), -\frac{y - b}{2f(x, y)} \right)$$

where $f(x, y)$ is given by (5). Generating a Julia set for the Mandelbrot system may now be described as follows: Beginning with $(x_0, y_0) = (0, 0)$, randomly select the value (x_{n+1}, y_{n+1}) each time to be either $F_1(x_n, y_n)$ with probability p_1, or $F_2(x_n, y_n)$ with probability p_2, where $p_1 + p_2 = 1$. After obtaining a sufficiently large number of iterates (x_n, y_n), graph all but the first few.

When functions are iterated this way, they may be described as an **iterated-function system.** In its most general sense, an iterated-function system of two variables is comprised of a collection of two or more functions:

$$F_1(x, y), \quad F_2(x, y), \quad \ldots, \quad F_k(x, y)$$

along with a set of probabilities:

$$p_1, \quad p_2, \quad \ldots, \quad p_k$$

whose sum is 1. With (x_0, y_0) given, each iterate (x_{n+1}, y_{n+1}) for $n = 0, 1, 2, \ldots$ is determined by first randomly choosing one of the functions $F_i(x, y)$, and then computing

$$(x_{n+1}, y_{n+1}) = F_i(x_n, y_n)$$

The mechanism for making these random choices is weighted so that each $F_i(x, y)$ is chosen with probability p_i.

This method of generating fractals has become quite popular in recent decades, primarily due to M. Barnsley. Unlike the two inverse functions of the Mandelbrot system, here the $F_i(x, y)$ need not necessarily be related to each other. Nearly any collection of functions could be used, although certain properties (that we shall not go into) would have to be satisfied to ensure convergence of the iteration process.

As an example, suppose we define the three linear functions

$$F_1(x, y) = \left(\frac{x}{2}, \frac{y}{2}\right), \quad F_2(x, y) = \left(\frac{2x + 1}{4}, \frac{y + 1}{2}\right) \quad \text{and} \quad F_3(x, y) = \left(\frac{x + 1}{2}, \frac{y}{2}\right)$$

and let $p_1 = p_2 = p_3 = 1/3$. Beginning with $(x_0, y_0) = (1, 1)$, if we then perform 10,000 iterations and graph all but the first 10 iterates, the fractal shown in Figure 5.20 results. This is known as the **Sierpinski triangle.**

FIGURE 5.20

FIGURE 5.21

On the other hand, using the four linear functions

$$F_1(x, y) = (0.85x + 0.04y + 0.075, -0.04x + 0.85y + 0.18)$$
$$F_2(x, y) = (0.2x - 0.26y + 0.4, 0.23x + 0.22y + 0.045)$$
$$F_3(x, y) = (-0.15x + 0.28y + 0.575, 0.26x + 0.24y - 0.086)$$
$$F_4(x, y) = (0.5, 0.16y)$$

with respective probabilities

$$p_1 = 0.77, \quad p_2 = 0.12, \quad p_3 = 0.10, \quad p_4 = 0.01$$

the fractal from Chapter 1 that is reproduced in Figure 5.21 results after graphing all but the first 10 of 20,000 iterates with initial value $(x_0, y_0) = (1, 1)$. This remarkably natural looking fractal is named the **Barnsley fern,** after its originator. Images such as this have spurred a great deal of interest in fractals from such diverse areas as computer graphics and the biological sciences.

Dimension of a Fractal

One of the most intriguing aspects of fractals is the non-integer or *fractional* dimensions that they appear to have. To compute the dimension of a fractal, the usual definition of dimension must be expanded. Although the dimension of an entire space is ordinarily defined as the positive whole number that corresponds to how many independent coordinates are needed to designate points in that space, other definitions are possible for a subset of a space.

One such definition involves first counting the minimum number of *boxes* $N = N(s)$ of *size* s that would be needed to contain or cover the subset, and then observing how N

increases as s approaches 0. This is sometimes called the **box-counting dimension.** For any set of points this dimension is defined as the value d that makes

$$N \text{ inversely proportional to } s^d \text{ or directly proportional to } \frac{1}{s^d}$$

as s approaches 0. This is equivalent to saying

$$d = \lim_{s \to 0} \frac{\ln N}{\ln(1/s)} \tag{6}$$

To verify that this definition makes sense, and in fact corresponds to our usual notion of dimension, suppose we compute d first for the closed interval [0, 1], which everyone would agree is one-dimensional. If boxes are viewed as being line segments along the real axis, and their sizes as the lengths of these line segments, then to cover the points of [0, 1] we would need at least

1	box of size	1	or
10	boxes of size	1/10	or
100	boxes of size	1/100	or

$$\vdots$$

This implies that we would always need at least

$$N = \frac{1}{s} \quad \text{boxes of size} \quad s$$

for any $s > 0$. From (6) this makes the dimension

$$d = \lim_{s \to 0} \frac{\ln(1/s)}{\ln(1/s)} = 1$$

which is the previously accepted dimension of [0, 1].

In the plane a box may be defined as a square, and the box size as the length of each side of that square. To cover the two-dimensional square defined by $0 \le x \le 3, 0 \le y \le 3$, we would therefore need a minimum of

9	boxes of size	1	or
900	boxes of size	1/10	or
90,000	boxes of size	1/100	or

$$\vdots$$

or in general

$$N = \frac{9}{s^2} \quad \text{boxes of size} \quad s$$

From (6) this means that

$$d = \lim_{s \to 0} \frac{\ln(9/s^2)}{\ln(1/s)} = \lim_{s \to 0} \frac{\ln 9 - 2 \ln s}{\ln 1 - \ln s} = \lim_{s \to 0} \frac{\ln 9 - 2 \ln s}{0 - \ln s}$$

$$= \lim_{s \to 0} \frac{\ln 9}{-\ln s} + \lim_{s \to 0} \frac{-2 \ln s}{-\ln s} = \frac{\ln 9}{\infty} + 2 = 0 + 2 = 2$$

Since $d = 2$, which is also the accepted dimension of the square $0 \le x \le 3, 0 \le y \le 3$, the two notions of dimension again correspond.

Since the box-counting dimension d appears to match our everyday understanding of dimension when applied to these simple subsets, it would be interesting to compute d for some of the fractals we have discussed. Consider first the Cantor set on the interval $[0, 1]$. It should be clear that the entire set could be contained within a single box of size $s = 1$. But, if the box size is instead $s = 1/3$, then at least two boxes are needed, namely, the intervals given in (1). Similarly, the four boxes of size $s = 1/9$ that are given in (2) could also cover the Cantor set. Another look at Figure 5.16 would be helpful in seeing this.

It therefore appears that covering the points of the Cantor set requires

$$1 \quad \text{box of size} \quad 1 \quad \text{or}$$
$$2 \quad \text{boxes of size} \quad 1/3 \quad \text{or}$$
$$4 \quad \text{boxes of size} \quad 1/9 \quad \text{or}$$
$$\vdots$$

which implies that in general at least

$$N = 2^n \quad \text{boxes of size} \quad s = \frac{1}{3^n}$$

are needed. Since letting $s \to 0$ corresponds to letting $n \to \infty$ here, then (6) becomes

$$d = \lim_{n \to \infty} \frac{\ln 2^n}{\ln 3^n} = \lim_{n \to \infty} \frac{n \ln 2}{n \ln 3} = \frac{\ln 2}{\ln 3}$$

Hence, the dimension of this Cantor set is $d = \ln 2/\ln 3 \approx 0.63$. A dimension between 0 and 1 is consistent with the idea that the Cantor set is substantially more than a countable set of points, which has a dimension of 0, but not quite a line segment whose dimension is 1. This *fractional* dimension is also partly responsible for B. Mandelbrot's choice of the term *fractal* to describe such sets.

To cover the Sierpinski triangle, it would take only

$$1 \quad \text{box of size} \quad 1 \quad \text{or}$$
$$3 \quad \text{boxes of size} \quad 1/2 \quad \text{or}$$
$$9 \quad \text{boxes of size} \quad 1/4 \quad \text{or}$$
$$\vdots$$

or in general

$$N = 3^n \quad \text{boxes of size} \quad s = \frac{1}{2^n}$$

Figure 5.22 shows some of these boxes. This time a calculation of the dimension would yield $d = \ln 3/\ln 2 \approx 1.58$, which indicates that the Sierpinski triangle amounts to something between a one-dimensional and a two-dimensional object.

Although the dimensions of many other simple fractal sets can be similarly computed using (6), determining the dimensions of more complex fractals, such as Julia sets for the Mandelbrot as well as most other nonlinear systems, or even the boundary of the Mandelbrot set itself, has proven a much more challenging task. This issue constitutes just one of many areas of current mathematical research involving fractals.

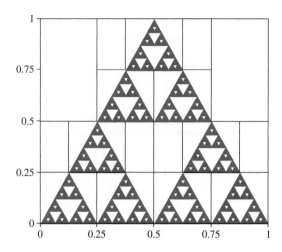

FIGURE 5.22

Exercises 5.4

1. Compute the total length of all intervals removed to create the Cantor set on the interval [0, 3], i.e., beginning with [0, 3] remove the middle thirds of all intervals that remain at each step.

2. Repeat the previous exercise for the general interval $[a, b]$ where $a < b$.

3. Other types of Cantor sets can be created on the interval [0, 1] other than the one discussed in this section. Find the total length of all intervals removed if the middle half of each remaining interval is removed at each step.

4. Repeat the previous exercise for the Cantor set in which the middle 3/4 of each remaining interval is removed at each step.

5. Compute the total area of all squares removed in constructing the Cantor set in the plane shown in Figure 5.17.

6. (a) Compute the area of the triangle whose corners are the points $(0, 0)$, $(1, 0)$ and $(1/2, 1)$. (b) Compute the total area of all triangles removed when constructing the Sierpinski triangle shown in Figure 5.22. (c) Compare the results of (a) and (b).

For each of the systems $(x_{n+1}, y_{n+1}) = F(x_n, y_n)$ *given in Exercises 7–9, find the multivalued inverse system* $(x_{n+1}, y_{n+1}) = F^{-1}(x_n, y_n)$.

7. $x_{n+1} = y_n^2 + 1$
   $y_{n+1} = x_n - 2$

8. $x_{n+1} = x_n^2 + y_n^2$
   $y_{n+1} = x_n$

9. $x_{n+1} = x_n^2 + y_n$
   $y_{n+1} = 2x_n - y_n$

10. Verify that $F^{-1}(x, y)$ given by (4) and (5) is the multivalued inverse of the Mandelbrot function $F(x, y) = (x^2 - y^2 + a, 2xy + b)$ by showing that $F(F^{-1}(x, y)) = (x, y)$.

11. Compute the dimension d of the fractal set described in Exercise 3.

12. Compute the dimension d of the fractal set described in Exercise 4.

13. Compute the dimension d of the scrambled set S of $x_{n+1} = f(x_n)$ for the function

$$f(x) = \begin{cases} 5x/2 & \text{if } x \le 1/2 \\ 5(1-x)/2 & \text{if } x > 1/2 \end{cases}$$

14. Repeat the previous exercise for the function $f(x) = \begin{cases} 6x & \text{if } x \le 1/2 \\ 6 - 6x & \text{if } x > 1/2 \end{cases}$

15. Suppose that when constructing the Cantor set on the square in the (x, y) plane defined by $0 \le x \le 1, 0 \le y \le 1$, the middle halves along both the x-axis and the y-axis are removed at each step, instead of the middle thirds. Compute the dimension d of the resulting fractal set.

16. Compute the dimension d of the Julia set for the system

$$x_{n+1} = \begin{cases} 3.5y_n & \text{if } y_n \le 1/2 \\ 3.5(1 - y_n) & \text{if } y_n > 1/2 \end{cases}$$

$$y_{n+1} = x_n$$

▆▆▆▆▆▆ Computer Projects 5.4

1. Generate the scrambled set for $x_{n+1} = 4.4x_n(1 - x_n)$ using at least 5000 iterations.

2. Generate the Julia set for the system $\begin{cases} x_{n+1} = 6y_n(1 - y_n) \\ y_{n+1} = x_n \end{cases}$ using at least 10,000 iterations.

3. Generate the Julia set of the Mandelbrot system for each of the following pairs of parameter values. In each case, perform at least 10,000 iterations using various probability distributions when choosing $+$ and $-$ in (4).
   (a) $a = -0.5, b = 0.3$; (c) $a = 0, b = 1$;
   (b) $a = -0.2, b = 0.7$; (d) $a = 0.2, b = 0.7$

For each of the iterated function systems given in Projects 4–9, generate and plot at least 10,000 iterates (x_n, y_n) using $(x_0, y_0) = (0, 0)$ and assuming that all probabilities p_i are equal. Skip the first 100 or so iterates when graphing.

4. $F_1(x, y) = (rx - 0.3y + 1, 0.8x + 0.3y)$ and $F_2(x, y) = (rx - 0.3y - 1, 0.8x + 0.3y)$ for (a) $r = 0.2$; (b) $r = 0.6$.

5. $F_1(x, y) = (0.4x - ry + 1, 0.2x + 0.7y)$ and $F_2(x, y) = (0.4x - ry - 1, 0.2x + 0.7y)$ for (a) $r = 0.1$; (b) $r = 0.5$.

6. $F_1(x, y) = (rx - sy + 0.25, sx^2 + ry^2)$ and $F_2(x, y) = (rx - sy - 0.25, sx^2 + ry^2)$ for (a) $r = 0.1, s = 0.9$; (b) $r = 0.85, s = 0.1$.

7. $F_1(x, y) = (rx^2 - ry^2 + 0.25, rx^2 + ry^2)$ and $F_2(x, y) = (rx^2 - ry^2 - 0.25, rx^2 + ry^2)$ for (a) $r = 1$; (b) $r = 1.35$.

8. $F_1(x, y) = (x/2, y/2)$, $F_2(x, y) = (x/2, (y + 1)/2)$ and $F_3(x, y) = ((x + 1)/2, y/2)$.

9. $F_1(x, y) = (x/2, (1 - y)/2)$, $F_2(x, y) = ((x + 1)/2, (2 - y)/2)$ and $F_3(x, y) = ((x + 2)/2, (1 - y)/2)$.

10. Fractal sets arise in many settings not discussed in this section. For example, when Newton's Root-Finding Method is applied to $f(z) = z^3 - 1$ where z is complex, the iterative equation becomes

$$z_{n+1} = z_n - \frac{z_n^3 - 1}{3z_n^2}$$

If separated into real and imaginary parts, this is equivalent to iterating the system

$$x_{n+1} = \frac{2x_n}{3} + \frac{x_n^2 - y_n^2}{3\left(x_n^2 + y_n^2\right)^2}$$

$$y_{n+1} = \frac{2y_n}{3} - \frac{2x_n y_n}{3\left(x_n^2 + y_n^2\right)^2}$$

where $z_n = x_n + iy_n$. Since $z = x + iy = 1$ is a root of $f(z)$, then $(x, y) = (1, 0)$ should be a stable fixed point of this system. However, two other stable fixed points exist as well, and all three have basins of attraction.

The boundary of the basin of attraction of $(1, 0)$ is a fractal. Show this by generating that basin of attraction, i.e., plot all points (x_0, y_0) for which (x_n, y_n) converges to $(1, 0)$. (*Note:* This requires more sophisticated computer programming than appears in the Appendix.)

Appendix

Here we describe two rather diverse alternatives for numerically and graphically investigating discrete dynamical systems. The first, *Microsoft Excel,* is unquestionably the most popular spreadsheet program in existence today. While spreadsheet programs are not ideal for performing very complex types of calculations, including some discussed in this text, their widespread availability and remarkable simplicity make them a natural choice for most users of personal computers. At the opposite extreme is *Mathematica,* perhaps the most powerful mathematical software presently available. Very lengthy and sophisticated computations can be performed using *Mathematica,* but its syntax is rather tedious and difficult to master. Between these two extremes lies a vast array of quantitatively-oriented software of varying complexity and mathematical sophistication. If neither of the methods discussed below proves suitable, the reader may wish to investigate those other alternatives.

Microsoft Excel

We first demonstrate how most of the tables and graphs that have appeared throughout this text might be generated using a spreadsheet. Since simplicity and brevity were primary goals in developing this set of instructions, the reader should seek out other sources for more detailed explanations, if needed. Such sources presently abound in the computer sections of most libraries and bookstores. Although the emphasis is on *Microsoft Excel,* the format of most instructions mentioned here should carry over quite readily to virtually all other spreadsheet programs. If discrepancies are found, one should consult the help files and/or manuals for the particular software being used.

As with most programs in a *Windows* environment, to start *Excel* one double-clicks on its icon. A blank spreadsheet will appear in a new window with cell A1 (column A, row 1) *selected* or *highlighted*. Other cells can be highlighted using the mouse or arrow keys. Either a number, a string of characters or a mathematical formula may be typed into a cell. Although we shall not go into such cosmetic issues here, character strings are useful when labeling tabulated data, or perhaps the entire spreadsheet. In what follows we give only those instructions necessary for the actual calculation of the quantities desired.

As a first example, suppose we wish to see the first 20 iterates of $x_{n+1} = 1.25x_n - 1$, starting with $x_0 = 5$. This is best accomplished by entering the initial subscript 0 into cell A1, and the initial value 5 into cell $B1$, i.e., type each value into its cell and press the

Enter key. Next, enter the formula

$$\texttt{=A1+1}$$

into cell A2, and the formula

$$\texttt{=1.25*B1-1}$$

into cell B2. Note that formulas always begin with =, blank spaces are generally inconsequential, and * means multiplication.

Once these formulas are entered, subscript number *1* and the value of x_1 should appear in cells A2 and B2 respectively. To generate the remaining values of n and x_n, the pair of cells A2,B2 must be *copied* and *pasted* into each of the pairs of cells A3,B3 through A21,B21. This can be accomplished all at once by first highlighting cells A2 and B2, by using the mouse to point to cell A2, and then pressing down and holding the mouse button while dragging the cursor to cell B2. After letting go, click the *copy* icon in the toolbar. Next, similarly highlight cells A3,B3, A4,B4, ..., A21,B21 by pointing to, pressing down and holding the mouse button on cell A3, and then dragging the cursor to cell B21. After letting go, this time click the *paste* icon in the toolbar. The desired quantities should appear in these cells: *2* in cell A3, the value of x_2 in cell B3, *3* in cell A4, the value of x_3 in cell B4, ..., *20* in cell A21, the value of x_{20} in cell B21. Note that the formulas in cells A2 and B2 were automatically updated as they were pasted.

To plot a time-series graph using these quantities, click the *Chart* icon in the toolbar, and choose *XY (Scatter)* and the sub-type that looks best, perhaps points connected by line segments. To indicate the *Source Data* click the *Series* tab and then the *X Values* box. Highlight (as before) cells A1 through A21 and click the *X Values* box again. Click the *Y Values* box, highlight cells B1 through B21 and click that box again. After clicking *Finish,* the graph should appear in the spreadsheet. The graph, as well as the tabulated data, can be formatted in numerous ways that we shall not go into. They can then be printed by clicking the *print* icon in the toolbar.

Just as easily, virtually any linear or nonlinear equation can be iterated. For example, one can instead iterate

$$x_{n+1} = x_n + \frac{1}{(n+1)^2} \quad \text{by entering} \quad \texttt{=B1+1/A2^2}$$

into cell B2, since / and ^ denote division and exponentiation respectively, or

$$x_{n+1} = 5x_n e^{-x_n} \quad \text{by entering} \quad \texttt{=5*B1*EXP(-B1)}$$

into that cell, since *EXP* denotes the exponential function. In each case cells A2,B2 must then be copied and pasted into cells A3,B3 through A21,B21. If desired, the initial value in cell B1 can be changed in a fairly straightforward manner, as can be the total number of iterations. Also, not all points in the table need be graphed. Only those points highlighted when choosing *X Values* and *Y Values* will be graphed.

In addition to *EXP,* most other special mathematical functions are available in *Excel,* including the natural logarithm *LN,* the absolute value *ABS* and the trigonometric functions *SIN, COS, TAN,* etc., whose arguments must always be enclosed in parentheses. Some functions can be applied to a range of cells. For example, *SUM(A1:A10)* computes the sum of the values in cells A1 through A10. The functions *MAX, MIN* and *AVERAGE* work similarly, giving the maximum, minimum and average respectively.

Next, suppose we wish to estimate the Lyapunov exponent of the solution of $x_{n+1} = 3.9x_n(1 - x_n)$ with $x_0 = 0.6$, using the first 1000 iterates. This can be accomplished by first entering the following data into the specified cells:

```
Cell A1:        0

Cell B1:        0.6

Cell C1:        0

Cell A2:        =A1+1

Cell B2:        =3.9*B1*(1-B1)

Cell C2:        =C1+LN(ABS(3.9*(1-2*B2)))/1000
```

Cells A2,B2,C2 should next be copied and pasted into cells A3,B3,C3 through A1001, B1001,C1001. Cell C1001 will then contain an estimate of the Lyapunov exponent. Here, once again the initial value in cell B1 can easily be changed, but if a different number of iterations is required, the *1000* appearing in cell C2 must reflect this new count. Note that in order to compute the Lyapunov exponent for a different equation, not only must the formula *=3.9*B1*(1−B1)* in cell B2 be modified, but also *3.9*(1−2*B2)* appearing in cell C2 must be changed as well, to make this expression once again correspond to the derivative of the function being iterated, evaluated at x_1, whose value is in cell B2. After any such change (other than to the initial value in cell B1), cells A2,B2,C2 must be copied and pasted as before.

For that same equation a histogram can be obtained showing the distribution of the first 1000 iterates over 50 intervals or *bins*. Begin by entering the following data into the specified cells:

```
Cell A1:        0

Cell B1:        0.6

Cell C1:        =MIN(B2:B1001)

Cell D1:        =MAX(B2:B1001)

Cell E1:        =$C$1+A1*($D$1-$C$1)/50

Cell A2:        =A1+1

Cell B2:        =3.9*B1*(1-B1)
```

Next, copy and paste cells A2,B2 into cells A3,B3 through A1001,B1001, and cell E1 into cells E2 through E50. (*Note:* Since C1 and D1 were referenced in cell E1, the actual values in cells C1 and D1, and not the formulas, will be used when E1 is copied and pasted into cells E2 through E50.) Click *Tools* on the toolbar and select *Data Analysis*. (If this option is not available, see below.) In the *Data Analysis* menu click *Histogram* and then *OK*. When supplying data ranges, make the *Input Range* cells B2 through B1001, the *Bin Range* cells E2 through E50, and the *Output Range* cell F1. Check the *Chart Output* box and then click *OK*. This creates the distribution in columns F and G, as well as the histogram. Although 50 bins are used here, this can be modified in a fairly straightforward manner, as can be the function, number of iterations, initial value, and so on.

If the *Data Analysis* option is not available in the *Tools* menu, one should select *Add-Ins* in that menu, and then check the *Analysis ToolPak* box. After clicking OK (and perhaps inserting the program disk) this option should appear in the *Tools* menu, and a histogram may then be created as described above. If it does not, a histogram can still be generated with the following steps.

First, enter the data into the appropriate cells as specified above, and then copy and paste cells A2,B2 into cells A3,B3 through A1001,B1001, and cell E1 into cells E2 through E50. Next, enter the data

```
Cell F1:      =TRUNC(50*(B1-$C$1)/($D$1-$C$1))
```

and copy and paste it into cells F2 through F1000. Finally, highlight cells G1 through G50 and type (without pressing *Enter*)

```
=FREQUENCY(F1:F1000,A1:A50)
```

While holding down the *Ctrl* and *Shift* keys, then press *Enter*. This creates the bin distribution in column G, which can then be graphed as a histogram by clicking the *Chart* icon and choosing a *Column* graph. Use cells E1 through E50 as the *Category (X) axis labels* and cells G1 through G50 as the *Values*.

To iterate a system of two or more equations, for example:

$$x_{n+1} = 1.1x_n - 0.2y_n + 50 \quad \text{and} \quad y_{n+1} = x_n + 0.8y_n - 35$$

with $x_0 = 500$ and $y_0 = 450$, begin by entering the following data into the specified cells:

```
Cell A1:      0
Cell B1:      500
Cell C1:      450
Cell A2:      =A1+1
Cell B2:      =1.1*B1-0.2*C1+50
Cell C2:      =B1+0.8*C1-35
```

Then copy and paste cells A2,B2,C2 into the desired number of cells below them. These formulas will again be updated as they are pasted, and the values of x_n and y_n will appear in columns B and C respectively. Any other linear or nonlinear system can be iterated this way, although additional columns will be needed for systems of three or more equations.

Several types of graphs are possible using such data. If two separate time-series graphs are desired, the one for x_n can be created as described earlier, using an *XY (Scatter)* graph in which columns A and B constitute the *X Values* and *Y Values* respectively. A separate time-series plot of y_n can then be created in essentially the same manner, except that columns A and C should be used instead. For larger systems, time-series plots of other columns can be handled similarly.

If these two graphs are to be plotted together in the same coordinate system, then the steps are similar, except that two different data series will be used within the same graph. After supplying the *X Values* and *Y Values* for *Series 1* using columns A and B as before, an additional data series can be obtained by first right-clicking on the graph and choosing

Source Data from the menu. Click on the *Series* tab and then on *Add*. For *Series 2* columns A and C should be used as the *X Values* and *Y Values* respectively. If any other variables are involved in the system, their time-series graphs may be plotted along with the first two by obtaining additional data series.

Of course, a solution-space graph of (x_n, y_n) is generally more useful for a system. Creating one is quite similar to what has been described thus far. The only difference is that this time column A is ignored, and columns B and C are used as the *X Values* and *Y Values* respectively on the single data series that is plotted. For systems of three or more variables, a solution space graph of any two of them by may be generated by designating their columns as the *X Values* and *Y Values* respectively.

To create fractal images using *Excel* on a personal computer, it must possess sufficient processing speed and main memory to support the extensive number of calculations and massive data arrays that are needed. If that is the case, then the Julia set of Mandelbrot's system for the parameter values $a = 0.3$ and $b = 0.04$ may be generated by first entering the following data into the specified cells:

```
Cell A1:      0

Cell B1:      0

Cell A2:      =(2*TRUNC(RAND()+0.75)-1)*(((A1-0.3)
                +((A1-0.3)^2+(B1-0.04)^2)^0.5)/2)^0.5

Cell B2:      =(B1-0.04)/(2*A2)
```

Here, *TRUNC* means *truncate to a whole number,* and *RAND* means *choose a random real number between 0 and 1*. This time, copy and paste cells A2,B2 into the next 10,000, 20,000, or perhaps 30,000 cells below them. Note that it may take some time to highlight all the cells when pasting. Next, create an *XY (Scatter)* graph with **no** lines or curves connecting consecutive points, using columns A and B as the *X Values* and *Y Values* respectively. To avoid transient dynamics, you may want to skip the first 100 or so points when graphing. After finishing the graph, it's best to adjust the size and shape of the points just plotted. This can be done by right-clicking on a point within the graph and choosing *Format Data Series* from the menu. Then choose the *Style* to be a triangle, a *Foreground* with no color, a black *Background* and a *Pointsize* of 2. Axes, labels, etc., may also be removed. After clicking on *OK* a reasonable rendering of the Julia set should appear. Julia sets for other parameter values may be generated by changing the 0.3 and 0.04 in Cells A2 and B2 to other quantities, and repeating the above steps.

The Sierpinski triangle can be created by first entering the following data into the specified cells:

```
Cell B1:      1

Cell C1:      1

Cell A2:      =TRUNC(3*RAND())

Cell B2:      =0.5*B1+A2/4

Cell C2:      =0.5*C1+A2*(2-A2)/2
```

After copying and pasting cells B2,C2 into as many cells below them as desired, the fractal is generated by graphing as above, except using columns B and C as the *X Values* and *Y Values* respectively. Other fractals can be created in a similar manner.

Mathematica

Rather than using a spreadsheet program, more sophisticated computer users may wish to take advantage of the computational powers of *Mathematica*. To do so one must start the *Mathematica* program by double-clicking on its icon, or by typing a command such as *mathematica* in a terminal window. The *Mathematica* prompt *In[1]:=* should appear, perhaps in a new window, and any valid *Mathematica* command may then be typed after it. However, to run a complex sequence of tasks, it is better to first save all instructions in a program file. The entire program can then be run by typing << followed by the name of that file. For example, one might say:

$$\text{In}[1]: = <<mathfile$$

where *mathfile* is the name of the *Mathematica* program file. This runs the program, and (assuming there are no errors) all screen output will appear, perhaps in a new window. This will be followed by a return to the updated prompt *In[2]:=*. Additional *Mathematica* commands may then be typed and/or other programs run in a similar manner. When finished one ordinarily exits *Mathematica* by either closing its window or typing *Quit* (with a capital *Q*) after the prompt.

Below are examples of *Mathematica* programs for accomplishing various tasks discussed in this book. Although they can be run as shown, some slight modifications will likely be necessary with regard to the functions, initial values, numbers of iterations, etc. After making appropriate modifications, each program should be typed into a file and saved, using some form of simple text editor (with no special typesetting characters), such as *Emacs* or *Notepad*. A Mathematica *Notebook* may also be used. The program can then be run as described above. Since simplicity and brevity were once again primary goals in developing these programs, the reader should consult one of the many sources presently available for more complete explanations of the *Mathematica* commands being used.

First suppose we wish to perform 20 iterations of $x_{n+1} = 1.25x_n - 1$ starting with $x_0 = 5$, and have the resulting table and time-series graph appear on the screen. With the equation written in the equivalent form $x_n = 1.25x_{n-1} - 1$, the following program file will accomplish this. Note that the multiplication operator * is optional, blank spaces and lines are generally inconsequential, but brackets, commas, and upper and lower case letters should be typed exactly as they appear:

```
Clear[x]

first=0

last=20

Array[x,last]

x[0]=5.0

x[n_]:=x[n]= 1.25 x[n-1] - 1
```

```
graph=Graphics[{Line[Table[{n,x[n]},{n,first,last}]],
        Table[Point[{n,x[n]}],{n,first,last}]},
            Axes->True,PlotRange->All]
Show[graph]
table=OutputForm[TableForm[Table[{n,PaddedForm[x[n],
        {16,6}]},{n,first,last}]]]
```

This program can be modified in various ways. For example: The tabulated values can be reformatted by changing the 16 and/or 6 to other values. If one wishes to see the table and graph of iterates 10 through 20 only, instead let *first=10*. The equation being iterated might be changed, for example to:

$$x_n = x_{n-1} + \tfrac{1}{n^2} \quad \text{using} \quad \texttt{x[n\_]:=x[n]=x[n-1]+1/n\^{}2,}$$

since / and ^ denote division and exponentiation respectively, or to

$$x_n = 5x_{n-1}e^{-x_{n-1}} \quad \text{using} \quad \texttt{x[n\_]:=x[n]=5x[n-1]Exp[-x[n-1]]}$$

Besides the exponential function *Exp,* other useful *Mathematica* functions include the natural logarithm *Log,* the absolute value *Abs,* and the trigonometric functions: *Sin, Cos, Tan,* etc., whose arguments must always be enclosed in square brackets.

The initial value can be changed to some other real quantity, such as *x[0]=0.0*. It's best to always use a decimal point when initializing *x[0]*. If more iterates are needed, 1000 for example, one should let *last=1000*. In this case, however, the length of the table may be a problem since it will likely scroll by on the screen too quickly. One may therefore wish to send this table to a file by inserting a line like

table>>*tablefile*

at the end of the program file. This will send the table to the text file called *tablefile* instead of the screen. It can later be viewed using a text editor after exiting *Mathematica*. Finally, if one wishes to print out the graph that was just generated, one should either type

PSPrint[graph]

(with *PSP* capitalized) after the *Mathematica* prompt when the program is finished, or put that instruction in the program itself just after the *Show[graph]* command.

To compute and display an estimate of the Lyapunov exponent using the first 1000 iterates of a solution, a program such as the following might be used:

```
Clear[x,d]
last=1000
Array[{x,d,s},last]
x[0]=0.6
x[n_]:=x[n]= 3.9 x[n-1](1-x[n-1])
d[n_]:=d[n]= 3.9(1-2 x[n])
```

```
Table[{x[n],d[n]},{n,1,last}]

s[0]=0.0

s[n_]:=s[n]=s[n-1]+Log[Abs[d[n]]]/last

Table[s[n],{n,1,last}]

s[last]
```

One can easily alter the number of iterations to perform, *last,* the equation being iterated, *x[n]*, and/or the initial point, *x[0]*. Note that if *x[n]* is changed, so too must the equation for *d[n]*, which should always correspond to the derivative of the function being iterated, evaluated at *x[n]*.

For the same equation a histogram or column-graph showing the distribution of the first 1000 iterates can be obtained from the following:

```
Clear[x]

last=1000

intervals=50

Array[{x,i},last]

x[0]=0.6

x[n_]:=x[n]= 3.9 x[n-1](1-x[n-1])

max=Max[Table[x[n],{n,1,last}]]

min=Min[Table[x[n],{n,1,last}]]

Array[{c,p},intervals]

Do[{c[n]=0,p[n]=min+n*(max-min)/intervals},

        {n,0,intervals}]

Do[{i[n]=IntegerPart[intervals*(x[n]-min)/(max-min)],

        c[i[n]]=c[i[n]]+1},{n,1,last}]

c[intervals-1]=c[intervals-1]+c[intervals]

lines={AbsoluteThickness[5],{Table[Line[{{p[n],0},{p[n],

        c[n-1]}}],{n,1,intervals}]}}

graph=Graphics[lines,Axes->True, PlotRange->All]

Show[graph]
```

The number of intervals used in this distribution can easily be increased or decreased from 50 by giving *intervals* a different value, and the width of the vertical bars in the graph can be adjusted accordingly by giving a different argument to *AbsoluteThickness*. The function, initial value and/or total number of iterations can also be changed as before, and the graph can again be printed using *PSPrint[graph]*.

A linear or nonlinear system involving x_n and y_n may be iterated using a program such as the one below:

```
Clear[x,y]

first=0

last=20

Array[{x,y},last]

x[0]=500.0

y[0]=450.0

x[n_]:=x[n]= 1.1 x[n-1] - 0.2 y[n-1] + 50

y[n_]:=y[n]= x[n-1] + 0.8 y[n-1] - 35

graph1=Graphics[{Line[Table[{n,x[n]},{n,first,last}]],
          Table[Point[{n,x[n]}],{n,first,last}]},
              Axes->True,PlotRange->All]

Show[graph1]

graph2=Graphics[{Line[Table[{n,y[n]},{n,first,last}]],
          Table[Point[{n,y[n]}],{n,first,last}]},
              Axes->True,PlotRange->All]

Show[graph2]

table=OutputForm[TableForm[Table[{n,PaddedForm[x[n],{8,4}],
          PaddedForm[y[n],{8,4}]},{n,first,last}]]]
```

In this case, two separate time-series graphs will be generated, but if one wishes to plot them together, the *Show[graph1]* command should be removed, and *Show[graph2]* should be replaced with *Show[graph1,graph2]*. On the other hand, if a solution-space graph of (x_n, y_n) is desired, the following will create one:

```
Clear[x,y]

first=0

last=50

Array[{x,y},last]

x[0]=500.0

y[0]=450.0

x[n_]:=x[n]= 1.1 x[n-1] - 0.2 y[n-1] + 50

y[n_]:=y[n]= x[n-1] + 0.8 y[n-1] - 35
```

```
graph=Graphics[{Line[Table[{x[n],y[n]},{n,first,last}]],
         Table[Point[{x[n],y[n]}],{n,first,last}]},
            Axes->True,PlotRange->All]
Show[graph]
table=OutputForm[TableForm[Table[{n,PaddedForm[x[n],{8,4}],
         PaddedForm[y[n],{8,4}]},{n,first,last}]]]
```

As before these programs can be easily modified to accommodate different equations, numbers of iterations, etc. Saving tables in files and printing graphs is also accomplished as described earlier.

With a few simple changes to these programs, it is also possible to iterate systems involving three or more variables. Adding variables to the tables and time-series graphs generated by the above programs can be accomplished in a fairly straightforward manner. However, with regard to solution-space graphs, one now has an alternative other than plotting one variable against another in a two-dimensional solution space, as we have been doing. In *Mathematica* three-dimensional graphs are possible, which means any three variables can be plotted together in solution space. The only real differences are that three coordinates are given when plotting points and lines, and that the function *Graphics3D* is used in place of *Graphics*. The following program demonstrates the idea:

```
Clear[x,y,z]
first=0
last=50
Array[{x,y,z},last]
x[0]=100.0
y[0]=500.0
z[0]=800.0
x[n_]:=x[n]= 0.5 x[n-1] - 0.4 y[n-1] + 0.2 z[n-1] + 300
y[n_]:=y[n]= 0.7 x[n-1] + 0.8 y[n-1] - 0.1 z[n-1] + 200
z[n_]:=y[n]= 0.1 x[n-1] - 0.2 y[n-1] + 0.9 z[n-1] + 250
graph=Graphics3D[{Line[Table[{x[n],y[n],z[n]},
        {n,first,last}]],
           Table[Point[{x[n],y[n],z[n]}],{n,first,last}]},
             Axes->True,
        AxesLabel->{"x","y","z"},PlotRange->All,
           Boxed->False]
```

```
Show[graph]

table=OutputForm[TableForm[Table[{n,PaddedForm[x[n],

    {8,4}],

        PaddedForm[y[n],{8,4}],PaddedForm[z[n],{8,4}]},

            {n,first,last}]]]
```

If one wishes to create some fractal images on a system that possesses adequate processing speed and main memory, the following will generate the Julia set of Mandelbrot's system for the parameter values $a = 0.3$ and $b = 0.04$, although it can be easily modified to accomodate other parameters:

```
Clear[x,y]

first=100

last=50000

Array[{x,y},last]

a=0.3

b=0.04

x[0]=0.0

y[0]=0.0

x[n_]:=x[n]=(2 Round[Random[Real]+0.25]-1)(((x[n-1]-a)

            +((x[n-1]-a)^2+(y[n-1]-b)^2)^0.5)/2)^0.5

y[n_]:=y[n]=(y[n-1]-b)/(2 x[n])

graph=Graphics[{PointSize[0.005],Table[Point[{x[n],

        y[n]}],{n,first,last}]}]

Show[graph]
```

Since we are not interested in the transient dynamics, the first 100 points are skipped when plotting the graph. The total number of points generated can be increased or decreased from 50,000 by changing the value of *last,* and with some trial and error, the size of these points, 0.005, can be adjusted accordingly to preserve the quality of the image. Similar modifications can be made to the following program, which generates the Sierpinski triangle:

```
Clear[x,y,r]

first=100

last=50000

Array[{x,y,r},last]
```

```
r[n_]:=r[n]=Round[3 (Random[Real])-0.5]

x[0]=1.0

y[0]=1.0

x[n_]:=x[n]=0.5 x[n-1]+0.25 r[n]

y[n_]:=y[n]=0.5 y[n-1]+0.5 r[n](2-r[n])

graph=Graphics[{PointSize[0.003],Table[Point[{x[n],
        y[n]}],{n,first,last}]}]

Show[graph]
```

Other fractals can be generated by altering the above equations for *x[n]* and *y[n]*. These graphs can once again be printed using *PSPrint[graph]*.

Answers to Odd-Numbered Exercises

Exercises 1.1

1. stochastic
3. deterministic
5. deterministic
7. deterministic
9. stochastic
11. stochastic
13. (a) and (c)
15. $r = 0.6$, $S_0 = 6$, $S_{n+1} = 0.6S_n$
17. $r = 0.7$, $S_0 = 5$, $S_{n+1} = 0.7S_n$
19. 3, 1.5, 0.75, 0.375, 0.1875
21. The least and greatest heights of the centerpoint as it oscillates

Exercises 1.2

1. asymptotic
3. ongoing
5. asymptotic
7. ongoing
9. asymptotic
11. asymptotic
13. (a) \$25,555.56; (b) 5.2 %
15. 12.5 years
17. (a) $P_{n+1} = 1.005P_n$, $P_0 = 8000$; (b) \$8040, \$8080.20, \$8120.60
19. (a) $S_n \to 0$ as $n \to \infty$; (b) all $S_n = S_0$ and so $S_n \to S_0$ as $n \to \infty$
21. (a) $v(t) \to -mg/k$ as $t \to \infty$; (b) -137.5

Exercises 2.1

1. (a) 1000, 2100, 3310, 4641, 6105.10; (b) all equal 100,000
3. (a) 21, 23, 19, 27, 11; (b) 4400, 4040, 4364, 4072.4, 4334.84
5. (a) $P_{n+1} = 1.01P_n + 150$, $P_0 = 800$; (b) $P_{n+1} = 1.005P_n + 225$, $P_0 = 470$
7. (a) 198; (b) 0.2
9. (a) $P_{n+1} = 1.05P_n - 75$; (b) $P_{n+1} = 1.03P_n - 500$
11. (a) $P_{n+1} = 1.15P_n$, $P_0 = 75,000$; (b) $P_{n+1} = 1.02P_n$, $P_0 = 125,000$
13. (a) $P_{n+1} = 1.07P_n + 175$, $P_0 = 5400$; (b) $P_{n+1} = 0.4P_n + 5000$, $P_0 = 7500$
15. (a) $P_{n+1} = 1.015P_n + 3500$, $P_0 = 275,000$; (b) $P_{n+1} = 1.015P_n - 6300$, $P_0 = 275,000$
17. (a) $T_{n+1} = (1-a)T_n$; (b) $T_{n+1} = 0.95T_n$

19. (a) $D_{n+1} = aD_n$; (b) $D_{n+1} = 0.96D_n$
21. (a) $\Theta_{n+1} = -a\Theta_n$; (b) $\Theta_{n+1} = -\sqrt{0.8}\Theta_n$

Exercises 2.2

1. (a), (c) linear
3. (a) $a_n = -2/n, b_n = 5/n$; (b) $a_n = n - 1, b_n = 1 - n$; (c) $a_n = 1/(n + 1), b_n = 1$;
   (d) $a_n = 1, b_n = n/3$
5. $x_{n+1} = (1 + n/10)x_n$; 1, 1.1, 1.32, 1.716, 2.4024
7. $x_{n+1} = x_n + (-1/2)^n$; 2, 1.5, 1.75, 1.625, 1.6875
9. $x_{n+1} = (-1)^n(2^n - x_n/2^n)$; 0, −2, 4.5, −7.4375, 16.4648
11. 1, 5, 14, 30, 55   13. $1/2, 7/6, 23/12, 163/60, 71/20$
15. $S_{n+1} = S_n + \frac{2(n+1)}{(n+1)^2+1}, N = 100$ 17. $S_{n+1} = S_n + \frac{(-1)^{n+1}}{2n+1}, N = 1000$
19. $S_{n+1} = S_n + \sqrt{20n + 10}, N = 50$ 21. $S_{n+1} = S_n + (-1)^n \frac{2n+2}{2n+1}, N = 50$
23. (a) $P_{n+1} = (1 + 1/300)P_n + 5n + 20, P_0 = 1000$; (b) $P_{n+1} = (1 + 1/300)P_n + 3^n$,
   $P_0 = 1000$
25. (a) $P_{n+1} = (1 + 1/2^n)P_n, P_0 = 100{,}000$; (b) $P_{n+1} = (2 - 1/2^n)P_n, P_0 = 100{,}000$
27. $T_{n+1} = 0.9T_n + 1 + 1/2^n, T_0 = 100$

Exercises 2.3

1. $P_{n+1} = \frac{2n+2}{3n+7} P_n$; $N = 100$; $\frac{2}{7}, \frac{4}{35}, \frac{24}{455}, \frac{12}{455}, \frac{24}{1729}$
3. $P_{n+1} = \frac{(-1)^{n+1}}{5^{n+1}+1} P_n$; $N = 20$; $-0.1667, -0.0064, 5.09 \times 10^{-5}, 8.13 \times 10^{-8}, -2.6 \times 10^{-11}$
5. $P_{n+1} = (-1)^n \frac{50(n+1)}{3^{n+1}} P_n$; $N = 10$; $\frac{50}{3}, -\frac{5000}{27}, -\frac{750{,}000}{3^6}, \frac{1.5 \times 10^8}{3^{10}}, \frac{3.75 \times 10^{10}}{3^{15}}$
7. $x_n = 1/(n + 1)$ 9. $x_n = (n + 1)(n + 2)/2$
11. $x_n = 10(1.5)^n$ 13. $x_n = (-1)^n 3000(1.75)^n$
15. (a) 0.5722 ft; (b) 3.75, 2.34375, 1.4648, 0.9155, 0.5722
17. (a) $859.47; (b) $898.46   19. (a) 9.633; (b) 1.125
21. (a) 19.7 min; (b) 6.6 min   23. (a) 18.4; (b) 46.9 mo
25. (a) 103; (b) 824 sec
27. (b) exponential growth, (c) exponential decay

Exercises 2.4

1. $a = 0.75, n = 11$; $S = 4 - 4(0.75)^{11}$
3. $a = -1.5, n = 13$; $S = 0.4(1.5)^{13} + 0.4$
5. 10   7. diverges
9. 2   11. $x_n = 4 - 2(0.75)^n$
13. all $x_n = 1$ 15. $x_n = 0$ for even n, $x_n = 2$ for odd n
17. (a) $231,206; (b) $2311.15   19. (a) 7096; (b) 14
21. (a) $94.70; (b) $114.89   23. (a) 6.4 tons; (b) 5 tons

25. $x_n = u_n + 20$; $u_n = u_0/2^n$; $x_n = 20 + (x_0 - 20)/2^n$
27. $x_n = u_n + 1.2$; $u_n = (-0.25)^n u_0$; $x_n = 1.2 + (-0.25)^n(x_0 - 1.2)$

Exercises 2.5

1. (a) \$343,587.64; (b) \$2910.47
3. (a) approx 96 weeks; (b) \$9600 deposited, \$586.24 interest
5. (a) \$17,168.64; (b) \$3854.38
7. (a) $S_{n+1} = 0.75S_n + 10{,}000$; (b) $S_n = 10{,}000(0.75)^n + 40{,}000$, $C_n = 60{,}000 - 10{,}000(0.75)^n$
9. (a) $N_{n+1} = 0.5N_n + 0.1$; (b) $N_n = 0.45(0.5)^n + 0.2$, $C_n = 0.8 - 0.45(0.5)^n$
11. $x_{n+1} = 0.55x_n + 0.2$; $x_n = (0.55)^n(x_0 - 4/9) + 4/9$, $y_n = 5/9 + (0.55)^n(y_0 - 5/9)$
13. $x_{n+1} = 1 - x_n$; $x_n = (-1)^n(x_0 - 0.5) + 0.5$, $y_n = 0.5 + (-1)^n(y_0 - 0.5)$
15. 33 trained, 27 untrained   17. 832 making profit, 168 not
19. 59% homeowners, 41% renters
21. $C_{n+1} = (0.75 - (-0.1)^{n+1} - (0.1)^{n+1})C_n + 100{,}000(0.15 + (0.1)^{n+1})$, $S_n = 100{,}000 - C_n$

Exercises 2.6

1. 3/4 unstable, no oscil   3. 8 stable, no oscil
5. 0 unstable with oscil   7. −4 stable, no oscil
9. 0 stable, no oscil   11. 0 stable with oscil
13. \$27.27 unstable with oscil   15. (a) 22,540; (b) 18,750
17. (a) $U_{n+1} = 0.91U_n + 200$; (b) \$2222.22
19. 40 trained, 20 untrained   21. 833 make profit, 167 do not
23. 80% homeowners, 20% renters
25. (a) $x = 1/3$, $y = 2/3$, stable; (b) $x = 1/2$, $y = 1/2$, not stable

Exercises 2.7

1. $f_x = 12x^3 + 5$, $f_y = -24y^2$ 3. $f_x = 8xy^5$, $f_y = 20x^2y^4$
5. $S_a = 110a - 30b$, $S_b = 10b - 30a$ 7. $S_a = 682a + 62b$, $S_b = 62a + 10b$
9. $(0, 0)$ 11. $(2, -1)$
13. minimum   15. neither
17. $5m + 3b = 8$, $3m + 3b = 5$; $y = 1.5x + 0.167$
19. $9m - b = -26$, $-m + 5b = 25$; $y = -2.386x + 4.523$
21. $x_{n+1} = 1.769x_n + 2.746$ 23. $P_{n+1} = 0.411P_n + 0.211$

Exercises 3.1

1. (a) $P_{n+1} = 2.75P_n(1 - P_n/50{,}000)$; (b) $P_{n+1} = 3.5P_n(1 - P_n/10^6)$
3. (a) $C = 25{,}000$; (b) $C = 200{,}000$

5. (a) $P_{n+1} = rP_n(1 - P_n/C)^2$; (b) $P_{n+1} = rP_n/(1 + P_n^2/C^2)$; (c) $P_{n+1} = rP_n(1 - P_n^2/C^2)$;
(d) $P_{n+1} = rP_ne^{-P_n^2/N^2}$

7. (a) $r = 6.5$, $N = 200$; (b) $r = e^2$, $N = 10^8$

9. (a) $I_{n+1} = I_n - 0.3I_n + 2I_n(1 - I_n/10{,}000)$; (b) $I_{n+1} = 3.5I_n(1 - I_n/10^6)$

11. (a) $I_{n+1} = I_n - 0.8I_n + \frac{1300}{999}I_n(1 - I_n/10^6)$; (b) $I_{n+1} = I_n - 0.6I_n + 0.5I_n(1 - I_n/400{,}000)$;
(c) $I_{n+1} = I_n + \frac{100}{99}I_n(1 - I_n/10{,}000)$

13. (a) $P_{n+1} = 3/P_n + 0.5P_n$; (b) $P_{n+1} = 1/P_n - 2P_n - 1$

15. (a) $P_0 = \sqrt{2}$, $P_1 = 2 + \sqrt{2}$; (b) $P_0 = 1 + \sqrt{2}$, $P_1 = 0$; (c) $P_0 = \infty$, $P_1 = 2$

17. (a) $P_0 = 2$, $P_1 = 1$; (b) $P_0 = 0$, $P_1 = 5$

19. $T_{n+1} = T_n - rT_n + \frac{s}{N}T_n^2(1 - T_n/N)^2$

21. $A_{n+1} = A_n - rA_n + s/A_n^k$ 23. $y = 2.945x - 0.6935x^2$

25. $y = 4.901/x - 0.0111x$

Exercises 3.2

1. (a) $f(x) = 2x - 7/x$; (b) $f(x) = 5x/(x + 8)$

3. (a) non-aut, $f(n, x) = \sqrt{x + n}$; (b) aut, $f(x) = e^x \sin x$; (c) non-aut, $f(x) = (x - 1)/(n - 1)$; (d) aut, $f(x) = x \ln x$

5. (a) $f'(C/2) = 0$, $f(C/2) = rC/4$; (b) $f'(N) = 0$, $f(N) = rN/e$

7. (a) max $(2000, 4000/9)$; (b) max $(2000, 8000/3)$

9. (a) max $(1000, 2000/e^2)$; (b) max $(1000, 6000/e^2)$

11. (a) $2, 1/2, 2, 1/2, 2$; (b) $5/8, 8, 5/8, 8, 5/8$; (c) $x_0, c/x_0$ repeat

13. $3/2, -1$ 15. $0, \pm 1$

17. $0, 3000$ 19. $-1, 6$

21. three fixed points   23. one fixed point

25. 0 stable for $-1 < x_0 < 1$, 1 unstable   27. 1 unstable

29. 0 stable for $-2 < x_0 < 2$, 2 unstable   31. 0 unstable, 1/9 stable for $x_0 > 0$

33. 1,3 unstable, 2 stable for $1 < x_0 < 3$ 35. 1 stable for $0 < x_0 < 2$, 2 unstable

Exercises 3.3

7. A,C stable, B unstable, no oscil   9. 0 stable, no oscil; 1 unstable, no oscil

11. 2 stable with oscil; $-1/2$ unstable with oscil

13. unstable, no oscil   15. unstable with oscil

17. 0 unstable, no oscil

19. 1 stable with oscil, -2 unstable with oscil

21. (a) 0 stable; (b) 0 unstable, 500 stable

23. (a) 0 stable; (b) 0 stable, $100(5 - \sqrt{5})$ unstable, $100(5 + \sqrt{5})$ stable

25. stable

27. $p = 0$, $y = 4x$; $p = 3/4$, $y = -2x + 9/4$

29. (a) $f'(1/2) = 0$; (c) $y = 1/2$; (d) fast convergence

31. (b) $f'(1) = 2$, $(f^{-1})'(1) = 1/2$

Exercises 3.4

1. 2.5, 2.05, 2.0006
3. 2, 1.6429, 1.5161
5. 3, 2.6233, 2.6180
7. 2 and -1
9. sequence alternates between 0 and 2
11. $f(x) = x - g(x)/g'(x) = 2(x+1)/3$; 2 stable fixed point
13. $f(x) = x - g(x)/g'(x) = (x^2 + 2x)/(x+3)$; 0 stable fixed point
15. (a) $g(t, y) = 2y^2 - y^3 + t$, non-aut; (b) $g(y) = 8y^2 \sin y$, aut; (c) $g(t, y) = 2y + t/3$, non-aut; (d) $g(y) = 2/y + 3/2$, aut
17. 0.3, 0.57, 0.813
19. 1.1, 1.1995, 1.2979
21. 1.1, 1.2331, 1.4156
23. 0.9, 0.8094, 0.7261
25. $y_{n+1} = 0.9y_n + 0.4$, $y_n = 4(1 - 0.9^n)$
27. $y(t) = 0$ unstable, $y(t) = 5/3$ stable
29. $y(t) = \sqrt{3}$ unstable, $y(t) = -\sqrt{3}$ stable
31. 68.5°, 80°

Exercises 3.5

1. (a) 0, 4
3. 0,1
5. $f^2(x) = x^4 + 2x^2 + 2$, $f^3(x) = x^8 + 4x^6 + 8x^4 + 8x^2 + 5$
7. 0 fixed point, $\pm 1/2$ 2-cycle
9. 0, 3 fixed points, $(1 \pm \sqrt{5})/2$ 2-cycle
11. ± 1 fixed points, no 2-cycle
13. stable
15. unstable
17. 1, 1/2 unstable
19. 1, 2 stable
21. (a) $300 \pm 100\sqrt{5}$ unstable; (b) 400, 600 stable
23. $-1/5, 3/5$ unstable
25. $f^{-1}(1) = -1$, $f^{-1}(-1) = 1$, $(f^{-1})'(1) \cdot (f^{-1})'(-1) = 1/9$
27. $a^n x + b(a^n - 1)/(a - 1)$
29. (a) $f^2(x) = x/(2x+1)$, $f^3(x) = x/(3x+1)$; (b) $f^n(x) = x/(nx+1)$; (c) only solution of $x/(mx + 1) = x$ is $x = 0$
31. (a) unstable; (b) stable
33. $1/11, -31/11, 29/11, 25/11, 17/11$ unstable 5-cycle

Exercises 3.6

1. $x_{n+1} = rx_n^2(1 - x_n)$, $x_n = P_n/C$
3. $x_{n+1} = rx_n/(1 + x_n^2)$, $x_n = P_n/C$
5. let $x_n = P_n/(rC)$
7. $p(c) = (\sqrt{4c + 1} - 1)/2$ exists for $c > 0$
9. $p(c) = e - c$ exists for $0 < c < e$
11. $0 < r < 1$
13. $0 < r < 24$
15. $r = 2$
17. $p(a) = (a - 1)^{1/3}$ exists for $a > 1$, stable for $1 < a < 5/3$
19. $p(a) = \sqrt{(2a - 1)/a}$ exists for $a > 1/2$, stable for $1/2 < a < 1$
21. $p_1(a), p_2(a) = (1 \pm \sqrt{4a - 3})/2$ exists for $a > 3/4$, stable for $3/4 < a < 5/4$
23. $p_1(a), p_2(a) = \pm\sqrt{a - 1}$ exists, stable for $a > 1$
25. (a) 2000; (b) 750
27. (a) -2; (b) $-\sqrt{6}$

Exercises 3.7

1. (a) 0 stable for $|r| < 0.2$; (b) $p(r) = (5r - 1)/(2r)$ exists for $r > 0.2$, stable for $0.2 < r < 0.6$; (c) $r_0 = 0.2$ from (a), (b) and $p(0.2) = 0$
3. (a) 0 stable for $|r| < 1$; (b) $p(r) = \ln r$ exists for $r > 1$, stable for $1 < r < e^2$; (c) $r_0 = 1$ from (a), (b) and $p(1) = 0$
5. (a) 0 stable for $|r| < 1$; (b) $p(r) = \sqrt{r - 1}$ exists and stable for $r > 1$; (c) $r_0 = 1$ from (a), (b) and $p(1) = 0$
11. (a) $p(a) = (\sqrt{4a + 1} - 1)/2$ exists for $a > -1/4$, stable for $-1/4 < a < 3/4$; (b) 2-cycle $p_1(a)$, $p_2(a) = (1 \pm \sqrt{4a - 3})/2$ exists for $a > 3/4$, stable for $3/4 < a < 5/4$; (c) $a_0 = 3/4$ from (a), (b) and $p(3/4) = p_1(3/4) = p_2(3/4)$
13. (a) fixed point 0 exists for all a, stable for $|a| < 1$; (b) 2-cycle $p_1(a)$, $p_2(a) = \pm\sqrt{a - 1}$ exists and stable for $a > 1$; (c) $a_0 = 1$ from (a), (b) and $p_1(1) = p_2(1) = 0$
17. bifurcation at $a = -1/4$ into unstable fixed point $(-1 - \sqrt{4a + 1})/2$ and stable fixed point $(-1 + \sqrt{4a + 1})/2$
19. bifurcation at $c = 2$ into unstable fixed point $(c + \sqrt{c^2 - 4})/2$ and stable fixed point $(c - \sqrt{c^2 - 4})/2$
21. (a) yes; (b) yes
31. stable fixed point 0 and unstable 2-cycle $\pm 1/\sqrt{a}$ exist for all $a > 0$
33. stable fixed points $\pm\sqrt{a - 1}$ exist for all $a > 1$ and no 2-cycle exists

Exercises 3.8

1. (a) yes; (b) yes   3. 3-cycle $1/7, 5/7, -3/7$
5. (a) fixed points: 4 and 11; 2-cycle: 3,8; 3-cycle: 1,9,5; (b) a 3-cycle implies chaos
7. (a) $3x \bmod 1$ is always between 0 and 1; (b) $L = \ln 3$
9. (a) $\frac{3}{2} \ln 2$; (b) $\frac{1}{3} \ln 1.2$
11. (a) let $I = [0, 1/3]$ and $J = [2/3, 1]$; (b) let $x_0 = 2/3$
13. (a) $f(-1) = -1$ and $f'(-1) = -2$; (b) for $n \leq 0$, $f(x_n) = 1 - 2|-1 + 2^n| = -1 + 2^{n+1} = x_{n+1}$, $f(x_1) = 1 - 2|1| = -1$ and for $n \geq 2$, $f(x_n) = f(-1) = -1$; (c) $\lim_{n \to \infty} x_n = \lim_{n \to \infty} -1 = -1$ and $\lim_{n \to -\infty} x_n = \lim_{n \to -\infty}(-1 + 2^n) = -1$
15. (a) let $I = [-2, 0]$ and $J = [0, 2]$; (b) let $x_0 = 0$
17. (a) let $I = [0, 2/3]$ and $J = [2/3, 1]$; (b) let $x_0 \approx 0.7295$
19. (a) let $I = [0, 1/\sqrt{3}]$ and $J = [1/\sqrt{3}, 1]$; (b) let $x_0 = 1/\sqrt{3}$

Exercises 4.1

1. (a) $P_1 = 450$, $Q_1 = 340$, $P_2 = 335$, $Q_2 = 588$
3. (a) $\begin{cases} P_{n+1} = 1.3P_n - 0.2Q_n \\ Q_{n+1} = 0.5P_n + 0.9Q_n \end{cases}$; (b) $\begin{cases} P_{n+1} = 2P_n + 3000 \\ Q_{n+1} = 1.2P_n + Q_n - 2000 \end{cases}$
5. (a) $\begin{cases} P_{n+1} = 1.2P_n - 0.2Q_n \\ Q_{n+1} = 0.3P_n + 1.3Q_n \end{cases}$; (b) $\begin{cases} P_{n+1} = P_n - 2Q_n \\ Q_{n+1} = 2P_n + Q_n \end{cases}$
7. (a) $\begin{cases} P_{n+1} = 1.35P_n - 0.1Q_n \\ Q_{n+1} = -0.05P_n + 0.75Q_n \end{cases}$; (b) $\begin{cases} P_{n+1} = P_n - Q_n - 100 \\ Q_{n+1} = 2Q_n + 150 \end{cases}$

9. (a) $\begin{cases} P_{n+1} = 2P_n - 0.5Q_n \\ Q_{n+1} = -1.6P_n + 1.5Q_n \end{cases}$; (b) $\begin{cases} P_{n+1} = 1.75P_n - 0.125Q_n \\ Q_{n+1} = -0.25P_n + 0.625Q_n \end{cases}$

11. $\begin{cases} P_{n+1} = 1.2P_n + 0.5Q_n \\ Q_{n+1} = P_n \end{cases}$, $P_0 = 1500, Q_0 = 1200$

13. (a) $P_{n+1} = 1.8P_n + 0.3P_{n-1}$; (b) $P_{n+1} = 0.7P_n + P_{n-1} + 1000$

15. (a) $P_{n+1} = 0.5P_n + 0.7P_{n-1}$; (b) $P_{n+1} = 0.4P_n + 4.4P_{n-1}$

17. (a) $P_1 = 9, D_1 = 10, S_1 = 15$; (b) $P_2 = 17, D_2 = 2, S_2 = 31$

19. $\begin{cases} P_{n+1} = d_2 P_n - a_2 S_n + h_2 \\ S_{n+1} = c_2 P_n + S_n - k_2 \end{cases}$

21. (a) $\begin{cases} a\sum x_i^2 + b\sum x_i y_i = \sum x_i z_i \\ a\sum x_i y_i + b\sum y_i^2 = \sum y_i z_i \end{cases}$; (b) $z = -0.724x + 1.90y$

Exercises 4.2

1. (a) $a = 1.25, b = -0.9, c = 1.5, d = 3$; (b) $a = 2.3, b = -0.5, c = -0.3, d = 0.3$

3. $x_{n+2} = 2x_n$ 5. $P_{n+2} = 6P_{n+1} - 11P_n$

7. $(40, 50)$ 9. $(500, 500)$

11. all points satisfying $x = y$ are fixed points

13. $x_n = a^n x_0, \; y_n = d^n y_0$

15. $x_n = a^n x_0 + \frac{a^n-1}{a-1}h, \; y_n = d^n y_0 + \frac{d^n-1}{d-1}k$

17. $(120, 160)$ sink   19. $(5000, 2800)$ saddle

21. sink   23. all points rotate by $90°$

Exercises 4.3

1. (a) $\begin{bmatrix} 9 \\ -6 \end{bmatrix}$; (b) $\begin{bmatrix} 2 \\ 0 \end{bmatrix}$; (c) $\begin{bmatrix} 1/2 \\ -1 \end{bmatrix}$; (d) $\begin{bmatrix} 7 \\ -6 \end{bmatrix}$; (e) $\begin{bmatrix} -2 \\ 16/3 \end{bmatrix}$

3. (a) both $\begin{bmatrix} 6.5 \\ 2 \end{bmatrix}$; (b) both $\begin{bmatrix} 8 \\ 0.25 \end{bmatrix}$ 5. (a) both $\begin{bmatrix} -13 \\ -9 \end{bmatrix}$; (b) both $\begin{bmatrix} -2 \\ 21 \end{bmatrix}$

7. (a) lengths $= \sqrt{65}, 2\sqrt{65}, 3\sqrt{65}, \sqrt{65}/2$; (b) lengths $= 2, 4, 6, 1$

9. (a) $\begin{bmatrix} 10 & 0 \\ 5 & -15 \end{bmatrix}$; (b) $\begin{bmatrix} 6 & 1 \\ -4 & -3 \end{bmatrix}$; (c) $\begin{bmatrix} 0 & -2 \\ 14 & -12 \end{bmatrix}$

11. (a) $\begin{bmatrix} 19 \\ 16 \end{bmatrix}$; (b) $\begin{bmatrix} 8 \\ 12 \end{bmatrix}$ 13. (a) both $\begin{bmatrix} 8 \\ -14 \end{bmatrix}$; (b) both $\begin{bmatrix} 31 \\ 43 \end{bmatrix}$

15. (a) $\begin{bmatrix} 8 & 2 \\ 19 & 1 \end{bmatrix}$; (b) $\begin{bmatrix} 9 & -3 \\ -10 & 0 \end{bmatrix}$; (c) $\begin{bmatrix} 4 & 0 \\ -1 & 9 \end{bmatrix}$; (d) $\begin{bmatrix} 8 & 0 \\ 7 & -27 \end{bmatrix}$

17. (a) both $\begin{bmatrix} 0 & 11 \\ -10 & 9 \end{bmatrix}$; (b) both $\begin{bmatrix} 6 & 6 \\ -8 & 9 \end{bmatrix}$; (c) $\begin{bmatrix} 4 & 2 \\ -4 & 3 \end{bmatrix} \neq \begin{bmatrix} 5 & 10 \\ -1 & 2 \end{bmatrix}$

19. both $\begin{bmatrix} 4 & 6 \\ -14 & 9 \end{bmatrix}$ 21. 15

23. 1

25. $\begin{bmatrix} 1 & -1 \\ -7/2 & 4 \end{bmatrix}$

27. $\begin{bmatrix} -10 & 1/2 \\ -4 & 0 \end{bmatrix}$

29. $x = 5/2, y = 4$

31. $x = -12, y = -7$

33. (a) both 0; (b) 8 and -2

Exercises 4.4

1. $-2, 3/2$

3. $1, 5$

5. (a) $\begin{bmatrix} 1500 \\ 4500 \end{bmatrix}, \begin{bmatrix} 2250 \\ 6750 \end{bmatrix}$; (b) $\left(\frac{3}{2}\right)^n \begin{bmatrix} 1000 \\ 3000 \end{bmatrix}$

7. $6, 7$

9. $(3 \pm \sqrt{13})/2$

11. source

13. source

15. saddle

17. sink

19. $0 < t < 1/2$

21. $r = 3, c \begin{bmatrix} 1 \\ 1 \end{bmatrix}; s = 1, c \begin{bmatrix} 1 \\ -1 \end{bmatrix}$

23. $r = 1 + \sqrt{2}, c \begin{bmatrix} \sqrt{2} \\ 1 \end{bmatrix}; s = 1 - \sqrt{2}, c \begin{bmatrix} \sqrt{2} \\ -1 \end{bmatrix}$

25. (a) $x_n = v_1(1.5)^n + u_1(-2)^n$, $y_n = v_2(1.5)^n + u_2(-2)^n$; (b) $x_n = \frac{20}{7}((1.5)^n - (-2)^n)$, $y_n = \frac{10}{7}(1.5)^n + \frac{4}{7}(-2)^n$; (c) no limit

27. (a) $x_n = v_1 6^n + u_1 7^n$, $y_n = v_2 6^n + u_2 7^n$; (b) $x_n = 300 \cdot 7^n - 50 \cdot 6^n$, $y_n = 300 \cdot 7^n - 100 \cdot 6^n$; (c) no limit

29. (a) cr, cs; (b) $1/r, 1/s$

Exercises 4.5

1. 2

3. 0.8

5. source

7. sink

9. source

11. sink

13. (a) $r = 3$; (b) $x_n = 1000(2 - n)3^n$

15. (a) $r = 3/4$; (b) $x_n = 6000(1 - n)(3/4)^n$

17. (a) $x_n = (v_1 + u_1 n)(-2)^n$, $y_n = (v_2 + u_2 n)(-2)^n$; (b) $x_n = (n - 1)(-2)^{n+1}$, $y_n = (1 - 2n)(-2)^n$; (c) no limit

19. (a) $P_n = (v_1 + u_1 n)(1/2)^n$, $Q_n = (v_2 + u_2 n)(1/2)^n$; (b) $P_n = 50(8 + n)(1/2)^n$, $Q_n = 100(6 + n)(1/2)^n$; (c) $(0, 0)$

21. (a) all $(x_n, y_n) = (1, 1)$; (b) $(x_n, y_n) = (n + 1, n)$, no limit; (c) all $(x_n, y_n) = (2, 2)$; (d) all $(x_n, y_n) = (-1, -1)$; (e) $(x_n, y_n) = (-n, 1 - n)$, no limit

Exercises 4.6

1. $-1 - 9i$
3. $-1 + 5i/2$
5. $-7 - 24i$
7. $-(3 + 4i)/5$
11. (a) -1; (b) $-i$; (c) 1; (d) i
13. $\pm i\sqrt{3}$
15. $(-1 \pm i\sqrt{3})/2$
17. $0, -1 \pm i\sqrt{2}$
19. $\pm 1, \pm i$
21. $4\sqrt{3} + 4i$
23. $\sqrt{18}e^{3i\pi/4}$
25. $e^{-i\pi/4}$
27. $\sqrt{12}e^{i\pi/3}$
29. 16
31. $-256i$

Exercises 4.7

1. $(1 \pm i)/2$
3. $1 \pm i\sqrt{3}$
5. sink
7. source
9. sink
11. sink
13. $|t| < \sqrt{3}/2$
19. (a) $x_n = 2^n(v_1 \cos(n\pi/2) + u_1 \sin(n\pi/2))$, $y_n = 2^n(v_2 \cos(n\pi/2) + u_2 \sin(n\pi/2))$; (b) $x_n = 2^n(\cos(n\pi/2) + 2\sin(n\pi/2))$, $y_n = 2^n(2\cos(n\pi/2) - \sin(n\pi/2))$; (c) no limit
21. (a) $P_n = (0.8)^n(v_1 \cos(n\pi/3) + u_1 \sin(n\pi/3))$, $Q_n = (0.8)^n(v_2 \cos(n\pi/3) + u_2 \sin(n\pi/3))$; (b) $P_n = 1000(0.8)^n(5\cos(n\pi/3) - \sqrt{3}\sin(n\pi/3))$, $Q_n = 2000(0.8)^n(\cos(n\pi/3) + \frac{5}{\sqrt{3}}\sin(n\pi/3))$; (c) $(0, 0)$
23. $|v_1|, |u_1|$
25. 8
27. 10
29. 5
31. 40

Exercises 4.8

1. $(4, 25)$, source
3. $(2400, 400)$, saddle
5. $(40/31, 56/31)$, sink
7. $(10, -12)$, source
9. $(0, 5000)$, sink
11. 8, sink
15. (a) $\begin{cases} x_{n+1} = \frac{2}{3}x_n + \frac{1}{6}y_n + \frac{1}{3} \\ y_{n+1} = -\frac{1}{3}x_n + \frac{1}{6}y_n + \frac{1}{3} \end{cases}$; (b) $(1, 0, 0)$, stable

17. (a) $\begin{cases} x_{n+1} = -\frac{1}{3}x_n + \frac{2}{3}y_n + \frac{1}{3} \\ y_{n+1} = \frac{2}{3}x_n - \frac{1}{3}y_n + \frac{1}{3} \end{cases}$; (b) $(1/2, 1/2, 0)$, unstable

19. (a) $\begin{bmatrix} 0.5 & 0 & 0.5 \\ 0 & 0 & 0.5 \\ 0.5 & 1 & 0 \end{bmatrix}$; (b) $\begin{cases} x_{n+1} = -0.5y_n + 0.5 \\ y_{n+1} = -0.5x_n - 0.5y_n + 0.5 \end{cases}$; (c) $(0.4, 0.2, 0.4)$, stable

21. (a) $\begin{bmatrix} 0 & 0 & 1 \\ 1 & 0 & 0 \\ 0 & 1 & 0 \end{bmatrix}$; (b) $\begin{cases} x_{n+1} = -x_n - y_n + 1 \\ y_{n+1} = x_n \end{cases}$; (c) $(1/3, 1/3, 1/3)$, unstable

23. $(31579, 26316, 42105)$, stable
25. $(0.2, 0.2, 0.6)$, stable

Exercises 5.1

1. $P_1 = 1359$, $Q_1 = 1392$, $P_2 = 1416$, $Q_2 = 1784$

3. (a) $r = 0.15$, $s = 1.75$, $t = 0.05$, $N = 10^5$; (b) $I_1 = 10{,}085$, $R_1 = 1075$

5. (a) $\begin{cases} P_{n+1} = 8P_n - 0.005 P_n Q_n - 120 \\ Q_{n+1} = 4Q_n - 0.002 P_n Q_n + 50 \end{cases}$; (b) $P_1 = 1239$, $Q_1 = 1442$

7. (a) $P_{n+1} = (r P_n + s P_{n-1})e^{-(P_n + P_{n-1})/N}$; (b) $\begin{cases} P_{n+1} = (r P_n + s Q_n)e^{-(P_n + Q_n)/N} \\ Q_{n+1} = P_n \end{cases}$

9. (a) $\begin{cases} P_{n+1} = r_1 P_n - s_1 P_n^2 Q_n^2 \\ Q_{n+1} = r_2 Q_n - s_2 P_n^2 Q_n^2 \end{cases}$; (b) $\begin{cases} P_{n+1} = r_1 P_n - s_1 \sqrt{P_n Q_n} \\ Q_{n+1} = r_2 Q_n - s_2 \sqrt{P_n Q_n} \end{cases}$

11. (a) $\begin{cases} P_{n+1} = P_n + a_1 D_n - b_1 \sqrt{P_n} + h_1 \\ D_{n+1} = D_n + c_1/P_n - k_1 \end{cases}$; (b) $\begin{cases} P_{n+1} = P_n + a_1 D_n - b_1 e^{P_n} + h_1 \\ D_{n+1} = D_n + c_1/P_n - k_1 \end{cases}$

13. $(2, 0.25)$

15. $(0, 0)$, $(200, 1600)$

17. $(0, 0)$, $(3000 \ln 2, 3000 \ln 2)$

19. $(-1, 2)$, $(1/3, -2)$

21. (p, p) where $p = C(r + s - 1)/(2r + 2s)$

23. $(x_n, y_n) = (x_0^{a^n}, b^n y_0)$

25. $(x_n, y_n) = (x_0, y_0)$ for even n, $(x_n, y_n) = (a/y_0, a/x_0)$ for odd n

27. (a) $\begin{cases} u_{n+1} = 2u_n + 8v_n \\ v_{n+1} = u_n \end{cases}$; (b) $(0, 0)$ is a source, $(u_n, v_n) \to \pm\infty$; (c) $(0, 0)$ is a sink, $(x_n, y_n) \to (0, 0)$

Exercises 5.2

1. $\begin{bmatrix} 2x & -2y \\ 2y & 2x \end{bmatrix}$

3. $\begin{bmatrix} 1 + \ln x & 1 + \ln y \\ e^y - y e^x & x e^y - e^x \end{bmatrix}$

5. sink, no rotation

7. source with rotation

9. $(0, 0)$ sink, $(10, -4)$ saddle

11. $(0, 0)$ saddle, $(3000, 3000)$ sink

13. $(5, 1.4)$ sink

15. $(0, 0)$ sink

17. $F^n(x, y) = ((x/y)^{2^{n-1}}, (y/x)^{2^{n-1}})$

19. $(0, 0)$ stable fixed point and $(1, 1)$ unstable fixed point; $(0, 1)$, $(1, 0)$ unstable 2-cycle

21. $(0, 0)$, $(2, 0)$ stable 2-cycle

23. $(2, 2)$, $(-1, 2)$, $(-1, -1)$, $(2, -1)$ unstable 4-cycle

25. $(0, 0)$, $(1, 0)$, $(0, 1)$ unstable 3-cycle

Exercises 5.3

1. $0 < a < 2$

3. $|a| < 4$

5. $1 < a < 3$

7. $|a| < 2\sqrt{5}$

9. (a) $0 \le r < 1$; (b) $0 \le s < 1$

11. $|a| < 1$, $|b| < 1$

13. $|ab| < 1$

15. (a) $p_a = (a - 1)/a$ exists for $a > 1$; (b) $(0, 0)$ stable for $|a| < 1$; (p_a, p_a) stable for $1 < a < 3$; (c) $\lim_{a \to 0}(p_a, p_a) = (0, 0)$

17. 8   19. 12
21. $A : 0 \le x \le 1/2,\ 0 \le y \le 1/2,\ B : 1/2 \le x \le 1,\ 0 \le y \le 1/2$

Exercises 5.4

1. 3   3. 1

5. 1   7. $\begin{cases} x_{n+1} = y_n + 2 \\ y_{n+1} = \pm\sqrt{x_n - 1} \end{cases}$

9. $\begin{cases} x_{n+1} = \pm\sqrt{x_n + y_n + 1} - 1 \\ y_{n+1} = \pm 2\sqrt{x_n + y_n + 1} - y_n - 2 \end{cases}$ 11. 1/2

13. $\ln 2/(\ln 5 - \ln 2)$ 15. 1

Bibliography

Abraham, R. H., Gardini, L. and Mira, C., *Chaos in Discrete Dyanamical Systems, A Visual Introduction in 2 Dimensions,* Springer-Verlag/TELOS, New York, 1997

Abraham, R. H. and Shaw, C. D., *Dynamics - The Geometry of Behavior, Parts 1-4,* Ariel, Santa Cruz, 1982–1988

Addison, P. S., *Fractals and Chaos, An Illustrated Course,* Instutute of Physics Publishing, Bristol, 1997

Adler, F. R., *Modeling the Dynamics of Life, Calculus and Probability for Life Scientists,* Brooks-Cole, Pacific Grove, CA, 1998

Alligood, K. T., Sauer, T. and Yorke, J. A., *CHAOS: An Introduction to Dynamical Systems,* Springer-Verlag, New York, 1997

Barnsley, M., *Fractals Everywhere,* Academic Press, San Diego, 1988

Beltrami, E., *Mathematics for Dynamic Modeling,* 2nd ed., Academic Press, New York, 1997

Beltrami, E., *Mathematical Models for Society and Biology,* Academic Press, New York, 2001

Devaney, R. L., *A First Course in Chaotic Dynamical Systems: Theory and Experiment,* Addison-Wesley, Reading, MA, 1992

Devaney, R. L., *An Introduction to Chaotic Dynamical Systems,* 2nd ed., Addison-Wesley, Redwood City, CA, 1989

Edelstein-Keshet, L., *Mathematical Models in Biology,* Birkhäuser Mathematics Series, McGraw-Hill, New York, 1988

Elaydi, S., *An Introduction to Difference Equations,* 2nd ed., Springer-Verlag, New York, 1999

Giordano, F. R., Weir, M. D. and Fox, W. P., *A First Course in Mathematical Modeling,* 3rd ed., Brooks-Cole, Pacific Grove, CA, 2003

Glass, L. and Mackey, M. C., *From Clocks to Chaos: The Rhythms of Life,* Princeton University Press, Princeton, NJ, 1988

Gleick, J., *Chaos, Making a New Science,* Viking, New York, 1987

Goodwin, R. M., *Chaotic Economic Dynamics,* Clarendon Press, Oxford, 1990

Hau, B. L., *Chaos II,* World Scientific, Singapore, 1990

Hénon, M., A two-dimensional mapping with a strange attractor, *Commun. Math. Phys.* **50,** pp 69-77, 1976

Holmgren, R. A., *A First Course in Discrete Dynamical Systems,* 2nd ed., Springer-Verlag, New York, 1996

Hirsch, M. W. and Smale, S., *Differential Equations, Dynamical Systems, and Linear Algebra,* Academic Press, New York, 1974

Jackson, E. A., *Exploring Nature's Dynamics,* Wiley, New York, 2001

Kalman, D., *Elementary Mathematical Models, Order Aplenty and a Glimpse of Chaos,* Math. Assoc. of Amer., Washington, DC, 1997

Kaplan, D. and Glass, L., *Understanding Nonlinear Dyamics,* Springer, New York, 1995

Kelly, W. G. and Peterson, A. C., *Difference Equations,* Academic Press, Boston, MA, 1991

Kulenović, M. R. S. and Merino, O., *Discrete Dynamical Systems and Difference Equations with Mathematica,* Chapman-Hall/CRC, Boca Raton, FL, 2002

Lakshmikantham, V. and Triggante, D., *Theory of Difference Equations,* Academic Press, Boston, MA 1988

Lay, D. C., *Linear Algebra and its Applications,* 2nd ed., Addison-Wesley, New York, 2000

Levy, H. and Lessman, F., *Finite Difference Equations,* Dover, New York, 1992

Li, T.-Y. and Yorke, J., Period three implies chaos, *Amer. Math. Monthly* **82,** pp 985-992, 1975

Lorenz, E. N., Deterministic nonperiodic flow, *J. Atmospher. Sci.* **20,** pp 130-141, 1963

Lorenz, H.-W., *Nonlinear Dynamical Economics and Chaotic Motion,* 2nd ed., Springer, Berlin, 1993

Lynch, S. *Dynamical Systems with Applications using Maple,* Birkhäuser, Boston, MA, 2001

Mandelbrot, B. B., *The Fractal Geometry of Nature,* Freeman, New York, 1982

Marotto, F. R., Snap-back repellers imply chaos in $\mathbf{R}^n$, *J. Math. Anal. Appl.* **63,** pp 199-223, 1978

Martelli, M., *Discrete Dynamical Systems and Chaos,* Pitman Monographs and Surveys in Pure and Applied Math. 62, Wiley, New York, 1992

May, R. M., Biological populations with nonoverlapping generations: stable points, stable cycles, and chaos, *Science* **186,** pp 645-647, 1974

Mira, C., *Chaotic Dynamics: From the One-Dimensional Endomorphism to the Two-Dimensional Diffeomorphism,* World Scientific, Singapore, 1987

Mooney, D. and Swift, R., *A Course in Mathematical Modeling,* Math. Assoc. of Amer., Washington, DC, 1999

Murray, J. D., *Mathematical Biology,* 2nd ed., Springer-Verlag, Berlin, 1993

Neuwirth, E. and Arganbright, D., *The Active Modeler - Mathematical Modeling with Microsoft Excel,* Brooks-Cole, Pacific Grove, CA, 2004

Peitgen, H.-O., Jürgens, H. and Saupe, D., *Chaos and Fractals, New Frontiers of Science,* Springer-Verlag, New York, 1992

Pesaran, M. H., and Potter, S. M., *Nonlinear Dynamics, Chaos and Econometrics,* Wiley, Chichester, 1993

Puu, T., *Nonlinear Economic Dynamics,* Springer, New York, 1997

Robinson, C. R., *Introduction to Dynamical Systems: Continuous and Discrete,* Prentice-Hall, U. Saddle River, NJ, 2003

Sandefur, J. T., *Discrete Dynamical Systems: Theory and Applications,* Clarendon, Oxford, 1990

Sandefur, J. T., *Elementary Mathematical Modeling: A Dynamical Approach,* Brooks-Cole, Pacific Grove, CA, 2003

Scheinerman, E. R., *Invitation to Dynamical Systems,* Prentice Hall, U. Saddle River, NJ, 1996

Schnoor, J. L., *Environmental Modeling,* Wiley, New York, 1996

Sedaghat, H., *Nonlinear Difference Equations, Theory with Applications to Social Science Models,* Kluwer, Boston, MA, 2003

Smale, S., Differentiable dynamical systems, *Bull. Amer. Math. Soc.* **73,** pp 714-817, 1967

Strogatz, S. H., *Nonlinear Dynamics and Chaos,* Addison-Wesley, Reading, MA, 1994

Taylor, H. M. and Karlin, S., *An Introduction to Stochastic Modeling,* Academic Press, San Diego, 1994

Thompson, J. M. T. and Stewart, H. B., *Nonlinear Dynamics and Chaos,* Wiley, Chichester, 1986

West, B. J., *Fractal Physiology and Chaos in Medicine,* Studies in Nonlinear Phenomena in Life Sciences - Vol. 1, World Scientific, Singapore, 1990

Index

A

Absolute error, 137, 138
Air-resistance, 4, 19, 133, 140
Alternating series, 32
Amortized loan, 60
Annuity, ordinary, 21, 22, 58
Annuity Savings Formula, 59
Annuity Savings Model, 22
Approximate solution
  differential equation(s), 134
    system of, 275, 302
  root of equation, 127–133, 154,
    175, 314
Asymptotic behavior/dynamics, 17, 69,
    See also Stability
Attractor, 106, *See also* Stability
  sink, 205, 271
  strange, 186, 296, 305
Autonomous differential
    equation(s), 136
  fixed point, 137
  system of, 275, 301, 302
Autonomous linear equation 28, 30
  fixed point, 71
  oscillation, 73
  solution
    homogeneous, 43
    non-homogeneous, 53, 54
  stability, 72
Autonomous linear system,
    See Linear system
Autonomous nonlinear equation,
    See Nonlinear equation
Autonomous nonlinear system,
    See Nonlinear system

B

Barnsley fern, 309
Barnsley, M., 308
Basin of attraction, 106, 121, 177, 179, 314
Bifurcation
  nonlinear equation
    fixed point, 166
    m-cycle, 171
    period-doubling cascade, 171–173
  nonlinear system
    Hopf, 291
    irrationally periodic cycle, 292
    fixed point, 290–294
    m-cycle, 293, 294
    period-doubling cascade, 290, 291
    period-multiplying cascade, 293–295
Birth rate, 23, 24
Bouncing Ball Model, 9
Bouncing ball, 6–9
Box-counting dimension, 309–311
Butterfly effect, 180

C

Cantor set, 303, 304, 311
Carrying capacity, 89, 267, 268
Change of variable, 55, 74, 155, 255
Chaos, 173, 176, 186
  nonlinear equation, 176–186
    butterfly effect, 180
    dense orbit, 178
    homoclinic orbit, 185
    horseshoe, 183
    infinite periodic cycles, 177
    Lyapunov exponent, 180–182

Chaos (*continued*)
    period three, 184
    scrambled set, 176, 177
    sensitive dependence, 179, 180
    snap-back repeller, 185
    transitive function, 178
  nonlinear system, 295–299
    homoclinic orbit, 298, 299
    horseshoe, 296–298
    snap-back repeller, 289, 290
    strange attractor, 296
Chronic illness, 260
Cobweb graph, 110–114
  and nonlinear dynamics
    chaos, 179, 183, 185
    fixed point, 119
    periodic cycle, 145, 149, 150
  and time series graph, 111, 113, 114
Coefficient matrix
  linear system, 219, 255
    eigenvalues, *See* Eigenvalue(s)
    eigenvectors, 220–225
  simultaneous equations, 214–216
Coin toss, 3, 4
Competing species, 191, 267
  Linear Competition Model I/II, 191, 192
  Nonlinear Competition Model, 267
Complex conjugate, 239
Complex eigenvalue(s), *See* Eigenvalue(s)
Complex number, 238
Complex plane, 240
Composition (notation), 147, 283
Compound interest, 13, 14, 21, 22, 58–61
  exponential growth of, 45
Compound Interest Formula, 14, 43
Compound Interest Model, 13
Computer graphics/animation, 194, 196, 207,
    302, 309
Computer software, 33
  Mathematica, 320–326
  Microsoft Excel, 315–320
Conjugate, complex, 239
Contagious Disease, 90, 91, 268, 269
  infection rate, 90, 99, 167, 268, 269, 288
  Nonlinear Infection Model I/II, 91
  Nonlinear Infection-Recovery
    Model, 269
  recovery rate, 90, 268, 269
Contamination, 27, 38, 48, 58
Continuous curve, 273, 295, 301
Continuous time, 5

Cooling, 27, 48
Credit card debt, 77

D

Death rate, 23
De Moivre's Formula, 240
Demand, 24, 91–93, 193, 194, 269
  Linear Price-Demand Model, 194
  Linear Price-Demand-Supply Model, 193
  Nonlinear Price-Demand Model, 270
Dense orbit, 178, 296
Dense set, 178, 302–304
Density dependence, 88, 89, 97, 101, 267, 268
Derivative, *See also* Differential equation(s)
  and bifurcation
    fixed point, 166
    m-cycle, 171
  and linearization, 124
  and Lyapunov exponent, 181, 182
  and Newton Root-Finding Method, 129, 133
  and oscillation, 119
  and stability
    differential equation, 137
    fixed point, 119, 157–162
    2-cycle, 146, 160–162
    m-cycle, 151
  partial, *See* Partial derivative(s)
Determinant, 215
  and eigenvalues, 226
  and inverse matrix, 215
  and nontrivial solutions, 215, 216
Deterministic process, 2, 4, 177, 307
Differential equation(s), 3, 5, 6, 133, 134
  autonomous, 136
    fixed point, 137
    system of, 275, 301, 302
  Euler method, 133–138, 275, 302
Dimension (of a fractal), 309–311
Direct numerical iteration, 8
Discrete dynamical system, 6
Discrete time, 5
Disease, *See* Contagious disease
Drum, 10
Dynamical system, 2
Dynamical systems approach, 75

E

Education, 65, 67
Eigenvalue Formula, 226

Eigenvalue(s), 220–222, 225, 226
  complex conjugate, 244, 245
    homogeneous solution, 246, 247
    Hopf bifurcation, 291
    rotation, 247–249, 256, 279, 292, 293
    stability, 250, 256, 279, 283, 288, 289
  one repeated real, 229, 230
    homogeneous solution, 231–233
    stability, 233, 256, 279, 283
  two real distinct, 223
    bifurcation, 290, 291
    homogeneous solution, 223–225
    stability, 223, 256, 279, 283,
      288, 289
Eigenvector(s), 220–225
Empirical model, 83, 84
Error, 137, 138
Equilibrium point, *See* Fixed point
Euler Method, 133–138
  for a system, 275, 302
Euler's Formula, 240
Exact solution, 12, *See also* Solution
Excel, Microsoft, 33, 315–319
Exponential function, 44
  growth/decay of, 44, 181, 199, 200

F

Falling object, 2–5, 19, 133, 140
Fibonacci Equation, 196
Filtering, 27, 38, 48, 58
Fixed point
  differential equation, 137
  linear equation, 71–74
  linear system, 204–206
    homogeneous, 233, 234, 250
    non-homogeneous, 255, 256
  nonlinear equation, 103–106, 119, 124
  nonlinear system, 270, 271,
    277–281, 291
  parameterized, *See* Parameterized
    family/system
Flu, 76
Formula for exact solution, 12, *See also*
  Solution
Fractal, 14, 176, 186, 302–311
  Barnsley fern, 309
  Cantor set, 303, 304, 311
  dimension, 309–311
  iterated-function system, 307–309
  Julia set, 305–307

Mandelbrot map/system, 14, 293–295,
  307, 308
Mandelbrot set, 294, 305
Sierpinski triangle, 308, 311
Frequency distribution, 179, *See also*
  Histogram

G

Generation, 23, 65, 261, 263, *See also*
  Overlapping generations
Geometric series, 50–52
Global stability, 106, 184, 271
Growth rate, 23, 24, 88, 89, 190–193,
  266–268

H

Harvesting, 24, 153, 163, 192, 193
Hénon map, 301
Histogram, *See also* Frequency distribution
  using Mathematica, 322
  using Microsoft Excel, 317, 318
Homeowners, 67
Homoclinic orbit
  nonlinear equation, 184, 185
  nonlinear system, 288, 289
Homogeneous linear equation, 39
  autonomous, 43, *See also* Autonomous
    linear equation
  non-autonomous, 40
Homogeneous linear system, 197, 219
  fixed point, 204–206
  rotation, 247–249
  solution
    complex eigenvalues, 246, 247
    one repeated eigenvalue, 231, 233
    two real eigenvalues, 223–225
  stability
    complex eigenvalues, 250
    one repeated eigenvalue, 234
    two real eigenvalues, 223
Homogeneous system of simultaneous
  equations, 215, 216
Hopf bifurcation, 291–293
Horseshoe
  nonlinear equation, 183
  nonlinear system, 296–298
Host, 190
Hyperbolic trajectory, 203, 206, 224,
  272, 280

I

Identity matrix, 213, 226–228, 264
Imaginary axis, 240
Imaginary number, 236, 237, 246
Imaginary part, 238, 246
Immigration, 24, 191–193
Income mobility, 261, 263
Infection, *See* Contagious Disease
Infection rate, 90, 99, 167, 268, 269, 288
    periodic, 99
Infinite series, 52, 53
Instability, *See* Stability
Interest rate, 11–13, 21, 22
Intermediate Value Theorem, 153
Interval of existence/stability, *See also*
      Bifurcation
    nonlinear equation, 157–162
    nonlinear system, 287–289
Invariant curve, 273, 295
Inverse
    equation, 77, 126, 184
    matrix, 214, 215
    system, 289, 307
Irrationally periodic cycle, 292, 293
Irrationally periodic rotation, 249, 292, 293
Iterate, iteration, iterative equation, 8
Iterated-function system, 307–309

J

Julia set, 305–307
    using Mathematica, 325
    using Microsoft Excel, 319

L

Learning, 93, 95
Least-squares regression, 78–83, 96,
      196, 197
L'Hopital's Rule, 51, 234
Li, T.-Y., 184
Linear combination, 220, 223, 231, 245, 246
Linear Competition Model I/II, 191, 192
Linear equation, 20, 30
    autonomous, *See* Autonomous linear
      equation
    non-autonomous
      homogeneous, 39, 40
      non-homogeneous, 39, 50
Linear growth, 46

Linear Immigration/Migration/Harvesting
      Model, 24
Linear model, *See also* Models
    equation, 9, 21–25, 27, 48, 76, 77
    Markov process,
      three-state, 256–264
      two-state, 62–68
    system, 190–194
Linear Overlapping-Generations Model I/II,
      192, 193
Linear Population Model, 23
Linear Prey-Predator Model I/II, 190, 191
Linear Price Model, 25
Linear Price-Demand Model, 194
Linear Price-Demand-Supply Model, 193
Linear regression, 78–84, 196, 197
Linear system, 189, 197
    homogeneous, *See* Homogeneous linear
      system
    non-homogeneous, *See* Non-homogeneous
      linear system
    of simultaneous equations, 213–216
Linearization, 124, 276–279
    matrix, *See* Matrix of partial derivatives
Loan, 22, 60
Loan Payment Formulas, 60
Loan Payment Model, 22
Local stability, 106, 271, 276–279,
      See also Stability
Logarithmic scale, 46
Logistic differential equation, 134
Logistic equation, 89, 97, 155, 156
Logistic family, 155, 156
    bifurcation, 165, 166
    chaos, 176–180, 183, 185
    intervals of stability, 158, 160, 161
    period doubling cascade, 169–173, 176, 177
Long-term behavior, 17, 69, *See also* Stability
Lorenz, E., 302
Lorenz system, 301, 302
Lyapunov exponent, 180–182
    using Mathematica, 321, 322
    using Microsoft Excel, 317

M

m-cycle, *See also* 2-cycle
    nonlinear equation, 147–152
    nonlinear system, 282–284
    parameterized, *See* Parameterized
      family/system

Maclaurin Series, 244
Mandelbrot, B., 14
Mandelbrot
    map/system, 14, 293–295, 307
    set, 294, 305
Market share, 67, 263
Markov process, 61–66, 76, 256–262
    regular, 76
    state/transition diagram, 61
    state vector, 257
    steady-state, 76, 257
    three-state, 256–262
        reducing to two variables, 259, 260
    transition matrix, 257
    two-state, 61–66, 76
        reducing to one variable, 63
Mathematica, 33, 315, 320–326
Mathematical model, 2,
        See also Models
Matrix, 211
    coefficient
        linear system, 219, 255
        simultaneous equations, 214
    determinant, 215
    eigenvalues, *See* Eigenvalue(s)
    eigenvectors, 220–225
    identity, 213, 214, 226, 264
    inverse, 214, 215
    multiplication, 212, 213
    norm, 218
    transition, 257–261
Matrix of partial derivatives, 278
    and dynamics, 279, 283, 290–292
May, R. M., 173
Mean Value Theorem, 123
Membrane, 25
Microsoft Excel, 33, 315–320
Migration, 24, 153, 191–193
Models
    Bio-medicine
        chronic illness, 260
        contagious disease, 76, 91, 269
        virus growth, 48
    Business
        market share, 67, 263
        profitability, 67
        unemployment level, 76
    Ecology
        competing species, 191–193, 267
        harvesting, 24, 153, 163, 191–193
        prey-predator, 190, 191, 266

single species, 23, 24, 89, 105, 110, 192,
        193, 267, 268
    Economics
        demand, 193, 194, 270
        price, 25, 92, 93, 193, 194, 270
        supply, 193
    Engineering
        filtering pollutants, 27, 38, 48, 58
        swaying skyscraper, 48
    Finance
        annuity savings, 22, 59
        compound interest, 13
        credit card debt, 77
        loan, 22, 60
        simple interest, 12
    Learning Theory
        learning a topic, 95
        stages of training, 64, 67, 264
    Physics
        beating drum, 10
        bouncing ball, 6, 9
        coin toss, 3, 4
        cooling object, 27, 48
        falling object, 3, 5
        radioactive decay, 48
        swinging pendulum, 10, 27, 48
        weather prediction, 301, 302
    Sociology
        education levels, 65, 67
        home ownership, 67
        income mobility, 261, 263
        spread of technology, 95
        urbanization, 61, 67, 257–259, 263
Modulus, 241, 242, 250, 278, 283, 291
Monotonic solution, 70, 73, 74, *See also*
        Oscillation

N

Negative of a vector, 210
Newton Root-Finding Method, 127–133, 154,
        175, 314
Newton's Law of Cooling, 27
Non-autonomous linear equation, 28–30
    homogeneous, 39, 40
    non-homogeneous, 39, 49, 50
Non-autonomous nonlinear equation, 98, 99
Non-homogeneous linear equation, 39
    autonomous, *See* Autonomous linear
        equation
    non-autonomous, 49, 50

Non-homogeneous linear system, 197, 254
  fixed point, 204–206
  rotation/stability, 255, 256
Nonlinear Competition Model, 267
Nonlinear equation, 87, 96, *See also*
    Parameterized family
  chaos, 176–186, *See also* Chaos
  fixed point, 104, 106, 119
  2-cycle, 141, 144, 146
  m-cycle, 147–151
  solution, 101, 109
Nonlinear Infection Model I/II, 91, 156, 157
Nonlinear Infection-Recovery Model, 269
Nonlinear model, *See also* Models
  differential equation(s), 3, 5, 6, 133, 134, 140
    system of, 275, 301, 302
  equation, 88–93, 95, 99, 105, 110, 153, 163
  system, 266–270, 274, 276
Nonlinear Overlapping-Generations
    Model, 268
Nonlinear Population Model I/II, 89, 97,
    155, 156
Nonlinear Prey-Predator Model, 266
Nonlinear Price Model I/II/III, 93
Nonlinear Price-Demand Model, 270
Nonlinear regression, 96
Nonlinear system, 265, 270, *See also*
    Parameterized system
  chaos, 295–299, *See also* Chaos
  fixed point, 270, 271, 278, 279
  of differential equations, 275, 301, 302
  periodic point/cycle, 282, 283
Norm of a matrix, 218
Normal equations, 82, 83
Numerical solution, 8, 32, 33
  Mathematica, 320–326
  Microsoft Excel, 315–320

O

One repeated real eigenvalue, *See* Eigenvalue(s)
One-parameter family, 155, 156, *See also*
    Parameterized family
Ordinary annuity, 21, 22, 59
Oscillation, *See also* Periodic point/cycle
  linear equation, 70, 73, 74
  nonlinear equation, 118, 119, 124
  periodic, 36, 99, 144, 145
Overlapping-generations, 192, 193, 267, 268
  Linear Overlapping-Generations Model I/II,
    192, 193

Nonlinear Overlapping-Generations
    Model, 268

P

Parameter, 21, 155, 287
  birth rate, 23
  carrying capacity, 89, 267, 268
  death rate, 23
  growth rate, 23, 88, 89, 190–193,
    266–268
  harvesting rate, 24, 153, 163, 191–193
  immigration rate, 24, 191–193
  infection rate, 90, 99, 268, 269, 288
  interest rate, 11–13, 21, 22
  migration rate, 24, 153, 191–193
  recovery rate, 90, 268, 269
Parameter space, 289–295
Parameterized family, 155
  fixed point, 157
    bifurcation, 166
    interval of existence/stability, 157–159
    period-doubling cascade, 171–73
  periodic point/cycle, 159, 160
    bifurcation, 169–171
    interval of existence/stability, 159–162
Parameterized system, 287–295
  fixed point, 287–293
    bifurcation, 290–295
    interval/region of existence/stability,
      287–289
  period-doubling cascade, 290, 291
  period-multiplying cascade, 293–295
Parasite, 190
Partial derivative(s), 80–82, 278
  and normal equations, 81, 82
  matrix of, *See* Matrix of Partial
    Derivatives
Partial product, 42
Partial sum, 31
Pendulum, 10, 27, 48
Period three (and chaos), 184, 296
Periodic oscillation, 36, 144, 145
  infection rate, 99
Periodic point/cycle, *See also* 2-cycle,
    m-cycle
  irrationally, 249, 292
  nonlinear equation, 141–152, 177
  nonlinear system, 282–285, 290–295
  parameterized, *See* Parameterized
    family/system

Periodic rotation, 247–249, 292
Phase plane graph, 202
Polar coordinates, 240, 292
Pollution, 27, 38, 48, 58
Population (single species), 23, 24, 88, 89,
    192, 193, 267, 268, *See also* Markov
    process
  carrying capacity, 89, 267, 268
  density dependence, 88, 89, 97, 101,
    267, 268
  differential equation, 134
  equation
    Linear Immigration/Migration/Harvesting
      Model, 24
    Linear Population Model, 23
    logistic, 89, 97, 155, 156
    Nonlinear Population Model I/II, 89, 97,
      98, 155, 156
    other, 105, 153, 163
  generation, 23
  growth rate, 23, 24, 88, 89, 192, 267, 268
  system
    Linear Overlapping-Generations Model
      I/II, 192, 193
    Nonlinear Overlapping-Generations
      Model, 267, 268
Predator/Prey, 190, 266
  Linear Prey-Predator Model I/II, 190, 191
  Nonlinear Prey-Predator Model, 266
Price, 24, 25, 91–93, 193, 194, 269, 270
  equation
    Linear Price Model, 25
    Nonlinear Price Model I/II/III, 92, 93
  system
    Linear Price-Demand Model, 194
    Linear Price-Demand-Supply
      Model, 193
    Nonlinear Price-Demand Model, 270
Probability distribution, 4
Product, 41–43
Profit, 67

Q

Qualitative behavior/dynamics, 15–18, 35, 68,
    75, 197, 201

R

Radioactive decay, 48
Random process, *See* Stochastic process

Real axis, 240
Real eigenvalue(s), *See* Eigenvalue(s)
Real part, 238
Recovery, 90, 91, 268, 269
Recursion, recursive equation, 9
Refining a model, 4, 5
Region of existence/stability, 289–295
Regression, 78–84, 96, 196
Regular Markov process, 76
Relative error, 137, 138
Rent (annuity), 59–61
Renters, 67
Repeated real eigenvalue(s),
    See Eigenvalue(s)
Repeller, 106, *See also* Stability
  fractal, 304–307
  snap-back, *See* Snap-back repeller
  source, 205, 271
Root, *See* Newton Root-Finding Method
Rotation
  linear system, 247–249, 251, 256
  nonlinear system, 279, 292, 293

S

Saddle, 206, 271, *See also* Stability
Scalar, 210
Scaling factor, 156
Scrambled set
  nonlinear equation, 176–179
  nonlinear system, 186, 296, 302–305
Second-order equation, 192, 267, 268
  converting to, 197, 198
  solving, 231–233, 245–247
Sensitive dependence on initial conditions,
    179–182, 296, 304
Separating variables, 134
Sierpinski triangle, 308, 311
  using Mathematica, 325, 326
  using Microsoft Excel, 319, 320
Simple interest, 11–12
  linear growth of, 46
Simple Interest Formula, 12
Simple Interest Model, 12, 21, 28
Simultaneous equations, 213–215
Single species, *See* Population
Sink, 205, 271, *See also* Stability
Skyscraper, 48
Smale, S., 183, 297
Snap-back repeller, 184, 185, 298, 299
Software, *See* Computer Software

Solution
  formula for exact, 12
  linear equation
    autonomous, homogeneous, 43
    autonomous, non-homogeneous, 54
    non-autonomous, homogeneous,
      39, 40
    non-autonomous, non-homogeneous,
      49, 50
  linear system, homogeneous
    complex eigenvalues, 245–247
    one repeated eigenvalue, 231–233
    two real eigenvalues, 223–225
  nonlinear equation, 101, 109
  nonlinear system, 275
  nontrivial, 215, 216
  numerical, *See* Numerical solution
  quantitative, vs. qualitative behavior, 15–17,
    68, 69
  trivial, 215
Solution space graph, 201
  and linear dynamics, 205, 206
  and nonlinear dynamics
    stability, 271–273, 292, 293
    chaos, 273, 295, 296, 299
  using Mathematica, 323–325
  using Microsoft Excel, 319
Solution vector, 219
  length, 222–224, 234, 250, 251
  rotation, 247–251, *See also* Rotation
Source, 205, 271, *See also* Stability
Spread of technology, 95
Stability, *See also* Parameterized
    family/system
  differential equation, 137
  linear equation, autonomous, 72, 74
  linear system, homogeneous, 205, 206
    complex eigenvalues, 250
    one repeated eigenvalue, 234
    two real eigenvalues, 223
  linear system, non-homogeneous, 205, 206,
    255, 256
  nonlinear equation
    fixed point, 106, 119, 124
    2-cycle, 143–146
    m-cycle, 148–151
  nonlinear system
    fixed point, 271, 279
    invariant curve, 273, 295
    irrationally periodic cycle, 292, 293
    m-cycle, 283

State (Markov process), 61
State diagram, 61
State vector, 257
Steady-state, 76, 257
Stochastic process, 2, 4, 61–66, 177, 256–262,
  307–309
Strange attractor, 186, 296, 305
Substitution, 55, 74, 155, 255
Sum, 30–32
Superstable, 126, 163
Supply, 24, 91, 92, 193, 194, 269
  Linear Price-Demand-Supply
    Model, 193
System
  discrete dynamical, 6
  dynamical, 2
  linear, 189, 197
    homogeneous, *See* Homogeneous linear
      system
    non-homogeneous, *See*
      Non-homogeneous linear system
  nonlinear, *See* Nonlinear system
  of differential equations, 275, 301, 302
  of simultaneous equations, 213–216

T

Taylor Series, 52, 239, 244
Telescoping product, 41
Temperature, 27, 48, 140
Term (of compounding), 13
Terminal velocity, 19, 140
Testing a model, 4, 5
Threshold, 75, 102, 110, 164, 173, 276
Time-series graph, 35
  and cobweb graph, 111, 113, 114
  and linear dynamics, 74, 199–202
  and nonlinear dynamics
    stability, 101, 102, 107
    chaos, 101, 102, 180
  using Mathematica, 320, 321, 323, 324
  using Microsoft Excel, 316, 318, 319
Topologically conjugate equations, 75
Trajectory, 201, 271
Training, 64, 67, 264
Transient dynamics, 17, 18, 33, 69
Transition diagram, 61
Transition matrix, 257
Transitive function, 178
Trivial Solution, 215
Two real eigenvalues, *See* Eigenvalue(s)

Two-cycle/2-cycle, *See also m*-cycle
  nonlinear equation, 141–147
  nonlinear system, 282–284
  parameterized, *See* Parameterized
    family/system

U

Uncoupled system, 197–199, 257,
      See also Second order equation
Unemployed, 76
Unimodal function, 97
Unimodal family, 169, 172, 173,
      177, 184
Unstable point/cycle, *See* Stability
Urbanization, 61, 67, 257–259, 263

V

Variation of parameters, 231
Vector, 208, *See also* Solution vector
  eigenvector, 220–225
  state, 257
Virus growth, 48

W

Wage earners, 261, 263
Weather prediction, 301, 302

Y

Yorke, J. A., 184